T0179281

Representing, Modeling, and Visualizing the Natural Environment

INNOVATIONS IN GIS

SERIES EDITORS

Jane Drummond
University of Glasgow, Glasgow, Scotland

Bruce Gittings
University of Edinburgh, Edinburgh, Scotland

Elsa João
University of Strathclyde, Glasgow, Scotland

INNOVATIONS IN GIS

Representing, Modeling, and Visualizing the Natural Environment

Edited by
Nick Mount
Gemma Harvey
Paul Aplin
Gary Priestnall

CRC Press
Taylor & Francis Group
Boca Raton London New York

CRC Press is an imprint of the
Taylor & Francis Group, an **informa** business

CRC Press
Taylor & Francis Group
6000 Broken Sound Parkway NW, Suite 300
Boca Raton, FL 33487-2742

First issued in paperback 2020

© 2009 by Taylor & Francis Group, LLC
CRC Press is an imprint of Taylor & Francis Group, an Informa business

No claim to original U.S. Government works

ISBN-13: 978-0-367-57739-1 (pbk)
ISBN-13: 978-1-4200-5549-8 (hbk)

Visit the Taylor & Francis Web site at
http://www.taylorandfrancis.com

and the CRC Press Web site at
http://www.crcpress.com

Contents

SECTION 1   Representing the Natural Environment

SECTION 2   Modeling the Natural Environment

SECTION 3 Visualizing the Natural Environment

Preface

The natural environment, including the ways in which humans interact with it, represents a complex and dynamic forum for scientific inquiry, and studies seeking to explore and predict characteristics and processes within this field are necessarily associated with a strong geospatial element. This volume identifies particular analytical challenges associated with the application of geographical information science (GIScience) in environmental contexts, and also serves to illustrate broader opportunities and themes relating to the use of geographic information systems (GIS) in other areas of science and social science.

This is the thirteenth volume in a series based on the Geographical Information Science Research UK (GISRUK) Conference Series. The GISRUK Conference Series, established in 1993, provides an interdisciplinary forum for the discussion and publication of GIS research and the promotion of research collaborations. Although GISRUK is a UK-based initiative, it attracts delegates from many countries and covers a diverse range of disciplines. This volume has been developed from research presented at the GISRUK 2006 Conference held at the University of Nottingham, and the GISRUK 2007 Conference held at the National University of Ireland, Maynooth, plus several invited keynote papers and research papers.

The contributions relate to the key themes of representing, modeling, and visualizing the natural environment in the context of GIS. The opening chapter provides an introduction to these themes and introduces each article included in the volume. The articles cover a range of theoretical, methodological, and empirical issues based on a diverse range of environmental applications of GIS, and provide innovative examples of the current state of the art. The editors wish to thank the contributors for their time and effort in preparing manuscripts for this volume, the reviewers for their invaluable comments on the manuscripts, and the GISRUK Steering Committee for its assistance with the selection of papers from GISRUK 2006 and 2007. We hope that the volume will provide researchers, students, and practitioners in both GIS and the environmental sciences with an overview of the opportunities for utilizing GIS for environmental applications and of some of the ongoing research challenges in this field.

The Editors

Nick Mount (BS, Bristol; PhD, Liverpool John Moores) is a lecturer in geographical information science at the University of Nottingham. His research interests extend across a broad range of GIS applications in sensitive natural environments, with a particular focus on the application of GIS in dynamic rivers, proglacial environments, and the estimation of spatial error in environmental, spatiotemporal data sets. His current focus is on the spatiotemporal representation and analysis of large dynamic rivers, including the Brahmaputra in India and Bangladesh. Prior to joining Nottingham in 2006, Mount was director of the applied GIS program at the University of London and a lecturer in geographic information systems at Charles Sturt University, New South Wales, Australia.

Gemma Harvey (BS, Liverpool; PhD, Nottingham) is a research associate in river management and restoration at the University of Nottingham. Her main research interests are in river restoration and the underlying science base, and flood risk management in the United Kingdom and China. She is interested in the application of multivariate- and geo-statistics, and GIS technologies in river science and management contexts to address complexity and dynamics across a range of spatiotemporal scales. Her current work includes the development of novel field techniques, as well as applications of geostatistical and multivariate techniques to the characterization of river habitat for management and restoration purposes, assessing the impacts of flood defense works on river habitat structure, and exploring changing flood risk in the United Kingdom and east China.

Paul Aplin (MA, Edinburgh; MS, Aberdeen; PhD, Southampton) is an associate professor in geographical information science at the University of Nottingham. His main research interests are in environmental remote sensing, specializing in land cover analysis, spatial scale investigation, and ecological applications. He is currently the chairman of the Remote Sensing and Photogrammetry Society, and the book series editor for the International Society of Photogrammetry and Remote Sensing.

Gary Priestnall (BS, Durham; PhD, Nottingham) is an associate professor in geographical information science at the University of Nottingham. His main research interests are in digital geographic representation, visualization, and location-aware computing. He is director of the master's program in GIScience at the University of Nottingham, was chair of the GIS Research UK 2006 conference at Nottingham in April 2006, and is Nottingham site manager for the collaborative Higher Education Funding Council for England (HEFCE)-funded Centre for Excellence in Teaching and Learning SPLINT (SPatial Literacy IN Teaching). He has edited a research volume titled *Chat Moss*, an art–geography collaboration focusing on landscape visualization, with artist Derek Hampson. He is focusing on landscape visualization in both teaching and research contexts and reviewing the state of the art in augmented and virtual reality in landscape portrayal.

Contributors

Paul Aplin
School of Geography
University of Nottingham
Nottingham, United Kingdom
paul.aplin@nottingham.ac.uk

Katy Appleton
School of Environmental Sciences
University of East Anglia
Norwich, United Kingdom
k.appleton@uea.ac.uk

Richard Aspinall
Macaulay Institute
Aberdeen, United Kingdom
r.aspinall@macaulay.ac.uk

Matthew E. Baker
Department of Geography &
  Environmental Systems
University of Maryland,
  Baltimore County
Baltimore, Maryland, USA
mbaker@umbc.edu

Darren J. Bender
Department of Geography
University of Calgary
Calgary, Alberta, Canada
dbender@ucalgary.ca

Stuart Blair
School of Geography
University of Leeds
Leeds, United Kingdom
stublair@gmail.com

Susanne Bleisch
Institute of Geomatics Engineering
University of Applied Sciences
  Northwestern Switzerland (FHNW)
Muttenz, Switzerland
susanne.bleisch@fhnw.ch

William E. Cartwright
School of Mathematical and
  Geospatial Sciences
RMIT University
Melbourne, Australia
w.cartwright@rmit.edu.au

Steve Carver
School of Geography
University of Leeds
Leeds, United Kingdom
s.j.carver@leeds.ac.uk

Sagi Dalyot
Mapping and Geo-Information
  Engineering
Technion–Israel Institute of
  Technology
Haifa, Israel
dalyot@technion.ac.il

Yerach Doytsher
Mapping and Geo-Information
  Engineering
Technion–Israel Institute of
  Technology
Haifa, Israel
doytsher@technion.ac.il

Matt Duckham
Department of Geomatics
University of Melbourne
Melbourne, Victoria,
  Australia
mduckham@unimelb.edu.au

Jason Dykes
giCentre, School of Informatics
City University London
London, United Kingdom
jad7@soi.city.ac.uk

Ronan Foley
Department of Geography
National University of Ireland
Maynooth, Ireland
ronan.foley@nuim.ie

Steffen Fritz
Forestry Program
International Institute for
  Applied Systems Analysis (IIASA)
Laxenburg, Austria
steffen.fritz@jrc.it,
fritz@iiasa.ac.at

John Fry
School of Biology and
  Environmental Science
University College Dublin
Dublin, Ireland
john.fry@ucd.ie

Alan Gilmer
Department of Environment &
  Planning
Dublin Institute of Technology
Dublin, Ireland
alan.gilmer@dit.ie

Ainhoa Gonzalez
Department of Environment &
  Planning
Dublin Institute of Technology
Dublin, Ireland
ainhoag@yahoo.com

Danni Guo
Kirstenbosch Research Center
South African National
  Biodiversity Institute
Cape Town, South Africa
guo@sanbi.org

Renkuan Guo
Department of Statistical Sciences
University of Cape Town
Cape Town, South Africa
renkuan.guo@uct.ac.za

Craig von Hagen
Food and Agricultural Organisation
Nairobi, Kenya
craigvonhagen@yahoo.co.uk

Glen Hart
Ordnance Survey, Research &
  Innovation
Romsey Road, Southampton,
  United Kingdom
glen.hart@ordnancesurvey.co.uk

Gemma Harvey
School of Geography
University of Nottingham
Nottingham, United Kingdom
gemma.harvey@nottingham.ac.uk

Fiona Hemsley-Flint
School of Life Sciences
Oxford Brookes University
Oxford, United Kingdom
fiona.flint@hotmail.com

Alejandro de las Heras
School of Environmental Sciences
University of East Anglia
Norwich, United Kingdom
a.heras@uea.ac.uk

Alastair Jardine
School of GeoSciences
University of Edinburgh
Edinburgh, United Kingdom
alij@mac.com

Andy Jones
School of Environmental Sciences
University of East Anglia
Norwich, United Kingdom
a.p.jones@uea.ac.uk

Iain R. Lake
School of Environmental Sciences
University of East Anglia
Norwich, United Kingdom
i.lake@uea.ac.uk

John Lee
Lovell Johns Ltd
  Long Honborough
Oxfordshire, United Kingdom
john.lee@lovelljohns.com

David J. Lieske
Department of Geography and
  Environment
Mount Allison University
Sackville, New Brunswick, Canada
dlieske@mta.com

Andrew Lovett
School of Environmental Sciences
University of East Anglia
Norwich, United Kingdom
a.lovett@uea.ac.uk

William Mackaness
School of GeoSciences
University of Edinburgh
Edinburgh, United Kingdom
william.mackaness@ed.ac.uk

Michel Massart
AGRIFISH Unit
Institute for the Protection and
  Security of the Citizen (IPSC)
  EC Joint Research Centre
Ispra, Italy
michel.massart@jrc.it

Colin J. McClean
Environment Department
University of York
York, United Kingdom
cjm8@york.ac.uk

Peter Mooney
Department of Computer Science
and
National Centre for Geocomputation
National University of Ireland
Maynooth, Ireland
peter.mooney@nuim.ie

Nick Mount
School of Geography
University of Nottingham
Nottingham, United Kingdom
nick.mount@nottingham.ac.uk

Thierry Nègre
AGRIFISH Unit
Institute for the Protection and
  Security of the Citizen (IPSC)
  EC Joint Research Centre
Ispra, Italy
thierry.negre@jrc.it

Tonny Oyana
Department of Geography
Southern Illinois University
Carbondale, Ilinois, USA
tjoyana@siu.edu

Gary Priestnall
School of Geography
University of Nottingham
Nottingham, United Kingdom
gary.priestnall@nottingham.ac.uk

Femke E. Reitsma
School of Geosciences
University of Edinburgh
Edinburgh, United Kingdom
femke.reitsma@ed.ac.uk

Felix Rembold
AGRIFISH Unit
Institute for the Protection and
  Security of the Citizen (IPSC)
  EC Joint Research Centre
Ispra, Italy
felix.rembold@jrc.it

Conrad E. S. Rider
School of GeoSciences
University of Edinburgh
Edinburgh, United Kingdom
c.e.s.rider@sms.ed.ac.uk

Peter Samson
Weardale Business Centre
Stanhope, United Kingdom
peter@northpenninesaonb.org.uk

Linda See
School of Geography
University of Leeds
Leeds, United Kingdom
l.m.see@leeds.ac.uk

Daniel Z. Sui
Department of Geography
Texas A&M University
College Station, Texas, USA
sui@geog.tamu.edu

John Sweeney
National Centre for Geocomputation
and
Department of Geography
National University of Ireland
Maynooth, Ireland
john.sweeney@nuim.ie

David G. Tarboton
Department of Civil and
  Environmental Engineering
Utah State University
Logan, Utah, USA
david.tarboton@usu.edu

Christien Thiart
Department of Statistical Sciences
University of Cape Town
Cape Town, South Africa
christien.thiart@uct.ac.za

Stewart Thompson
School of Life Sciences
Oxford Brookes University
Oxford, United Kingdom
sthompson@brookes.ac.uk

Nigel Waters
Department of Geography and
  Geoinformation Science
George Mason University
Fairfax, Virginia, USA
nwaters@gmu.edu

Adam Winstanley
Department of Computer Science
and
National Centre for Geocomputation
National University of Ireland
Maynooth, Ireland
adam.winstanley@nuim.ie

A-Xing Zhu
State Key Laboratory of
  Resources and Environmental
  Information Systems
Institute of Geographical
  Sciences and Natural Resources
  Research
Chinese Academy of Sciences
Beijing, China
and
Department of Geography
University of Wisconsin
Madison, Wisconsin, USA
axing@lreis.ac.cn, azhu@wisc.edu

1 Introduction to Representing, Modeling, and Visualizing the Natural Environment

Gemma Harvey, Nick Mount,
Paul Aplin, and Gary Priestnall

CONTENTS

Geographic information systems (GIS) provide a range of opportunities for exploring the complexity of the natural environment, offering tools and software packages that undergo continual innovation and development in response to new analytical demands for solutions to spatial problems. Geographic information science (GIScience) provides the academic framework for these technologies, addressing the theoretical and scientific issues associated with the use of GIS. This volume explores issues associated with the application of GIS and GIScience to geographic problems associated with the natural environment, particularly addressing the subjects of representation, modeling, and visualization within GIS.

This chapter provides an introduction to the term *natural environment*, describing its usage in the context of this volume. This is followed by an overview of how GIS can be used to explore various aspects of the natural environment. Finally, the three main parts of the volume are introduced, focusing in turn on the key themes of representation, modeling, and visualization of the natural environment.

1.1 THE NATURAL ENVIRONMENT

Common perceptions of *naturalness* in relation to the environment are often associated with remote, wilderness landscapes that experience no influence from humans. In contrast, *urban areas* or *built environments*, particularly the process of urbanization, are often considered a threat to the natural landscape [1]. As of 2008, though, more than half of the world's population lived in urban areas [2], and indirect environmental impacts like atmospheric pollution influence even the most remote wildernesses. A range of environments of differing degrees of naturalness can therefore be perceived, creating a spectrum from pristine wilderness environments that experience no direct human interaction to built urban environments resulting from prolonged and intense human influence. A large range of environments is found in between these two extremes, and in terms of naturalness, most can at best be considered *seminatural*. The chapters in this volume fall at various points along this spectrum of environmental naturalness, but generally focus on environments with minor or moderate human influence and interaction. The chapters range from those dealing with wilderness areas and remote landscapes, physical landscape processes, and ecological applications (Chapters 4, 7, 12, 13 19, 20, and 21), to human interactions with the rural environment through agriculture (Chapters 5, 6, and 22), environmental pollution and degradation issues (Chapters 14 and 15), and the engagement of communities and other stakeholders in the management of their local environment (Chapters 8, 9, and 18).

1.2 GIS FOR REPRESENTING, MODELING, AND VISUALIZING THE NATURAL ENVIRONMENT

The natural environment, in its various forms, is a complex and dynamic entity, presenting a variety of analytical challenges to researchers, environmental managers, and the public. The acquisition of environmental data, for instance, often requires costly sampling efforts, particularly for the most remote or hostile environments where our knowledge and understanding is most lacking. Time and cost limitations associated with data collection and processing have been alleviated in recent years, at least in part, by technological advancements within the fields of computer science and remote sensing, and through the dissemination of information and software tools via the Internet. The huge increase in the amount of data available to the various stakeholders of the natural environment, however, presents is own set of challenges associated with the ways in which that environment is represented, modeled, and visualized. For instance, discrepancies in data quality (including precision, accuracy, uncertainty, and metadata availability) and data structure (including spatial and temporal resolutions and associated issues of scale) between various information sources present significant data processing and analysis challenges for researchers and environmental managers.

GIS offer a range of sophisticated tools and techniques for coping with the challenges outlined above, and for exploring and predicting the complexity and dynamism of the natural world and the ways in which human beings interact with it [3–5]. Recent developments, for instance, include the application of GIS technology to the assessment of habitat and prediction of species distributions and biodiversity patterns [6,7], the exploration of the impacts of climate change on the natural environment [8,9], and the monitoring and prediction of natural hazards [10]. This volume explores some of the theoretical, empirical, and methodological issues associated with the representation, modeling, and visualization of the natural environment within GIS; offers state-of-the-art tools and techniques for addressing these; and suggests future directions for the research community.

Engagement with environmental data is not, however, restricted to the realm of the academic researcher or environmental manager. GIS are increasingly employed within education, business, government, and recreational contexts. For instance, GIS-based mapping products are now widely used within the media [11], and the availability of free, interactive Web-based GIS applications such as Google Earth has assisted the transfer of GIS technology to schools [12]. Public participation GIS (PPGIS) within the context of environmental issues has developed in response to the increasing public awareness of environmental pressures such as climate change, pollution, habitat degradation, and flooding. Web-based GIS allow members of the public to model and visualize the natural environment and potential changes to it, offering a means of engaging the public in environmental issues and involving them in decision-making processes undertaken by local and national governments [13,14]. Web-based GIS also allow members of the public to interact with the environment at a more informal level, for instance, through planning journeys and recreational activities (e.g., Multimap [15]).

1.3 INNOVATIONS IN REPRESENTATION, MODELING, AND VISUALIZATION OF THE NATURAL ENVIRONMENT

This volume focuses on three important and intimately linked themes associated with the exploration and prediction of the natural environment using GIS: representation, modeling, and visualization. Digital representations of geographic attributes are powerful tools for understanding the real world and communicating spatial data [16]. *Representations* are both supported by and contribute to *models*, which provide the set of constructs for describing and representing parts of the real world digitally [17]. *Visualizations* offer a flexible medium for analyzing and interacting with real and artificially created environments [17,18]. This volume deals with key topics such as the representation of concepts and relationships within a GIS, techniques for dealing with uncertainty, scale and sampling issues, and GIS for public participation. These issues are explored in relation to aspects of the natural environment such as land cover and land use, biodiversity, environmental policy, modeling flow and terrain, distribution of environmental pollutants, and economic, social, cultural and emotional interactions between people and the landscape.

The volume is organized into three sections: "Representing the Natural Environment," "Modeling the Natural Environment," and "Visualizing the Natural Environment." Each section begins with two keynote papers that provide broad commentaries on the part's theme and highlight linkages between the different sections. In each part, keynote papers are followed by a series of invited and research articles.

1.4 REPRESENTING THE NATURAL ENVIRONMENT

The first section of the book explores a broad range of challenges and opportunities for representing the natural environment within GIS, encompassing the development of ontologies, data aggregation and disaggregation, the integration of public perceptions of the environment with the study of physical attributes, and the development of Web-based GIS applications within the context of environmental data. The keynote paper by Duckham (Chapter 2) discusses some classical GIScience issues within the context of the natural environment, focusing on three themes: context and meaning (including ontological and extensional approaches), uncertainty (including accuracy, precision, and vagueness), and dynamism. This is complemented by a second keynote paper by Waters (Chapter 3) that considers the role of GIS in representing the surface of the earth within both environmental research and formal and informal educational contexts, considering issues associated with methodological approaches, interpretation, and understanding. The issue of ontologies in representation is then addressed by the research of Hemsley-Flint et al. in Chapter 4, introducing a methodology for the establishment of semiformal ontologies that are both easily developed and understood by humans, and may be readily translated into formal, machine-readable ontologies.

Today's wealth of spatial information provides many opportunities for geographical research, but large discrepancies in data quality, resolution, and coverage can represent significant challenges for analysis. Furthermore, consideration of human interactions with the natural environment can necessitate more flexible approaches. Such issues are highlighted in Chapters 5, 6, and 7, and potential solutions are offered within the context of representing and analyzing land cover and land use characteristics. In Chapter 5, McClean presents and evaluates a methodology for the spatial disaggregation of administrative-level land use data to finer spatial units appropriate for environmental research. The chapter discusses the impact of missing data and misrepresentations, but overall the method is considered to perform well and has the potential for application to other geographical data sets. Fritz et al. then compare representations of agricultural land use data from four different information sources in Chapter 6, employing Boolean and fuzzy approaches to facilitate spatial comparisons and highlighting issues of precision and uncertainty. In Chapter 7, Blair et al. explore a range of methods for identifying areas within the North Pennines that may be suitable for rewilding programs. The work illustrates the flexibility offered by a fuzzy set approach, and additionally by a composite approach developed by the authors to allow integration of human perceptions of wildness in addition to the study of physical environmental parameters.

The section is then concluded with two chapters that deal with the use of Web-based GIS for representing environmental data. In Chapter 8, Mooney and Winstanley suggest that future directions in the field will be underpinned by a move from desktop to Web-based GIS, and discuss the representation of environmental data for delivery and dissemination through this medium. The authors outline key technological developments, emphasizing the importance of high-quality metadata in Web-based applications. An evaluation of a web-based GIS approach to involving the public in environmental policy is then provided by Gonzalez et al. in Chapter 9. Overall, the

Web-based tool was received positively, an important finding in light of the growing importance of e-participation, but the authors also discuss potential barriers to such approaches. These barriers, including levels of computer literacy, access to technology, perceptions of GIS, and issues associated with confidentiality and political will, should be considered in future work.

1.5 MODELING THE NATURAL ENVIRONMENT

The second section of the book deals with scaling and uncertainty in modeling the natural environment, offering a series of modeling approaches that can be used as means of overcoming data processing challenges associated with variations in data and model structures between GIScience and the physical sciences; maintaining precision and accuracy while merging spatial datasets; and coping with problems associated with sampling bias and low data coverage. In Chapter 10, Aspinall provides a keynote paper that explores three of the challenges associated with modeling the natural environment: representation of environment; representation of process (including process dynamics and spatial processes); and representation of time. The chapter describes some of the key developments and approaches used in each of these areas, and identifies how improvements to GIS capabilities in relation to these challenges will have the potential to address some key questions in environmental science and management. The second keynote paper by Zhu (Chapter 11) focuses on the effects of neighborhood size and the modifiable areal unit problem on the processing of spatial data for modeling the natural environment, suggesting procedures for selecting appropriate neighborhood sizes and minimizing scale incompatibilities between geographic variables.

In Chapter 12, Tarboton and Baker present a general method for recursive flow analysis that integrates multiple inputs and a class of algebraic rules into the calculation of flow-related quantities. The authors illustrate the potential of the approach using example functions applicable to hydrologic and environmental modeling, and emphasize the potential for integrating GIS with hydrological modeling despite differences in data structures and modeling approaches between the disciplines. In Chapter 13, Dalyot and Doytsher introduce a new hierarchical approach for merging digital terrain models, combining the monitoring of global geometric discrepancies with local scale matching and merging to avoid overlooking localized topographic trends, which can often occur in conventional height averaging approaches. Chapters 14 and 15 offer some model-based solutions to problems associated with sampling bias and resolutions commonly encountered within the context of environmental data. In Chapter 14, de las Heras and Lake use a combination of linear, spatial autocorrelation and geographically weighted regression (GWR) models to explore tree-cover change in Brazil, identifying the benefits of GWR in coping with data complexities such as spatial variation, spatial autocorrelation, and heteroskedasticity, but without the fragmentation of the data set required by other approaches. An integrated approach is taken by Guo et al. in Chapter 15, combining the gray differential equation model GM(1,1) with ordinary kriging to predict soil dioxin patterns. The authors suggest that such an approach can be used to derive meaningful patterns from spatial data sets that are poorly sampled and characterized by spatial covariance.

1.6 VISUALIZING THE NATURAL ENVIRONMENT

Recent technological advances have strengthened capabilities for using GIS to create 2-D and 3-D visualizations of the natural environment, and widened the access to such tools to the general public. This volume's final section explores the development of such technologies within the domains of environmental research, recreation, and land management; providing comparisons of different techniques and identifying challenges associated with interpretation and understanding, quality, and privacy.

In Chapter 16, Cartwright provides a keynote paper that explores some of the challenges for the visualization community in providing tools for the interrogation, analysis, and visualization of geographical data associated with the increasing availability of vast quantities of information. The second keynote paper by Sui (Chapter 17) reviews the development of wiki cartography and implications for the visualization community, discussing the development of key enabling technologies and the resulting increase in Map 2.0 products used predominantly for recreational purposes. Some concerns, such as the deprofessionalization of mapping tasks, data standards, and quality assurance and privacy, are also explored.

In Chapter 18, Lovett et al. review the state of the art in GIS-based 3-D visualization of rural environments, using illustrations from recent research into landscape change and energy crops. The chapter notes that significant advancements in the field have led to improved levels of detail in the photorealism of both still and real-time visualizations, but that some significant differences between visualization systems exist. The authors suggest that future challenges will relate to interactivity (see also Chapters 21 and 22) and the representation of uncertainty and nonvisual phenomena in landscape visualizations. In Chapter 19, Lieske and Bender provide an illustration of how GIS-based visualizations can be used to improve understanding of spatial trends in the natural environment. Geostatistical techniques are applied to visualize breeding distributions of bird species, highlighting the patchy spatial trends commonly observed throughout the natural environment, and the importance of applying the correct analytical technique for the scale of interest (see also Chapter 11).

In Chapter 20, Jardine and Mackaness employ a risk model to assist the explicit visualization of risk for hill walkers, as an alternative to leaving individuals to interpret risk from standard map information. While the risk map necessarily contains more information, which can often hinder interpretation, with some training, participants were able to cope with the additional information and findings suggesting that the risk map did influence their choice of route. Bleisch and Dykes then examine a Web-based approach to planning hikes through 3-D visualizations in Chapter 21, combining the use of photoreaslistic 3-D maps with abstract information, finding that the 3-D maps assisted hikers in gaining an overview of the hike area, but the abstract information was important for extracting detailed information from the maps. The development of GIS-based visualizations for non-experts is also explored by Rider and Reitsma in Chapter 22, this time focusing on the development of a visualization tool for pasture management that incorporates a biophysical growth model, an agent-based sheep model, and a physical model of management options. The authors describe the development and testing of the model and make suggestions for further work, including explorations of usability among farmers.

1.7 SUMMARY

This introductory chapter describes the key concepts around which the volume is focused, and introduces the chapters in the rest of the volume. The use of the term *natural environment* is described along with the positioning of the volume's chapters along a spectrum of environmental naturalness. Some of the key technical, theoretical, and social issues associated with the application of GIS technologies to the exploration of the natural environment are identified, and a brief introduction to the concepts of representation, modeling, and visualization within GIS is offered. The final three sections of this chapter introduce the research chapters that make up each of the volume's main three sections, highlighting the key GIS, GIScience, and environmental issues addressed and emphasizing the linkages between them. We hope readers find the volume an interesting and instructive insight into the range of opportunities for applying GIS to the study of the natural environment.

REFERENCES

1. Niemelaè, J., Ecology and urban planning, *Biodiversity and Conservation*, 8, 119, 1999.
2. United Nations Population Fund (UNFPA), State of World Population 2007: Unleashing the Potential of Urban Growth, 2007, http://www.unfpa.org/swp/2007/presskit/pdf/sowp2007_eng.pdf
3. Skidmore, A., *Environmental Modelling with GIS and Remote Sensing*, Taylor & Francis, Boca Raton, FL, 2002.
4. Kanevski, M., *Analysis and Modelling of Spatial Environment Data*, Marcel Dekker, New York, 2004.
5. Dikau, R. and Saurer, H., *GIS for Earth Surface Systems: Analysis and Modelling of the Natural Environment*, Gebruder Borntraeger, Berlin, 1999.
6. Joy, M. K. and Death, R. G., Predictive modeling and spatial mapping of freshwater fish and decapod assemblages using GIS and neural networks, *Freshwater Biology*, 49, 1036, 2004.
7. Romero-Calcerradda, R. and Luque, S., Habitat quality assessment using weights-of-evidence based GIS modeling: The case of the Picoides tridactylus as a species indicator of the biodiversity value of the Finnish forest, *Ecological Modelling*, 196, 62, 2006.
8. Eatherall, A., Modelling climate change impacts on ecosystems using linked models and a GIS, *Climatic Change*, 35, 17, 1997.
9. Clarke, M. E., Rose, K. A., Levine, D. A., and Hargrove, W. W., Predicting climate change effects on Appalachian trout: Combining GIS and individual-based modeling, *Ecological Applications*, 11, 161, 2001.
10. Gogu, R. C., Dietrich, V. J., Jenny, B., Schwandner, F. M., and Hurni, L., A geo-spatial data management system for potentially active volcanoes—GEOWARN project, *Computers and Geosciences*, 32, 29, 2006.
11. Sui, D. Z. and Goodchild, M. F., Are GIS becoming new media? *International Journal of Geographical Information Science*, 15, 387, 2001.
12. Patterson, T. C., Google Earth as a (not just) geography education tool, *Journal of Geography*, 106, 145, 2007.
13. Lovett, A. and Appleton, K., *Innovations in GIS 12: GIS for Environmental Decision Making*, Taylor & Francis, Boca Raton, FL, 2007.
14. Kingston, R., Carver, S., Evans, A., and Turton, I., Web-based public participation geographical information systems: An aid to local environmental decision-making, *Computers, Environment and Urban Systems*, 24, 109, 2000.

15. Multimap, www.multimap.com, accessed April 2008.
16. Kang-tsung, C., *Introduction to Geographic Information Systems*, McGraw-Hill, New York, 2003.
17. Longley, P. A., Goodchild, M. F., Maguire, D. J., and Rhind, D. W., *Geographic Information Systems and Science*, John Wiley & Sons, Chichester, UK, 2005.
18. MacEachren, A. M. and Kraak, M.-J., Research challenges in geovisualization, *Cartography and Geographic Information Science*, 28, 3, 2001.

Section 1

Representing the Natural Environment

2 Keynote Paper
Representation of the Natural Environment

Matt Duckham

CONTENTS

OVERVIEW

Different representations of the natural environment form the foundation of any spatial information system for natural resource management. However, the development and selection of appropriate representations of the natural environment present an array of difficulties to the researcher or system designer. Many of these difficulties relate to classical problems in geographic information science (GIScience). In the context of the natural environment, these problems can be classified into three categories. The first of these categories is context and meaning, where the meaning of terms used to describe the natural environment may be ill-defined, or vary between specific information communities. The second category is uncertainty in information, in particular to imprecision and inaccuracy. Any measurement of the natural environment is subject to imperfection, which must be taken into account when developing different representations. The third category is dynamism, since the natural environment is constantly changing, presenting substantial problems to any attempt to represent it. Effective representation of dynamism must go beyond mere snapshot models of change over time, to explicit modeling of processes and events. This chapter explores these three key issues, using as an example the automated fusion of heterogeneous land cover data sets.

2.1 INTRODUCTION

Different representations of the natural environment are at the foundation of any spatial information system for natural resource management. However, the development and selection of appropriate representations of the natural environment present an array of difficulties to the researcher or system designer. Many of these difficulties relate to classical problems in geographic information science (GIScience), but in the context of the natural environment lead to their own specific challenges.

A typical example of a classical problem in GIScience that presents particular problems to the representation of the natural environment is *vagueness*. Vagueness concerns boundary indeterminacy. A vague concept has no clearly defined boundary [1]. For example, the spatial concept "southern England" is vague since there exist some locations that are definitely in "southern England" (like Southampton), some that are definitely not in "southern England" (like Manchester), and others for which it is indeterminate whether they are in "southern England" or not (like, say, Bristol).

Vague spatial concepts and relations are especially prevalent in connection with the natural environment because the infinite and intergrading variety found in the natural environment defies crisp delineation. Geomorphological features, for example, are usually vague in that is it not possible to draw crisp lines between those locations that are part of the feature and those that are not. Basic geomorphological terms like "mountain," "valley," or "estuary" are all vague. Similarly, terms connected with land cover and land use are also typically vague, exemplified by "shrubland," "wetland," or "forested upland."

An important idea in studying vagueness is that the indeterminacy found in everyday vague spatial concepts cannot be solved simply by more precise definitions, nor is it a result of unscientific or "sloppy" thinking. Vagueness is an unavoidable feature of human cognition, and persists even if we try to refine our definitions to eliminate boundary indeterminacy. However, the more precise a definition of a vague concept becomes, the less meaningful it becomes and further away it gets from human natural language. For example, it is possible to conceive of a precise definition of "forest," such that all boundary indeterminacy is eliminated (cf. Bennett [2]). However, such a definition becomes very long and complex, requiring so many special cases and fine distinctions that having written it, it becomes practically useless, requiring an inordinate amount of time and effort to apply to real environments. More important, it arguably bears little or no relation to the concepts labeled "forest" that are actually used in different information communities such as by ecologists, foresters, and recreational walkers.

Vagueness is but one difficulty facing the representation of the natural environment. In this chapter these problems are classified into three categories, which together encompass the primary problems of representation of the natural environment. The first of these categories is *context and meaning*. Vagueness is an example problem in the category, being related to the meaning of terms used to describe the natural environment, but several other related problems exist. The second category is *uncertainty in information*, since any measurement of the natural environment is subject to imperfection, and in particular to imprecision and inaccuracy. The third category is *dynamism*, since the natural environment is constantly changing, presenting substantial problems to any attempt to represent it.

2.2 CONTEXT AND MEANING

Given the rich variety of both features and uses of the natural environment, it is no surprise that different information communities have highly heterogeneous terms, descriptions, and interpretations for the natural environment. Achieving integration between these different perspectives requires representations of the natural environment that allow for the different contexts and meanings of key terms. However, as we have already seen, the terms used are often vague and defy the formulation of precise descriptions.

Later in this book, the chapter by Hemsley-Flint et al. (Chapter 4) addresses exactly this issue by examining the development of formal specifications (ontologies) for the terms used by an information community for data about the natural environment. The task of developing such ontologies is formidable and presents a range of hurdles, some of which Chapter 4 focuses on overcoming. However, once developed, such formal specifications provide a framework for representation of the natural environment that enables the subsequent integration of heterogeneous data sets. These ontologies avoid the trap of simply providing precise definitions of vague concepts (as discussed earlier, inevitably leading to vacuous or useless results) by focusing instead on the relationships between different vague terms. For example, while it may not be possible or useful to precisely define terms like "river" or "sea," an ontology that represents the (spatial) relationships between these terms (e.g., that "a river flows into a sea") may still help to provide an underlying framework for integrating heterogeneous data sets, which in turn may use different or more specialized terms.

Instead of attempting to formally define the relationships between terms in an ontology, an alternative approach is to attempt to automatically infer these relationships based on evidence contained within existing data sets. This is the approach adopted by Duckham and Worboys [3,4], with the aim of automated fusion of heterogeneous geographic data sets about the natural environment. Figure 2.1 contains a simplified example of this automated geographic information fusion process. In Figure 2.1, each data set is represented as an extensional component (the mapped spatial data) and an intensional component (the ontology for the categories represented in that spatial data). On the left-hand side of Figure 2.1, the intension for data set A contains the categories "Forest" and "Built-up area," while the extension contains two regions, one of each category. Similarly, on the right-hand side of

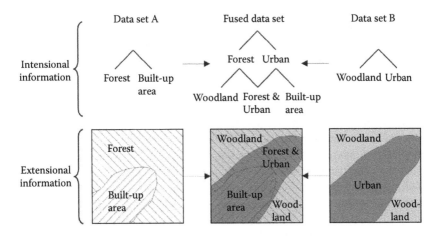

FIGURE 2.1 Simplified example of the automated geographic information fusion process.

Figure 2.1, data set B contains the intensions "Woodland" and "Urban" along with a map of the spatial extensions of the "Woodland" and "Urban" categories. Thus, Figure 2.1 might represent the situation where two different environmental agencies have both mapped the same geographic region using different ontologies.

The classic ontology-based approach to fusing such data sets would be to focus solely on the intensional component and to develop formal descriptions of how the terms represented in those intensions relate to one another. Ideally, if successful, this will result in formal descriptions than enable an information fusion system to infer that "Woodland" is a special case (subtype) of "Forest," and "Built-up area" is a special case (subtype) of "Urban" (as shown in the fused intensional component in the center of Figure 2.1).

However, Duckham and Worboys [3,4] show how similar results can be obtained automatically by using the extensional components of the data sets, ignored by conventional ontology-based approaches, in concert with simple machine learning techniques. Referring back to Figure 2.1, because all locations that are classified as "Built-up area" in data set A are classified as "Urban" in data set B, it is possible to infer automatically that the category "Built-up area" is a subclass of the category "Urban." Similarly, because all locations that are classified as "Woodland" in data set B are classified as "Forest" in data set A, "Forest" subsumes "Woodland" in the integrated ontology. At the core of the approach is a process of inferring semantic relationships from spatial relationships. This process is an example of inductive inference: reasoning from specific cases to general rules. If all instances of a category in one data set are spatially included in all instances of a different category in another data set, we may wish to infer that the first category is a special case (subsumed by) the second category. Crucially, this inference process is easily automated, because it does not rely on an understanding of the meaning of the different categories, merely on an analysis of the relationships between their spatial extents.

Similarly, in Figure 2.1 a new class, "Forest & Urban," can be created to represent those regions that are classified as "Forest" in data set A and "Urban" in data set B. In other words, although there exists no subsumption relationship between "Forest" and "Urban," we have inferred that these categories overlap, on the grounds that their extensions overlap. Although the example in Figure 2.1 is highly simplified, it is important to highlight that the process illustrated is more than a simple overlay. The data sets have been fused, in the sense that we have gained (a small amount of) new information about the (subsumption) relationships between the categories represented in each of the input data sets.

The advantages of such an approach, when compared with conventional ontology-driven approaches to information fusion (cf. Hemsley-Flint et al., Chapter 4, this volume), are primarily simplicity and automation. Taking advantage of abundant extensional information to drive an automated inference process alleviates the need for complex processes for capturing and storing domain knowledge and human expertise as formal ontologies. However, the approach is not without disadvantages, such as the need to move beyond simple taxonomies (hierarchies of concept connected by subsumption relationships, as used in the intensional components of the data sets in Figure 2.1) in developing richer representations of the natural environment. To date, extensions of this approach to automatic inference of nontaxonomic relationships are yet to be explored.

2.3 UNCERTAINTY

Uncertainty is an endemic feature of all geographic information. Vagueness is closely connected with uncertainty, but is primarily concerned with uncertainty in concepts (intensions). Different forms of uncertainty affect the measurements and observations (extensions) in data, primarily inaccuracy and imprecision.

Inaccuracy concerns a lack of correctness in information. For example, classifying a region as natural vegetation when it is in actuality cropland (cf. Fritz et al., Chapter 6, this volume) is an example of inaccuracy. Conversely, imprecision concerns a lack of detail in information. Imprecision is closely related to granularity, the existence of grains or clumps in a data set. Spatial granularity is an unavoidable feature of spatial information, since there must always be a limited "resolution" (in this context, meaning the "ability to discern fine detail") for any spatial data set. Granularity and imprecision are classical topics in GIScience since the level of granularity in spatial information fundamentally affects the features that are apparent. For example, in the specific case of representing the natural environment, Fisher et al. [5] provide examples of how landscapes can exhibit different fundamental morphological characteristics at the same location at different granularities (e.g., a location that exhibits channel features at fine granularity, may become part of a ridge at coarser granularities).

The chapters in this volume by McClean, Fritz et al., and Blair et al. (Chapters 5, 6, and 7, respectively) address different aspects of dealing with uncertainty in the representation of the natural environment. Blair et al. investigate inaccuracy in land cover classification by comparing the different spatial extents of regions classified as "wilderness" using different common definitions of the term. Fritz et al. use a similar methodology to investigate the accuracy of classification of agricultural land in Africa, with the addition that the data sets compared are at different granularities. McClean presents an approach to the disaggregation of administrative-level land use data to scales that are more appropriate to biodiversity studies. An interesting feature of these approaches is that none have access to the "true" spatial distribution of the phenomena they are investigating. Instead, all must compare and contrast related data about the natural environment, and make inferences about the likely characteristics of the true distribution from any discrepancies in observed data.

Returning to the example of information fusion, inaccuracy and imprecision in data are important considerations in developing automated approaches to geographic information fusion. In automated geographic information fusion, inaccuracy degrades the reliability of the inductive inference process, potentially leading to ontological relationships being inferred between classes that are, in reality, unrelated. If, in our example land cover data set B, part of the "Urban" region has been misclassified as "Woodland" such that it overlaps the "Built-up area" in data set A, then this will lead to the incorrect inference that "Woodland" and "Built-up area" are semantically overlapping (Figure 2.2). Note that the inaccuracy has produced a fused ontology that is not particularly informative, in the sense that we have gained no new information about the relationships between the categories in the input data sets (i.e., we could have achieved the same results using a simple overlay of the two data sets).

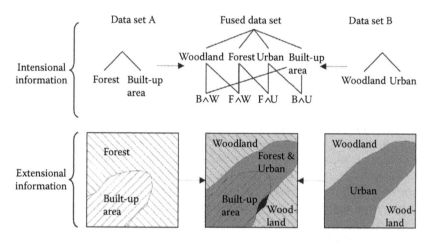

FIGURE 2.2   Inaccuracy in input data sets (black region indicates sliver polygon).

Like inaccuracy, heterogeneous levels of granularity degrade the reliability of the automated information fusion process. For example, imagine that land cover data set A has been collected at a coarser level of spatial granularity than data set B. Then it will be likely that the detailed features found in data set B will simply not be represented in data set A (such as small pockets of "Woodland" within the predominately "Urban" area that are represented in data set B, but have no correspondent in data set A). As a result, a naïve automated fusion process may again incorrectly infer that "Woodland" and "Built-up area" are semantically overlapping, as in Figure 2.3 (similar to the effects of inaccuracy in Figure 2.2). As for inaccuracy, the fusion product in Figure 2.3 is not particularly informative, as it is essentially a simple overlay of the data.

However, the spatial structure of geographic information does allow for inferences between information sources at different levels of detail (e.g., Duckham et al. [6]). Duckham and Worboys [3,4] explore various techniques for addressing the challenges presented to automated geographic information fusion by inaccuracy and imprecision, the simplest of which is to use the spatial characteristics of these imperfections to eliminate them from the inference process.

2.4 DYNAMISM

The natural environment is highly dynamic, subject to constant change. Reflecting this dynamism is a key challenge in representations of the natural environment and has long been a focus of research in GIScience more generally. Traditional GIS-based representations of the geographic environment are purely static. The limitations of such static representations have led to a range of different approaches to incorporating dynamism. Worboys and Duckham [7] identify three main classes of spatiotemporal models. Stage 1 models are collections of time-stamped static snapshots of geographic environments. Stage 1 models are limited in that they do not represent change explicitly, only implicitly as a product of the differences between two

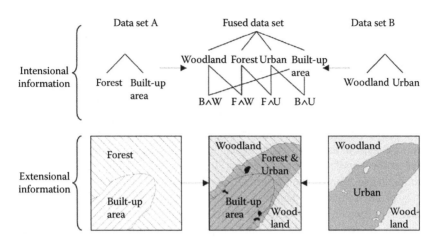

FIGURE 2.3 Granularity in input data sets (black regions in fused spatial data shows fine-grained pockets of Woodland).

snapshots at different times. Stage 2 models address this limitation by cataloging the sequence of changes that occur to objects in the geographic environment as object "lifelines." More recent research has focused on stage 3 models, which go further than stage 2 models by providing explicit representations of events and processes and their interactions (e.g., Galton [8], Grenon and Smith [9], and Worboys [10]).

Incorporating dynamism into representations of the natural environment constitutes one of the major ongoing challenges in GIScience. It is perhaps not surprising that none of the following chapters in this first book part are focused directly on the representation of change in the natural environment (although Fritz et al., Chapter 6, grapple indirectly with this issue having data sets captured at different times). Similarly, the work on automated geographic information fusion summarized earlier has yet to be extended to account for dynamism in spatial information. However, work on such an extension is currently in progress. The need for such extensions is particularly pertinent when considering emerging technologies, such as geosensor networks.

A geosensor network is a wireless network of miniaturized, sensor-enabled computers used to monitor geographic environments [11]. Geosensor networks offer the capability to capture spatial and temporal information about the environment at a much finer spatial and temporal granularity than has previously been possible. The information generated by geosensor networks complements more traditional data sets, such as remotely sensed data, or land cover mapping. These traditional mapping sources typically cover much wider spatial extents than geosensor networks can hope to cover (at least in the short- to medium-term future) and can maintain higher levels of accuracy and reliability, but lack the ability to monitor fine-grained dynamic changes over hours, minutes, or even seconds.

The complementary characteristics of these data sets (fine-grained, dynamic data from emerging geosensor networks, and course-grained, but reliable information over wide extents from more traditional sources) make the development of new techniques for integrating these heterogeneous data sources a new challenge for automated information fusion research. The goal is to be able to update traditional spatial data sources using highly dynamic information from geosensor networks. By taking advantage of the best characteristics of both information sources, the resulting data sets should ideally be more reliable than geosensor network data, and provide more information about dynamic changes in the environment than traditional data sources.

In order to be able to begin to apply automated fusion techniques, it is vital to be able to distinguish between discrepancies due to inaccuracy and imprecision, and those due to actual changes in the environment. The problem is related to classical problems of update and revision in spatial databases: determining whether an apparent change reflects a change in the world (requiring an update), or simply a change in our information about the world (requiring a revision) [7]. Conceptually, an important tool in distinguishing update from revision is through the development of stage 3, process-oriented models of environmental change. Changes that match an expected environmental process over time are likely to reflect actual changes in the world, requiring updates. Conversely, those changes that do not fit expected environmental processes are more likely to reflect improvements in our knowledge about the world, requiring revisions.

2.5 SUMMARY

Emerging new technologies are fueling a renewed need for rich, flexible representations of the natural environment. At the same time, the problems faced by researchers developing and applying those representations relate directly to classical problems in GIScience: context and meaning of geographic terms; uncertainty in geographic data; and dynamism in geographic phenomena. As a result, geographic information scientists engaged in developing new representations of the natural environment need to pay close attention to the existing research in our literature. As a maturing research topic, GIScience contains a substantial body of knowledge, techniques, and tools that have direct application to the problems posed by new technologies.

Awareness of the importance of our natural environment and the irreplaceable resources it provides is finally becoming ubiquitous in society, when in the past it was restricted to special interest groups and activities. GIScience has a key role to play in contributing to the development of new ideas and technologies that support sustainable management of the natural environment. However, only by building on the existing body of expertise in the field can GIScience hope to realize its potential wider contribution to society in this domain.

REFERENCES

1. Keefe, R. and Smith, P. *Vagueness: A Reader*, MIT Press, Cambridge, MA, 1997.
2. Bennett, B., What is a forest? On the vagueness of certain geographic concepts, *Topoi*, 20(2), 189, 2001.
3. Duckham, M. and Worboys, M. F., An algebraic approach to automated information fusion, *International Journal of Geographical Information Science*, 19(5), 537, 2005.
4. Duckham, M. and Worboys, M., Automated geographic information fusion and ontology alignment, in *Spatial Data on the Web: Modelling and Management,* Belussi, A., Catania, B., Clementini, E., and Ferrari, E., Eds., Springer, Berlin, 2007, chap. 6.
5. Fisher, P., Wood, J., and Cheng, T., Where is Helvellyn? Fuzziness of multiscale landscape morphometry, *Transactions of the Institute of British Geographers*, 29(1), 106, 2004.

6. Duckham, M., Lingham, J., Mason, K., and Worboys, M., Qualitative reasoning about consistency in geographic information, *Information Sciences*, 176(6), 601, 2006.

7. Worboys, M. F. and Duckham, M., *GIS: A Computing Perspective,* 2nd ed., CRC Press, Boca Raton, FL, 2004.

8. Galton, A., *Qualitative Spatial Change,* Oxford University Press, Oxford, UK, 2000.

9. Grenon, P. and Smith, B., SNAP and SPAN: Towards dynamic spatial ontology, *Spatial Cognition and Computation*, 4, 69, 2004.

10. Worboys, M., Event-oriented approaches to geographic phenomena, *International Journal of Geographical Information Science*, 19, 1, 2005.

11. Nittel, S., Stefanidis, A., Cruz, I., Egenhofer, M., Goldin, D., Howard, A., Labrinidis, A., Madden, S., Voisard, A., and Worboys, M., Report from the first workshop on geo sensor networks, *ACM SIGMOD Record*, 33(1), 2004.

Representing Surfaces in the Natural Environment— Implications for Research and Geographical Education

Nigel Waters

CONTENTS

OVERVIEW

Representing surfaces is fundamental to research concerning the natural environment. The state of the art as addressed in current textbook literature is discussed. This is followed by an historical account that describes the development of the three primary ways of representing surfaces, namely, as triangulated irregular networks, as digital elevation models, and as interpolated, contoured surfaces. New representations of surfaces as global spheres are described along with their role in the teaching of grade-school geography. Current research from the neurosciences on the ways in which individuals encode spatial data are recounted for the first time in the geographical literature and the implications of this research for learning to think spatially are discussed. The chapter concludes with an exploration of the future of surface representations.

3.1 INTRODUCTION

The natural environment may be characterized in various ways. Researchers need to represent the surface of the earth, the environment that exists above it, and the vegetation and wildlife that populate the environment. This chapter addresses how researchers represent the surface of the earth and why this is important for studying and learning geography.

Surface representation, in its most basic form the elevation of the land or an ice surface or indeed some abstracted climate variable, is fundamental to much subsequent research. For example, by knowing the height of points on the earth's surface the researcher can determine critical points; assess the complexity of the landscape; build surface representations; infer aspect, slope, vegetation, microclimate; and, if they are a habitat modeler, determine prey and predator distributions and build habitat models to gain an understanding of landscape processes [1] (see Chapter 7, this volume). It is a chain of logic that has a certain elegance.

The second half of this chapter will consider how surface representations can enhance the learning of geography. A short section on how human and physical geographic approaches to surface representation have cross-fertilized will precede the final section, which anticipates the future. Throughout the chapter it will be important to realize how closely representation is linked to visualization, and how both representation and visualization inform those approaches used in modeling the natural environment. Occasionally, discussions in the literature have made these links explicit (see the remarks by Mitas and Mitasova [2] on the role of interpolation in modeling). Even the words digital elevation model (DEM) and digital terrain model (DTM) conflate a surface representation procedure with a model and, by implication, with a method for visualization. Weibul and Heller [3] argue that the modeling process includes surface representation (generation and manipulation), visualization, and application (analysis). Yuan et al. [4] suggest that representation may occur at three levels—data models, formalization, and visualization—and so again the concepts of representation, modeling, and visualization within a GIS are hopelessly entangled between common English usage and language that has been adopted by computer and mathematical scientists.

3.2 REPRESENTING THE SURFACE OF THE EARTH

3.2.1 THE EXISTING LITERATURE: A BRIEF AND CRITICAL REVIEW

Characterization of the earth's surface has been researched extensively. Pity the poor student of digital terrain models, for the literature has exploded in recent years. Five volumes are of note [5–9]. Li et al. [7] opine that, finally, the more than decade-long absence of a textbook on terrain modeling following the publication of Petrie and Kennie's text [10] has been comprehensively resolved. Unfortunately, their observation on the lack of literature in the years following 1996 ignores the publication of the first edition of David Maune's *DEM Users Manual* [11], a benchmark reference work that is widely used in the industry, and Wilson and Gallant's [12] volume, *Terrain Analysis*.

The relative contemporaneity of the more recent texts has meant that they do not reference one another and so our putative student of terrain modeling, on reading one of these tomes, may remain unaware of the others. El-Sheimy et al. [6] aggravate this problem of inadequate referencing by citing texts that have been superseded by new editions published fifteen or more years later. Relying on references to Davis' first edition [13] for models that summarize the features of a surface or, similarly, Clark [14] for kriging explanations to characterize and represent the covariance structure of a DEM is unfortunate. Davis has provided a greatly expanded third edition of his classic text [15], and Clark similarly has published a comprehensively lengthened version of her seminal work along with extensive online resources, software, and data sets [16,17].

Maune [9] offers the most exhaustive review of the literature on representing the surface of the earth through the use of DEMs. The second edition of his widely cited resource opens with an introduction to 3-D surface representations. Confusion over terminology among DEMs, DTMs, and DSMs (digital surface models) is resolved, although it is noted that in many countries these terms are used interchangeably and this is largely the approach that Maune himself adopts and is essentially the attitude employed by Weibul and Heller [3] in their review. This acronym soup is further complicated by the use of the acronym DTED (digital terrain elevation data) by the U.S. National Geospatial-Intelligence Agency (NGA). The NGA uses the DTED acronym for the data collected in 2000 by the Shuttle Radar Topography Mission (SRTM), which obtained the most complete, relatively high resolution (90 m), and near global coverage of the earth's elevation data to date [18].

3.2.2 Representing Surfaces: The Three Primary Options

Three surface representation methods will be considered in detail here: the triangulated irregular network, the DEM, and isolines. However, it must be noted that there are other possibilities, including the use of voxels for fully three-dimensional graphics, and also representations by LiDAR (light detection and ranging) point clouds. Voxel-based approaches are of particular interest to geologists, atmospheric scientists, and oceanographers, but the awareness of voxels extends beyond research on the physical environment. Medical scientists, among others, use these representations extensively. Software packages that are designed to represent volumetric data, such as ScienceGL, are often marketed primarily for medical applications, though GIS applications are also prominent [19]. The management of extremely large LiDAR point clouds for surface representation is discussed by Cothren [20], and software for handling this type of surface representation is available (e.g., Ref. 21).

3.2.2.1 The Triangulated Irregular Network

Surface elevation may be represented in various ways but most commonly either as a grid, as a triangulated irregular network (TIN), or as contours. The TIN method is credited to Peucker (now Poiker; see Peucker [22] and the discussion in Mark [23]). Mark [23] attributes the original idea for representing the surface as a set of triangles to Bengtsson and Nordbeck [24]. According to Mark, it was Peucker's contribution to develop this as a topologically integrated data structure. However, Bengtsson and Nordbeck did consider topology because, using hardware with the limited capacities of the time, they faced storage problems and had to "connect-up" isarithms stored separately [24]. Thus the concepts of topology and connectivity were explicitly integrated into their software system. Surprisingly, Mark, in his historical review of the development of the TIN approach to surface representation, fails to mention Warntz's [25] conceptual breakthrough in deriving the critical points of a surface that came a few years earlier than Peucker's primary contributions. Even more interesting is the fact that Warntz extended this conceptualization to a "surface" that lacked any physical representation but had applications in economic geography as well as physical geography (specifically climatology) [26].

The merits of the TIN versus the grid-based approaches have been examined by Mark [27] and by Kumler [28]. Wang and Lo [29] have reviewed the earlier work and have conducted their own experiments. They concluded, in conflict with Kumler, that TINs are superior in terms of the accuracy of their surface representation, but these differentials decreased as the number of sample points increased. Wang and Lo's results were further qualified by their understandable admission that the results apply only to the software employed (Arc/Info 6.0) and might vary depending on which software and algorithms were used. For their experiments, the algorithms involved were proprietary to ESRI [30] and therefore could not be described in detail.

More recently, Smith and Mark [31] have addressed the question of whether mountains exist at all, arguing that geographers, geographic information scientists, and the public at large use both object- and field-oriented views of the world in determining what is and what is not a mountain. The critical points approach would appear to favor an object-oriented view of the world. However, as Warntz and Waters [26] demonstrated, a surface could be represented either as a set of peaks (mountains) or pits (depressions), but the latter would scarcely capture the imagination. They are not the "quintessential geographic things" to which Smith and Mark [31] so eloquently refer. Smith and Mark argue for a field-based ontology (see Chapter 4, this volume, for a discussion of ontologies from the perspective of the domain expert), and yet it is ironic that a true systems approach to geomorphology would indeed adopt an object-based approach, a "stocks and flows" representation as implemented in system dynamics models. The inability to accommodate these two approaches to representing the natural environment is perhaps why the system dynamics view of the world has never been fully and successfully spatialized (for a somewhat incomplete attempt, see "Appendix I—Spatial Dynamics" in Ford [32]). Development of a comprehensive ontology for geoscientific data is provided by Brodaric and Gahegan [33].

3.2.2.2 Digital Elevation Models

DEMs are gridded surfaces that reflect the limitations of the interpolation methods used to create them, and the spatial distribution of sampling points. Such interpolation methods include inverse distance weighted (IDW) algorithms [15], natural neighbor interpolation (NNI) [34], kriging [16], and splines, among others.

Mitas and Mitasova [2], in an authoritative review of interpolation methods, categorize the various surface representation techniques as follows: the local neighborhood approach, where existing points influence the surface up to a given distance (e.g., IDW, NNI, and TIN-based algorithms to produce smooth surfaces for the flat faces of the triangles); the geostatistical approach (e.g., kriging in its various forms); and the variational approach that requires that the surface honor the data points and should, in addition, be as smooth as possible (splines epitomize this method).

Each technique may have certain features that make it particularly attractive. Natural neighbors take advantage of any increased sampling in areas of high variability. Geostatistical approaches utilize the covariance structures of the data and allow for the examination of the strength and range of these properties, together with variations in sampling that are captured by the nugget effect, directional biases, and covariation with associated attributes (namely, cokriging). Algorithms available in commercial GIS software, such as the Geostatistical Analyst in ArcGIS, and in the Idrisi and Surfer software packages for variogram modeling, tend to be complementary. As such, procedures and variogram estimation strategies available in one package may not be included in the others and vice versa. That kriging is being subjected to innovative new approaches for surface interpolation is amply demonstrated by Goovaerts [35] for area-to-point estimates.

Geostatistical approaches are also being used to represent uncertainty in the data. Gotts [36] provides a contemporary review of sequential Gaussian simulation, a technique that permits an assessment of how the uncertainty in the data varies spatially. A tutorial on this intriguing method of spatial exploratory data analysis is available from the γ Statios Web site [37], and downloadable public domain software at the γ gslib Web site [38]. Figure 3.1 from Gotts displays porosity values for one possible realization of the top surface of a single facies, mature, petroleum reservoir.

The United States Geological Survey's (USGS) recently established Center of Excellence in Geographic Information Science (CEGIS) has launched various research projects, including one focused on the use of fractal and variogram analysis to determine the effects of scale and resolution on data integration for the National Map and National Spatial Data Infrastructure [39].

Mitas and Mitasova [2] note that many interpolation methods are application specific, and that for any given application, the interpolation process may be specially modified. Of particular interest are interpolation methods for data on the sphere where the interpolation functions are dependent on angle rather than distance. Representation of surfaces on the sphere is considered in Section 3.3.

3.2.2.3 Interpolation to a Contour or Isoline

The automatic generation of isarithms or isolines was of concern to Bengtsson and Nordbeck [24] (see Section 3.2.2.1), and remains a method of characterizing the

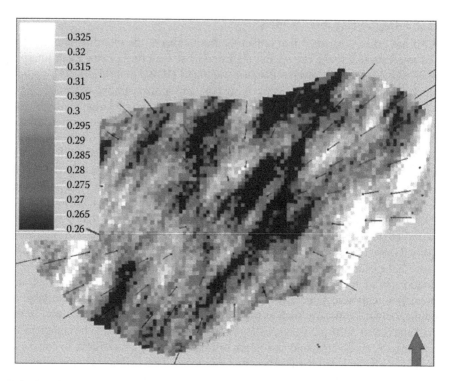

FIGURE 3.1 Percentage porosity values for a sequential Gaussian simulation in a single facies, mature, petroleum reservoir (Source: Gotts [36]; used with permission.)

surface that is dominated by local interpolation approaches [40]. The converse problem of producing a digital representation from a set of contours is of equal interest and has produced various solutions that again favor local interpolation [41].

Dahlberg [42] has compared computer-based contouring algorithms and hand-drawn solutions for the hydrocarbon exploration industry, and has shown that the computer algorithms behave analogously to a conservative geologist resisting the urge to play a hunch. Wren [43], by contrast, makes a plea for the objectivity of the computer algorithm. Mulugeta [44], apparently unaware of Wren's study, agrees with Dahlberg, citing the improved appearance of hand-drawn maps that emphasize regional patterns while admitting that the computer-generated surfaces had an accuracy that "equals or surpasses that of manually drawn maps" [44]. He concludes by advocating some combination of the manual editing process with an automated contouring algorithm. The potential for approaching the contouring problem through the use of expert systems has been addressed by Maslyn [45], Waters [46], and Dutton-Marion [47]. It is to be hoped that there will remain a role for expert interpretation of model-based output. Controversial new estimates of Antarctic ice mass loss [48], while touted as improvements that exploit new sensor technology for ice surface estimation not available when estimates were made in the past [49,50], might well benefit from expert intervention into the modeling process.

Carrara et al. [51] review the literature evaluating procedures for generating digital elevation data (DED) from contour lines, and then conduct their own experiments
on four data sets using five evaluation criteria: DED for spots close to the original contour lines should vary by less than 5%; DED falling between two contour
lines should have values falling within this range; DED values should vary linearly
between contour lines; DED in areas of low relief should reflect this morphology;
and DED defining unrealistic morphological features (artifacts) should represent less
than 0.2% of the data. In the future, the authors suggest that new procedures for generating extremely high resolution DEMs will include the use of softcopy photogrammetric methods, and this is indeed the approach used by Delparte [52] to produce a
high resolution, 5 m DEM to represent the terrain in Rogers Pass, Glacier National
Park, British Columbia, for the modeling of avalanche runout paths. Interestingly,
this was produced using expert, manual identification of the critical points to improve
the accuracy of the final result (see Mulugeta's conclusions [44], Molander [53], and
more recently McGlone [54] for a discussion of the strengths and weaknesses of
such automated methods). Nevertheless, and as Delparte notes, the future belongs to
LiDAR; however, at present these data are not widely available and their acquisition
for specific applications on an ad hoc basis is expensive [55]. Flood [56] argues for an
integration of photogrammetry, including existing hardware and personnel skill sets
and expertise, with LiDAR imaging in his article on lidargrammetry.

3.3 REPRESENTING SURFACES ON A SPHERE

3.3.1 TOBLER'S CALL TO ARMS

Tobler [57] has commented with dismay on how most of the early GIS packages were
designed merely to represent small parts of the earth's surface without representing
the earth's curvature. Considering that a version of his paper was originally presented
in 1992, his comments on representing GIS data on a sphere were prescient. Even
more discerning was Lukatela's earlier development of the Hipparchus GIS [58] that
remains the only GIS conceptualized from the outset so as to represent the surface of
the earth as a sphere. Tobler boldly asserts that not only should the representation of
the earth change, but that we should develop a truly geographic analysis system that
abandons the notion that the earth is flat. That researchers in other disciplines, such
as operations research, have long had solutions for location problems on a sphere
[59]—indeed their own geographic analysis systems—is somewhat embarrassing.

In an era of globalization, it is ironic that Friedman's [60] book, *The World Is
Flat*, has become a best seller. Despite its popularity, the book has been widely criti-
cized [61] and runs counter to current concerns over globalization and planetary
processes such as global warming. These concerns, editorials such as Tobler's, and
the development of new software and hardware technologies have produced innovative representations of the surface of the globe. The most obvious examples are
Google Earth and Microsoft's Virtual Earth programs. New interest in these technologies and the world as a single entity has led to the development of the *Digital
Earth* journal, published for the first time in January 2008 by Taylor & Francis and
the International Society for Digital Earth [62]. Grossner et al. [63] have argued that

a useful representation of a digital earth is one of the grand challenges for the GIS community to address in the immediate future.

3.3.2 VIRTUAL GLOBE REPRESENTATIONS IN SCIENTIFIC RESEARCH

Virtual globe sessions have been organized at leading scientific conferences such as the American Geophysical Union for some time [64], and a list of those conferences that occurred in 2006, 2007, and early 2008 is available [65]. Representing the surface of the spherical earth has proved extremely useful for climatologic studies, and the American Meteorological Society organized a special session on this topic at its annual meeting in New Orleans in January 2008 [66]. One paper presented at this special session showed how radar beam propagation could be affected by terrain (beam occultation) and other variables (e.g., wind power generators) leading to flawed predictions of precipitation events [67]. Images representing these problems may be downloaded from the Wx Analyst Web site [68] (also see Figure 3.2).

The implementation of hardware solutions has resulted in the National Oceanic and Atmospheric Administration's (NOAA's) Science-on-a-Sphere educational program [69]. This sphere is essentially a room-sized globe that uses computers and video projectors to display planetary data to enhance the understanding of terrestrial processes. It remains to be seen whether representing the physical environment in this fashion has any additional pedagogic value. This topic is discussed next. Intuitively this would seem to be the case, but studies similar to those conducted for various approaches to the teaching of GIS [70] should be implemented as soon as

FIGURE 3.2 Three-dimensional occultation pattern overlaid with radar reflectivity on January 4, 2008. (Source: Shipley et al. [64]; used with permission.)

possible to determine if the investment in these kinds of technologies has an acceptable cost–benefit ratio.

3.3.3 Digital Earth in the Schools

Spatial representations such as Google Earth have been advocated for teaching geography in schools. Specifically, Patterson [71] has used Google Earth in seventh-grade classroom exercises. He suggests that the work of Solem and Gershmel [72] provides evidence that online resources increase students' comprehension of skills and concepts while also increasing their knowledge of geographic issues. Among Google Earth's advantages, Patterson cites first its entertainment value, a quality advocated by Greenspan [73]. Second he notes the ability to use this freeware at any location, including the child's home, as long as there is a computer with an Internet connection. Patterson lists the third advantage as the support of an online community that may be accessed at the Google Earth Community Web site [74] (see also the Google Earth Projects Web site [75]), a resource that includes KMZ files for illustrating aspects of the geography of the environment. Such assets allow the worldwide community of teachers to build resources in a collaborative manner. They can volunteer information at their community level that can then be shared globally. Volunteered geographic information is a new way to represent local knowledge [76]. There are numerous Web sites that provide online support for the neophyte using Google Earth. One of the more popular may be found at the Google Maps Mania Web site [77], a site that provides links to online tutorials to assist students in creating their own Google Earth content and representations. Finally, the ability to represent features, such as the Grand Canyon, from a variety of geographical perspectives can aid students' comprehension of the physical characteristics and processes that created these landforms.

According to Patterson [71], the main disadvantage of Google Earth and, presumably closely related technologies such as Microsoft's Virtual Earth and ESRI's ArcExplorer [78], is their inability to carry out basic GIS operations that permit spatial analysis. Perhaps they might be conceived as some form of "minimal GIS" that Marsh et al. [79] have recently advocated, although one suspects that those authors are unlikely to be satisfied by this technology regardless of its attraction for students. Patterson concludes his discussion of the usefulness of Google Earth to represent the world and to teach students about geography with a demand for scientifically designed studies to determine its effectiveness in the educational process. If the arguments of Lynch et al. [80] are to be believed, the development of an effective practice that integrates representations of the earth into a true e-learning environment is likely to be a complex process.

Traditional teaching of geography has relied on two-dimensional representations of the earth's surface. GeoWall attempts to move beyond this method of representing geographical features and surfaces by using stereo images that allow for the projection of three-dimensional representations [81]. Other new ways to represent the natural environment include computer-assisted virtual environments (CAVE). Although much of this research has been associated with the reconstruction of buildings, archaeological features, and urban areas, new work is extending these approaches into reconstructions of the physical environment [82].

3.4 LEARNING TO UNDERSTAND SPATIAL REPRESENTATIONS

3.4.1 PRINCIPLES OF SURFACE REPRESENTATION

Morse [83] provided two guiding principles for the comprehension and subsequent analysis of computer-generated data: proportional effect and least effort. Proportional effect, the first principle, requires that the size and identity, namely, the relevant attributes of the data, have to be encoded. Common approaches are to do this using position (e.g., location on a map; a pseudo representation of a third dimension such as height), length, size, angle, color, brightness, texture, time (in animation), or symbols. The various ways of portraying surfaces and spatial data are commonly covered in cartography texts [84], by graphic design specialists such as Tufte [85–87], or on Web sites such as that maintained by Cindy Brewer, a cartographic professor at Penn State University [88].

The second of Morse's [83] principles, that of least effort, requiring ease of perception and interpretation, includes optimal scaling (i.e., allowing meaningful distinctions without requiring unnecessary detail), display integration, and minimization of stimulus load (although too little stimulus may be as undesirable as too much). Finally, conceptual and task compatibility must also conform to the viewer's expectations and this may vary with the experience of the viewer.

The National Research Council's report "Learning to Think Spatially" states that "spatial representations are powerful tools that can enhance learning and thinking" [89]. The report argues that this is achieved, first, because spatial representations are a powerful way to encode new information. Second, the report states that generating images of existing information allows for a greater degree of recall. Third, spatial representations are claimed to enhance problem solving in some but not all instances. Each of these claims will be considered in turn.

3.4.2 LEARNING AND ENCODING NEW DATA FROM SPATIAL REPRESENTATIONS

There is evidence to suggest that representing new data, both spatial and nonspatial, as a map is an effective way to encode the data [90,91], and GIS researchers have long asserted that spatial is special [92,93]. Current research into the brain suggests that it is even more special than the authors of the National Research Council report might have suspected. Research at the Medical Research Council (MRC) Centre for Synaptic Plasticity at the University of Bristol has focused on "how, where and why the brain modifies synaptic strength during normal function" [94]. This work includes research into place cells, neurons in the hippocampus that fire when the subject is in a particular location [95]. The neuroscience research completed to date is even more intriguing and, so far, largely ignored by geographers.

Muller [96] states that, in a seminal paper, O'Keefe and Dostrovsky [97] discovered place cells. The latter two researchers showed the importance of place cells by demonstrating that they will fire whenever a rat returns to a familiar location known as a place field. The fact that place cell activity appears goal oriented may have important implications for habitat modeling. Directional bias also occurs in linearly constrained environments. Existing research has largely been conducted in lab environments involving animals such as rats and cats, and in virtual environments for humans [98]. It remains to be seen whether this research translates to natural environments, how it varies for wildlife (for example, do wildlife corridors generate the same directional biases and how does the acquisition of spatial information in the dark occur and differ among species), and how persistent are the spatial representations. For the habitat modeler, it is intriguing to note that there are several representations of the animal's environment: (a) the real world, (b) the representation in the internal hippocampus and other parts of the brain, and (c) the researcher's attempt to replicate what is important in the habitat and in such devices as resource selection functions. The extent to which these are one and the same is important but moot.

Besides place cells, neuroscientists have identified head direction cells [96] that fire when an animal (usually a rat) looks in a certain direction, spatial view cells for primates (monkeys) observing objects in an environment [99], and grid cells where the neuron firing patterns have a distinct topographical structure with strong spatial autocorrelation properties [100]. Again this research raises questions for spatial representations of the natural environment. First, to what extent does it translate to human observations of the environment? Second, what spatial representations are most suited to learning about the environment? Perhaps more specifically, is the best representation for learning a large wall that can be viewed in three dimensions (such as the GeoWall discussed earlier), a physical sphere, or a three-dimensional virtual environment? It might be a good idea to perform the research before schools, universities, and museums invest large sums of money on one or another of these strategies.

Burgess et al. [98] reviewed neuroscience research into spatial memory, compared studies across species and within species, and introduced new research involving virtual reality representations of a town. They described and contrasted the egocentric and allocentric spatial frameworks. In the former, the framework moves with the observer and locations of objects are not fixed; but in the latter, the locations of objects do not change as the observer moves through the environment. Memory tests associated with two-dimensional landscape scenes, for example, do not discriminate between the two frameworks, but a virtual reality environment allows for an allometric spatial framework to be examined. Indeed Arthur et al. [101] concluded, following a set of experiments where subjects were asked to reproduce the spatial layout of a virtual environment, that "interaction with a virtual environment was indistinguishable from interaction with real objects at least within the constraints of the present procedure." Burgess et al. and Arthur et al. reported gender differences. Both reported superior performances from males in terms of the accuracy of spatial reconstructions, and Burgess et al. also discuss strategy differences between males and females, with males making use of geometric and landmark information to aid learning and recall, whereas females relied more exclusively on landmarks. Inexplicably these results are contradicted in the work of some geographers [102].

3.4.3 IMPROVING RECALL THROUGH SPATIAL REPRESENTATIONS

Returning to the discussion in the National Research Council Report "Learning to Think Spatially" [89], the authors of that report argue for the use of spatial representations as an aid to memory. The argument is that by teaching children to learn to think spatially, they will then be able to use spatial representations as an aid to memory. Others have argued in a similar vein [103], but it is likely that a spatial representation only aids the memory when it is a highly familiar environment and it is possibly little better than a mnemonic. Indeed it may work the other way around; a mnemonic may aid recall and retention of a spatial arrangement.

3.4.4 SPATIAL REPRESENTATIONS AND PROBLEM SOLVING

When problem solving it is often, though not always, better to represent the problem spatially, or so argues the report "Learning to Think Spatially" [89]. A virtual representation of a dangerous environment such as avalanche terrain (see Delparte [52], and Figure 3.3) can provide recreational, backcountry skiers with an understanding of the terrain that they are proposing to enter. This is even more effective if it is merged or mashed up in a Google Earth setting and combined with an effective warning system that is based on current weather conditions [52].

3.4.5 CROSS-FERTILIZATION BETWEEN THE GEOGRAPHY OF THE HUMAN AND NATURAL ENVIRONMENTS

Human geographers and geoinformation scientists at large have considered many of the problems that have been addressed here. Thus the problem of occultation noted by Shipley et al. [67] with respect to wave beam propagation has been analyzed by ReMartinez [104] when considering the reach of radio stations in the mountainous terrain of Peru (Figure 3.4). Such models were required to determine the influence of radio stations on the indigenous electorate and involved the integration of wave propagation models, DEMs, and socioeconomic data showing population distributions and languages spoken [105].

The rich literature in human geography on how place matters, and such classic texts that describe the geography of the city in terms of landmarks and visual cues

FIGURE 3.3 Avalanche exposure map. (Source: Delparte [52]; used with permission.)

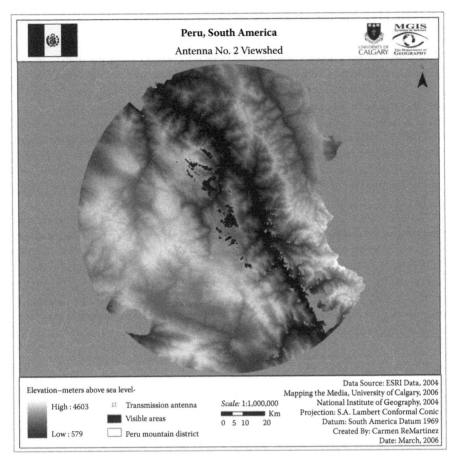

Peru, South America
Antenna No. 2 Viewshed

Elevation–meters above sea level-

High : 4603   ⊠ Transmission antenna

     Visible areas

Low : 579   Peru mountain district

Scale: 1:1,000,000
Km
0 5 10   20

Data Source: ESRI Data, 2004
Mapping the Media, University of Calgary, 2006
National Institute of Geography, 2004
Projection: S.A. Lambert Conformal Conic
Datum: South America Datum 1969
Created By: Carmen ReMartinez
Date: March, 2006

FIGURE 3.4   Represention of a Radio Antenna Viewshed in the Peruvian Andes. (Source: ReMartinez [104], used with permission).

[106], validated by work in neuroscience (see Section 3.4.2), also authenticates and extends research that seeks to produce an understanding of the natural environment. Virtual environments do generate synaptic responses that assist in an understanding of the real world, endorsing the research of all those who seek to provide more interesting and accurate representations of natural environments.

GIS specialists have long been interested in human cognition of the spatial environment. Early summaries of this literature have been provided in both versions of the core curriculum for GIS developed at the National Center for Geographic Information and Analysis (in the original version by Suchi Gopal, "Spatial Cognition," and in the revised version by Daniel Montello, "Human Cognition of the Spatial World" [107]). This is a large and growing literature in GIS but it has yet to be linked effectively with that in neuroscience discussed earlier. Perhaps the best hope for such an integration lies in the work of Barkowsky [108] and his MIRAGE model that seeks to reconstruct mental images of space based on topological properties, orientations, and shapes.

3.5 THE FUTURE OF SURFACE REPRESENTATIONS

Goodchild [93] notes that in existing DEMs "time is ignored because elevation is assumed to be a static property." Many applications require that temporal changes in the natural environment be easily represented to aid visualization, analysis, and understanding. The theoretical basis of including time in GIS is now beginning to be understood [109], and temporal animations are now facilitated in new versions of commercial GIS software such as ArcGIS 9.3 [110]. ArcGIS 9.3 will include an image server making it still easier to incorporate new data and surface representations, but understanding how the data were produced and the implications of the error incorporated into every surface portrayal will remain paramount, and *caveat emptor* will still be critical for the user whether a teacher or researcher.

The research agenda for the portrayal of geographic data in the future may be discerned in the updates to McMaster and Usery [111] (see, for example, Buckley et al. [112]); in the newly published *Handbook of Geographic Information Science* [113]; in state-of-the-art developments in cognate disciplines; in specialized software that can be integrated with GIS packages; and in the priorities set by such agencies as the National Geospatial-Intelligence Agency [114]. It is hoped that this chapter has covered the primary issues that have been and will be of concern in surface representation during the coming years.

ACKNOWLEDGMENTS

I would like to acknowledge Jeff Gotts, Scott Shipley, Donna Delparte, and Carmen ReMartinez for permission to reproduce figures from their unpublished work.

REFERENCES

1. Alexander, S. M., Waters, N. M., and Paquet, P. C., A Probability-based GIS Model for Identifying Focal species Linkage Zones across Highways in the Canadian Rocky Mountains, in *Applied GIS and Spatial Analysis*, Stillwell, J. and Clarke, G., Eds., Wiley, Chichester, 2004, chap. 13, 233.
2. Mitas, L. and Mitasova, H., Spatial Interpolation, in *Geographical Information Systems: Principles and Technical Issues*, Vol. 1, Longley, P. A., Goodchild, M. F., Maguire, D. J., and Rhind, D. W., Eds., Wiley, New York, 1999, chap. 34.
3. Weibul, R. and Heller, M., Digital Terrain Modelling, in *Geographical Information Systems: Principles and Applications*, Vol. 2, Goodchild, M. F., Maguire, D. J. and Rhind, D. W., Eds., Wiley, New York, 1991.
4. Yuan, M., Mark, D. M., Egenhofer, M. J., and Peuquet, D. J., Extensions to Geographic Representations, in *A Research Agenda for Geographic Information Science: Section E—Nutritional Disorders*, McMaster R. B. and Usery, E. L., CRC Press, Boca Raton, FL, 2005, chap. 5.
5. Rana, S., *Topological Data Structures for Surfaces*, Wiley, Chichester, 2004.
6. El-Sheimy, N., Valeo, C., and Habib, A., *Digital Terrain Modeling*, Artech House, Boston, 2005.
7. Li, Z., Zhu, Q., and Gold, C., *Digital Terrain Modeling: Principles and Methodology*, CRC Press, Boca Raton, FL, 2005.

8. Peckham, R. J. and Gyoso, R., *Digital Terrain Modelling: Development and Applications in a Policy Support Environment,* Lecture Notes in Geoinformation and Cartography, Springer, New York, 2007.
9. Maune, D. F. (Ed.), *Digital Elevation Model Technologies and Applications: The DEM Users Manual* (2nd ed.), The American Society for Photogrammetry and Remote Sensing, Maryland, 2007.
10. Petrie, G. and Kennie, T. (Eds.), *Terrain Modelling in Survey and Civil Engineering,* Whittles Publishing, Caithness, Scotland, 1990.
11. Maune, D. F. (Ed.), *Digital Elevation Model Technologies and Applications: The DEM Users Manual* (1st ed.), The American Society for Photogrammetry and Remote Sensing, Bethesda, MD, 2001.
12. Wilson, J. P. and Gallant, J. G., *Terrain Analysis: Principles and Applications,* Wiley, New York, 2000.
13. Davis, J. C., *Statistics and Data Analysis in Geology* (2nd ed.), Wiley, New York, 1986.
14. Clark, I., *Practical Geostatistics,* Elsevier Applied Science, New York, 1979, available at: http://www.kriging.com/PG1979/PG1979_pdf.html
15. Davis, J. C., *Statistics and Data Analysis in Geology* (3rd ed.), Wiley, New York, 2002.
16. Clark, I., *Practical Geostatistics 2000,* Ecosse North American, Ohio, 2000.
17. What is Kriging?, http://www.kriging.com/, accessed April 2008.
18. Shuttle Radar Topography Mission, http://www2.jpl.nasa.gov/srtm/, accessed April 2008.
19. Science GL, http://www.sciencegl.com/, accessed April 2008.
20. Cothren, J. *Managing Very Large LIDAR Point Clouds in an Enterprise Database,* GIS for Local Government Conference, Penn State University, October 2005.
21. Q Coherent Software, http://www.qcoherent.com/, accessed April 2008.
22. Peucker, T. K., *Computer Cartography,* Resource Paper 17: Commission on College Geography, Association of American Geographers, Washington DC, 1972.
23. Mark, D., The history of geographic information systems: Invention and re-invention of triangulated irregular networks (TINs), *Proceedings of GIS/LIS '97,* ACSM/ASPRS, Falls Church, VA, October 1997, Available at: http://www.ncgia.buffalo.edu/gishist/GISLIS97.html
24. Bengtsson, B.-E. and Nordbeck, S., Construction of Isarithms and Isarithmic Maps by Computer, *BIT Numerical Mathematics,* 4, 87, 1964.
25. Warntz, W., The Topology of a Socioeconomic Terrain and Spatial Flows, *Papers of the Regional Science Association,* 17, 47, 1966.
26. Warntz, W. and Waters, N. M., Network Representations of Critical Elements of Pressure Surfaces, *Geographical Review,* 65, 476, 1975.
27. Mark, D., Computer Analysis of Topography: A Comparison of Terrain Storage Methods, *Geografiska Annaler A,* 57, 179, 1975.
28. Kumler, M. P., An Intensive Comparison of Triangulated Irregular Networks (TINs) and Digital Elevation Models (DEMs), Monograph 45, *Cartographica,* 31(2), 1, 1994.
29. Wang, K. and Lo, C.-P., An Assessment of the Accuracy of Triangulated Irregular Networks (TINs) and lattices in ARC/INFO, *Transactions in GIS,* 3, 161, 1999.
30. ESRI, www.esri.com, accessed April 2008.
31. Smith, B. and Mark, D. M., Do Mountains Exist? Towards an Ontology of Landforms, *Environment and Planning B,* 30, 411, 2003.
32. Ford, A. T., *Modeling the Environment: An Introduction to System Dynamics Modeling of Environmental Systems,* Island Press, Washington DC, 1999.
33. Brodaric, B. and Gahegan, M., Representing Geoscientific Knowledge in cyberinfrastructure: Some Challenges, Approaches and Implementations, in *Geoinformatics: Data to Knowledge,* Sinha, A. K., Ed., Geological Society of America, Special Paper, 397, 1, 2006.

34. Gold, C. M., Surface Interpolation, Spatial Adjacency and GIS, in *Three Dimensional Applications in GIS,* Raper, J., Ed., Taylor & Francis, London, 1989, 21.

35. Goovaerts, P., Kriging and Semivariogram Deconvolution in the Presence of Irregular Geographic Units, *Mathematical Geosciences*, 40, 101, 2008.

36. Gotts, J. W., *Geostatistical Modelling of Porosity in a Single Facies Sandstone Reservoir,* unpublished MGIS Project, Department of Geography, University of Calgary, Alberta, Canada, 2007.

37. Gaussian Simulation for Porosity Modeling, γ Statios, http://www.statios.com/Resources/08-sgsim.pdf, accessed April 2008.

38. γ gslib, http://www.gslib.com/, accessed April 2008.

39. Center of Excellence for Geospatial Information Science (CEGIS), http://cegis.usgs.gov/projects.html#fractal_and_variogram, accessed April 2008.

40. Watson, D. F., *Contouring: A Guide to the Display and Analysis of Spatial Data,* Pergamon, New York, 1992.

41. Auerbach, S. and Schaeben, H., Surface Representations Reproducing Given Digitized Contour Lines, *Mathematical Geology*, 22, 723, 1990.

42. Dahlberg, E. C., Relative Effectiveness of Geologists and Computers in Mapping Potential Hydrocarbon Exploration Targets, *Mathematical Geology*, 7, 373, 1975.

43. Wren, A. E., Contouring and the Contour Map: A New Perspective, *Geophysical Prospecting*, 23, 1, 1975.

44. Mulugeta, G., Manual and Automated Interpolation of Climatic and Geomorphic Statistical Surfaces: An Evaluation, *Annals of the Association of American Geographers*, 86, 324, 1996.

45. Maslyn, R., Gridding Advisor: An Expert System for Selecting Gridding Algorithms, *Geobyte*, 2, 42, 1987.

46. Waters, N. M., Expert Systems and Systems of Experts, in *Geographical Systems and Systems of Geography: Essays in Honor of William Warntz*, Coffey, W. J., Ed., Department of Geography, University of Western Ontario, London, Ontario, 1988, chap 12.

47. Dutton-Marion, K.E., *Principles of Interpolation Procedures in the Display and Analysis of Spatial Data: A Comparative Analysis of Conceptual and Computer Contouring,* unpublished PhD thesis, Department of Geography, University of Calgary, Calgary, Alberta, 1988.

48. Rignot, E., Bamber, J. L., van den Broeke, M. R., Davis, C. Li, Y., van de Berg, J., and van Meijgaard, E., Recent Antarctic Ice Mass Loss from Radar Interferometry and Regional Climate Modelling, *Nature Geoscience*, 1, 106, 2008.

49. Giovinetto, M. B., Waters, N. M. and Bentley, C., Dependence of Antarctic Surface Mass Balance on Temperature, Elevation, and Distance to Open Ocean, *Journal of Geophysical Research–Atmospheres*, 95(D4), 3517, 1990.

50. Giovinetto, M. B. and Zwally, J., Spatial Distribution of Net Surface Accumulation on the Antarctic Ice Sheet, *Annals of Glaciology*, 31, 171, 2000.

51. Carrara, A., Bitelli, G., and Carla, R., Comparison of Techniques for Generating Digital Terrain Models from Contour Lines, *International Journal of Geographical Information Science*, 11, 451, 1997.

52. Delparte, D., *Avalanche Terrain Modeling in Glacier National Park, Canada,* unpublished PhD thesis, Department of Geography, University of Calgary, Alberta, Canada, 2007.

53. Molander, C. W., Photogrammetry, in *Digital Elevation Model Technologies and Applications: The DEM Users Manual* (1st ed.), Maune, D. F., Ed., The American Society for Photogrammetry and Remote Sensing, Maryland 2001, 121.

54. McGlone, J. C., Photogrammetry, in *Digital Elevation Model Technologies and Applications: The DEM Users Manual* (2nd ed.), Maune, D. F., Ed., The American Society for Photogrammetry and Remote Sensing, Maryland, 2007, 119.

55. Fowler, R. A., Samberg, A., Flood, M. J., and Greaves, T. J., Topographic and Terrestrial LiDAR, in *Digital Elevation Model Technologies and Applications: The DEM Users Manual* (2nd ed.), Maune, D. F., Ed., The American Society for Photogrammetry and Remote Sensing, Maryland, 2007, 199.

56. Flood, M., LiDARgrammetry, *Geoworld*, 19 (2), 2006, available at: www.geoplace.com.

57. Tobler, W., Global Spatial Analysis, *Computers, Environment and Urban Systems*, 26, 493, 2002.

58. Lukatela, H., Hipparchus Geopositioning Model: An Overview, AUTO-CARTO 8, Baltimore, March 1987, available at: http://www.geodyssey.com/papers/hlauto8.html.

59. Wesolowsky, G. O., Location Problems on a Sphere, *Regional Science and Urban Economics*, 12, 495, 1982.

60. Friedman, T. L., *The World Is Flat*, Farrar, Strauss and Giroux, New York, 2005.

61. Aronica, R. and Ramdoo, M., The World Is Flat?: A Critical Analysis of New York Times Bestseller by Thomas Friedman, Meghan-Kiffer Press, Tampa, FL, 2006.

62. Digital Earth Journal, http://www.digitalearth-isde.org/, accessed April 2008.

63. Grossner, K.E., Goodchild, M. F. and Clarke, K. C., Defining a Digital Earth System, *Transactions in GIS*, 12, 145, 2008.

64. Virtual Globes at AGU, http://conferences.images.alaska.edu/agu/2006/index.htm, accessed April 2008.

65. Virtual Globes in Science, http://conferences.images.alaska.edu/, accessed April 2008.

66. Atmospheric Special Interest Group, http://www.gis.ucar.edu/sig/index.html, accessed April 2008.

67. Shipley, S. T., Steadham, R. M., and Berkowitz, D. S., Comparison of Virtual Globe Technologies for Depiction of Radar Beam Propagation Effects and Impacts, 24th IIPS Conference in New Orleans, January 2008, available at: http://ams.confex.com/ams/pdfpapers/135325.pdf

68. Wx Analyst Virtual Globe Radar Project, http://wxanalyst.com/radar/, accessed April 2008.

69. NOAA Science on a Sphere, http://sos.noaa.gov/index.html, accessed April 2008.

70. Doering, A. and Veletsianos, G., An Investigation of the Use of Real-Time, Authentic Geospatial Data in the K-12 classroom, *Journal of Geography*, 106(6), 217, 2007.

71. Patterson, T. C., Google Earth as a (Not Just) Geography Education Tool, *Journal of Geography* 106, 146, 2007.

72. Solem, M. and Gershmel, P., Online Global Geography Modules Enhance Undergraduate Learning, *Newsletter: Association of American Geographers*, 40(8), 11, 2005.

73. Greenspan, B., Mapping Play: What Cybercartographers Can Learn from Popular Culture, in *Cybercartography: Theory and Practice*, Taylor, D. R. F., Ed., Elsevier Science, New York, 2006, chap 13.

74. Google Earth Community, http://bbs.keyhole.com, accessed April 2008.

75. Google Earth Projects, http://www.mi-perm.ru/gis/earth/index.htm, accessed April 2008.

76. Goodchild, M. F., Citizens as Sensors: the World of Volunteered Geography, *GeoJournal*, 69, 211, 2007.

77. Google Maps Mania, http://googlemapsmania.blogspot.com, accessed April 2008.

78. Ball, M., Digital Reality: Comparing Geographic Exploration Systems, *Geoworld*, 18(1), 2006, available at: www.geoplace.com.

79. Marsh, M., Golledge, R., and Battersby, S. E., Geospatial Concept Understanding and Recognition in G6–College Students: A Preliminary Argument for Minimal GIS, *Annals of the Association of American Geographers*, 97, 696, 2007.

80. Lynch, K., Bednarz, B., Boxall, J., Chalmers, L., France, D., and Kesby, J., E-Learning for Geography's Teaching and Learning Spaces, *The Journal of Geography in Higher Education*, 32, 135, 2008.

81. GeoWall, www.geowall.org, accessed April 2008.
82. Levy, R. M., Virtual Dieppe, http://www.ucalgary.ca/evds/levy#VirtualDieppe, accessed April 2008.
83. Morse, A., Some Principles for the Effective Display of Data, in Proceedings of the 6th Annual Conference on Computer Graphics and Interactive Techniques, Chicago, Illinois, August 8–10, 1979, 94.
84. Slocum, T. A., McMaster, R. B., Kessler, F. C., and Howard, H. H., *Thematic Cartography and Geographic Visualization* (2nd ed.), Prentice Hall, New Jersey, 2003.
85. Tufte, E., *Envisioning Information,* Graphics Press, Cheshire, CT, 1990.
86. Tufte, E., *The Visual Display of Quantitative Information* (2nd ed.), Graphics Press, Cheshire, CT, 2001.
87. Tufte, E., *Beautiful Evidence,* Graphics Press, Cheshire, CT, 2006.
88. ColorBrewer, http://www.personal.psu.edu/cab38/ColorBrewer/ColorBrewer.html, accessed April 2008.
89. National Research Council, Learning to Think Spatially, National Academies Press, Washington DC, 2006.
90. Waters, N., New Software Organize Information Spatially, *GEOWorld,* 12(5), 34, 1999.
91. Skupin, A. and Fabrikant, S. I. Spatialization, in *The Handbook of Geographic Information Science,* Wilson, J. P. and Fotheringham, A. S., Eds., Blackwell Publishing, Oxford, UK, 2008, chap. 4.
92. Goodchild, M. F., What's Special about Spatial?, 2002, available at: http://www.csiss.org/aboutus/presentations/files/goodchild_qmss_oct02.pdf
93. Goodchild, M. F., The Nature and Value of Geographic Information, in *Foundations of Geographic Information Science,* Duckham, M., Goodchild, M. F., and Worboys, M. F., Eds., Taylor & Francis, New York, 2003, available at: http://www.geog.ucsb.edu/%7Egood/papers/374.pdf
94. Medical Research Council (MRC) Centre for Synaptic Plasticity, University of Bristol, http://www.bristol.ac.uk/depts/Synaptic/research/res2.html, accessed April 2008.
95. Neural Basis of Spatial Memory, MRC Centre for Synaptic Plasticity, University of Bristol, http://www.bristol.ac.uk/depts/Synaptic/research/projects/memory/spatialmem.htm, accessed April 2008.
96. Muller, R., A Quarter of a Century of Place Cells, *Neuron,* 17, 813, 1996.
97. O'Keefe, J. and Dostrovsky, J., The Hippocampus as a Spatial Map: Preliminary Evidence from Unit Activity in the Freely-Moving Rat. *Brain Research,* 34, 171, 1971.
98. Burgess, N., Maguire E. A., and O'Keefe, J., The Human Hippocampus and Spatial and Episodic Memory, *Neuron,* 35, 625, 2002.
99. Georges-Francois, P., Rolls, E. T., and Robertson, R. G., Spatial View Cells in the Primate Hippocampus: Allocentric View Not Head Direction or Eye Position or Place, *Cerebral Cortex,* 9(3), 197, 1999.
100. Hafting, T., Fyhn, M., Molden, S., Moser, M. B., and Moser, E. I., Microstructure of a Spatial Map in the Entorhinal Cortex, *Nature,* 436, 801, 2005.
101. Arthur, E. J., Hancock, P. A., and Chrysler, S. T., The Perception of Spatial Layout in Real and Virtual Worlds, *Ergonomics,* 40, 69, 1997.
102. Monetllo, D. R., Lovelace, K. L., Golledge, R. G., and Self, C. M., Sex-Related Differences and Similarities in Geographic and Environmental Spatial Abilities, *Annals of the Association of American Geographers,* 89, 515, 1999.
103. Hale-Evans, R., *Mind Performance Hacks,* O'Reilly Media, Sebastopol, CA, 2006.
104. ReMartinez, C., *A GIS Broadcast Coverage Prediction Model for Transmitted FM Radio Waves in Peru,* unpublished MGIS Project, Department of Geography, University of Calgary, Alberta, Canada, 2006.

105. Waters, N. M., Hansen, C., Sun, H., Gao, J., and ReMartinez, C., Mapping Media Influence on the Electoral Process in Peru, Paper presented at the Annual Meeting of the ESRI Users Conference in August 2006, San Diego, CA, 2006, Available at: http://gis. esri.com/library/userconf/proc06/papers/papers/pap_1539.pdf

106. Lynch, K., *The Image of the City*, MIT Press, Cambridge, MA, 1960.

107. Gopal, S. and Montello, D., GIS and Spatial Cognition, Online Lecture Notes, University of British Columbia, Available at: http://www.geog.ubc.ca/courses/klink/gis.notes/ ncgia/u73.html, accessed April 2008.

108. Barkowsky, T., *Mental Representation and Processing of Geographic Knowledge: A Computational Approach* (Lecture Notes in Artificial Intelligence, Vol. 2541), Springer, Berlin, 2002.

109. Peuquet, D., *Representations of Space and Time*, Guilford Press, New York, 2002.

110. Dangermond, J., Opening Address, ESRI Federal User Conference, Washington DC, February 20, 2008.

111. McMaster, R. B. and Usery, E. L. (Eds.), *A Research Agenda for Geographic Information Science*, CRC Press, Boca Raton, FL, 2005.

112. Buckley, A., Gahegan, M., and Clarke, K., UCGIS Geographic Visualization Research Priorities, Revisited, 2006, Available at: http://www.ucgis.org/priorities/research/ 2006research/chapter_11_update_new.pdf.

113. Deng, Y., Wilson, J. P., and Gallant, J. C., in *The Handbook of Geographic Information Science*, Wilson, J. P. and Fotheringham, A. S., Eds., Blackwell, Malden, MA, 2008, chap. 23.

114. National Geospatial Intelligence Agency, Priorities for GeoInt Research at the National Geospatial Intelligence Agency, The National Academies Press, Washington DC, 2006, available at: http://www.nap.edu/catalog.php?record_id=11601#toc.

4 Developing Ontologies from a Domain Expert Perspective

Fiona Hemsley-Flint, Glen Hart,
John Lee, and Stewart Thompson

CONTENTS

OVERVIEW

Ontologies are increasingly being developed and used within geographic information science and other related study areas to provide a means of more effective data integration between heterogeneous data sets. Most ontology research concerns the development of fully formal, computer-readable representations. This chapter presents a novel approach that puts the domain expert at the forefront of the development process. A method is described that enables domain experts to develop semiformal ontologies. These are more structured and less ambiguous than natural language; can be transformed into a formal, computer-readable representation; and, most important, are simple to understand and create by nonontology experts. Ontologies created by domain experts are evaluated against those created by an ontology expert. The

41

results demonstrate that the proposed method can successfully be implemented by domain experts to create viable semiformal ontologies.

4.1 INTRODUCTION

There is an increasing recognition of the role ontologies play in assisting in the integration of disparate data sets, especially within the geographic domain (e.g., de Bruijn [1] and Fonseca et al. [2]). The spatial aspect of data can be an important underlying factor in data integration. For example, if two organizations want to integrate their data regarding pesticide use and water quality, the common linkage between these two data sets will be geographic. Therefore, there needs to be a common understanding of not only the data sets themselves but the geographic features that they are associated with.

An ontology provides an explicit representation of the concepts and relations of a domain of interest within a formal, structured, and well-defined vocabulary from a particular viewpoint. By representing data and knowledge this way, others can fully understand the meaning of the domain and see how it can be incorporated with their own domain/data/knowledge. One of the aims of ontologies is that they can be handled and implemented by computers to carry out complex queries over disparate data sets without the user being aware of the underlying processes, but being provided with the results they require. Although the completely automatic interaction between ontologies is unlikely to be achievable (except in very special circumstances), such querying is possible when ontologies are semiautomatically merged or prepared.

There are currently two views of what an ontology is. In hard computer science terms, an ontology is a completely formal description of a domain usually expressed in formal logic, with the majority of ontologies being developed using a form of description logic as expressed by the Web Ontology Language (OWL) [3]. Such an approach provides the necessary mathematical rigor to enable inferences to be made by machines but at the cost of the accuracy with which the domain may be modeled. In particular, there can be no uncertainties or fuzziness in definition. The other view is more philosophical, where there is a general acceptance that there will be variation in interpretation. Geography is littered with such examples: my forest may be your woods. Within the task of geographic data integration such discussion about these variances are both routine and necessary. Indeed, the need to support geographic information systems (GIS) with ontologies has long been recognized in, for example, the work of Fonseca and Egenhofer [4]. Such work recognizes the need to bridge between geographic reality on one side and the restrictions of GIS implemented using current computational technologies on the other.

This chapter introduces a new methodology that enables experts of a domain to establish semiformal ontologies that can be understood by other humans and can also be easily translated into a formal ontology that is machine-readable. Although the ultimate aim of the research is to understand how geo-ontologies may be reused by experts in other domains, such as ecology, many of the examples in this chapter are taken from early work that concentrated on species descriptions. The following two sections provide an overview of semiformal ontologies and establish domain experts as an important resource in the ontology development process. Sections 4.4

and 4.5 describe the methodology that has been developed and the results of an evaluation measuring the presence of matching terms between ontologies of the same domain produced by different domain experts. Future directions and the summary and conclusion of this chapter can be found in Sections 4.6 and 4.7.

4.2 SEMIFORMAL ONTOLOGIES

Most of the literature concerning ontology development concentrates on creating formal ontologies for machine understanding and system interoperation. As such, the ontologies are developed using a formal structure such as description logic and represented in a particular computer language (e.g., OWL [3]). Uschold and Gruninger [5] define an ontology as "a shared and common understanding of a domain that can be communicated between people and across application systems." This is an often-quoted definition, yet the communication "between people" is not usually taken into account in ontology development, and resulting ontologies are generally not comprehensible to humans. Gruber [6] proposes that a knowledge-level definition of terms within the ontology should exist independently of a specific ontology language. Blazquez et al. [7] support this view and have introduced a set of intermediate representations that specify an ontology at the knowledge level. These intermediate representations consist of a glossary of terms and a number of concept classification trees that define the hierarchy of the concepts within the ontology. Although they remove some of the dependence on a particular computer language, these techniques have some limitations. First, there is still the assumption that it will be an ontology expert that is responsible for developing the ontology. Second, the representations seem to focus on a hierarchical representation of the concepts, which may not always be relevant for every domain that needs to be modeled; there are many other types of relationships that may need to be represented through the course of ontology development.

Semiformal ontologies allow a domain to be represented in a structured and less ambiguous way than natural language, which can be understood by humans and translated into a formal computational representation as required. These types of ontologies, also referred to as the conceptual stage (analogous to the knowledge level discussed earlier), are already recognized as an important stage in ontology development by others [8,9]. As well as being understandable to humans they are also useful because they remove any computer language or system barriers that are inherent in formal ontologies. There is also the opportunity for nonontology experts to begin to understand and develop ontologies at the semiformal level, thereby removing some of the potential costs involved in embarking upon ontology development.

4.3 THE ROLE OF THE DOMAIN EXPERT

A domain expert is someone who holds a large amount of implicit knowledge about the domain of interest (i.e., the domain an ontology is being developed for). As the number and range of domains that are implementing ontologies increases, the more important it will become for the expert(s) in that domain to have input in the development process. At present, the responsibility of developing an ontology lies with the ontology expert, with little or no interaction with a domain expert. Some literature does acknowledge the role that domain experts might play in the ontology development process, although this is mainly only as part of the knowledge acquisition phase [10]. However, the domain expert can potentially have a much bigger part to play in ontology development. He or she could create a semiformal, structured ontology, which can then be translated by an ontology expert into a fully formal, machine-readable ontology as required. This approach gives more control to the domain expert and can essentially enhance the ontology by ensuring that from the outset the concepts within the ontology are relevant and correct. It also reduces the amount of time that the ontology expert will need to spend on a project, as a large component of the work will have been undertaken before he or she needs to be involved.

4.4 METHODOLOGY

4.4.1 BACKGROUND

A methodology has been developed that enables domain experts to create a semiformal ontology—one that is intermediate between the imprecise, disjoint, and often implicit representations of a domain that exist in traditional forms of documentation, and the formal, computer-readable representation of this knowledge. The output consists of three components: semistructured sentences, a semantic network, and a knowledge dictionary. Each component is an intermediate representation, which together form a semiformal ontology with structure and minimal ambiguity.

Several small, trial ontologies were created to assist in the development of the methodology. These ontologies were for different freshwater plants and animals including mayfly (order *Ephemeroptera*) and starfruit (*Damasonium alisma*). Examples from these ontologies will be used in this chapter to illustrate the different stages of the method.

4.4.2 PURPOSE AND SCOPE

The first step that must always be carried out in ontology development is the definition of the purpose and scope of the ontology. The purpose establishes the reasons for building the ontology, what uses it will be put to, and, therefore, what needs to be contained within the ontology. Defining the scope of the ontology gives clear boundaries as to what should and should not be included in the ontology. The scope will also determine the granularity of the ontology—the scale at which the ontology should be developed. For example, when describing a particular species of animal, you may only want to concentrate on a particular anatomical level, for example, to only represent the fact that the animal has brown fur, rather than going on to explicitly describe an individual hair in terms of its molecular structure. The purpose and scope of the trial ontologies were to define the characteristics of particular organisms associated with freshwater habitats to allow identification of these organisms and species records to be associated with them.

4.4.3 KNOWLEDGE GATHERING

It is likely that the domain expert will need to supplement his or her own knowledge with other sources of information, such as reference texts. This tacit knowledge will enable him or her to more fully understand and augment this information, as opposed to an ontology expert who will not know all the specialized terms and nuances related to a particular domain. For the trial ontologies, a combination of textbooks, field guides, and reliable Internet resources (e.g., relevant government organization's Web sites) were used as text resources.

4.4.4 SEMISTRUCTURED SENTENCES

The development of the three different representations is an iterative process. However, the semistructured sentences provide the initial and main step toward identifying the key concepts and relationships that need to be incorporated in the ontology. A semistructured sentence aims to remove the ambiguity associated with natural language resulting in clear, concise sentences that can still be understood by humans but are structured enough that they can be easily translated into a more formal, machine-readable representation. The ideal outcome of the semistructured sentences is that they form triples consisting of subject, object, and predicate so that the subject and object generally translate into concepts and the predicate becomes the relationship between them. Although this is not possible in every case, making the sentences fit this structure greatly helps the construction of the other components of the ontology.

To establish the semistructured sentences, words considered superfluous (i.e., not directly related to the main concept in the sentence) are removed from the text resources. Such words are mainly prepositions, pronouns, and adverbs. For example, the original sentence "Starfruit is so named because of the distinctive fruits this plant bears which appear as six-sided stars" [11], would become "Starfruit named because of fruits plant bears which appear as six-sided stars."

Because this is still not particularly concise, the sentence can be rearranged and reworded to become two sentences in the triple form: "Starfruit bears fruit" and "Fruit looks like six-sided star," where "starfruit," "fruit," and "six-sided star" would be concepts and "bears" and "looks like" the relationships.

4.4.5 SEMANTIC NETWORK

The semantic network provides a more structured and visual representation of the ontology and consists of nodes (circles) that represent the concepts with arcs (lines) between them to represent the relationships. An example section of the semantic network for starfruit is shown in Figure 4.1. The concepts and relationships are derived from the semistructured sentences, although some adjustment might be required for the sentences to more readily fit into the semantic network framework or to maintain consistency. For example, the relationship "bears" from the semistructured sentence "starfruit bears fruit" has been changed to "has part" so that it is consistent with the other concepts that are parts of the starfruit concept. In some cases, it is not always

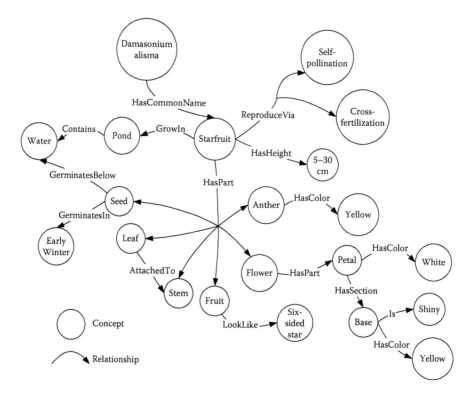

FIGURE 4.1 Subsection of the semantic network for the starfruit ontology.

possible to represent a semistructured sentence in the semantic network as it is too complex. Such cases should be noted so that the information is not lost.

4.4.6 KNOWLEDGE DICTIONARY

All the concepts and relationships included in the semantic network (and therefore the semistructured sentences) must also be contained in a knowledge dictionary. This representation provides the definitions of all the terms used in the ontology, along with any synonyms related to them. Defining all the terms in this way removes any ambiguity that might be associated with particular words and ensures any potential users of the ontology can fully understand the context in which concepts and their relationships are being used. The definitions can come from the domain expert themselves, any text sources being used (for example, some texts contain glossaries), or in some cases reference material (for example, WordNet, an online lexical resource) [12]. Table 4.1 shows a section of the knowledge dictionary for some of the concepts in the starfruit ontology. Most of the definitions are derived from WordNet, although some have been defined by the author.

Knowledge dictionaries are good for providing a baseline for agreement within a domain and, therefore, for identifying differences in terms or definitions across domains. However, they may also unwittingly introduce another level of ontological

TABLE 4.1

Example Section of the Starfruit Knowledge Dictionary: Defining the Concepts

Concept	Synonym(s)	Definition
Cross-fertilization		Fertilization by the union of male and female gametes from different individuals of the same species.
Flower	Bloom, blossom	Reproductive organ of angiosperm plants, especially one having showy or colorful parts.
Fruit		The ripened reproductive body of a seed plant.
Pond	Pool	A small, self-contained water body containing standing water.
Six-sided star		A plane figure with six radiating points.

complexity through the use of definitions obtained from other sources such as WordNet and the choices that are made from inclusion. Awareness of this additional complexity is therefore necessary lest unintended definitions are used in the mistaken belief that they fulfill the needs of the current domain.

4.5 RESULTS AND DISCUSSION

4.5.1 Developing the Method

While developing the trial ontologies used in establishing the methodology, several issues were found relating to the difficulty of representing dynamic and spatial domains and to developing some of the different representations in general. As mentioned in Section 4.4.2, defining the scope of the ontology can help determine its granularity, the level of detail that needs to be represented. However, for the mayfly ontology, it was found that while using the order level of taxonomy there were several concepts that were more difficult to represent than if the species level was used. For example, the number of molts a mayfly undergoes is determined by species, and therefore representing this for the order of mayfly introduced a degree of uncertainty. The dynamic nature of both the starfruit and mayfly also proved difficult to represent as part of the semantic network; it was not easy to show how their life cycles involve a change between one life stage and another. For starfruit, another aspect that was complicated to represent as part of the semantic network was how the surrounding environment can affect the appearance of the plant; for example, the leaves have different growth forms depending on the level of the water in which the plant is growing.

4.5.2 Evaluation of the Method

Three domain experts were asked to use the aforementioned methodology to develop ontologies for either mayfly or starfruit. Two domain experts were supplied with the text sources used previously to create the trial ontologies; the other used his own text resources. The domain experts were also expected to use their own knowledge as part of the development process. The resulting ontologies were compared with the trial ontologies developed by the current authors, which were viewed as the benchmark ontologies (that is not to say these ontologies are viewed as the ideal, but simply a basis on which the other ontologies can be compared). The number of matching terms between the benchmark and each domain expert's ontology were used to calculate precision and recall metrics that are widely used in text analysis and ontology evaluation [13,14]. The precision value (P) is the number of matching terms (T_m) as a proportion of the total number of terms in the domain expert's ontology (T_{de}):

$$P = (T_m / T_{de})$$

Recall (R) is the number of matching terms as a proportion of the total number of terms in the benchmark ontology (T_b):

$$R = (T_m / T_b)$$

Precision and recall can take values between 0 and 1. For precision, a value of 1 shows that all the terms contained in the domain expert's ontology matched with the benchmark ontology. The closer the precision value gets to 0, the fewer number of terms within the domain expert's ontology were matched with the benchmark. For the recall metric (also referred to as "coverage" by Guarino [14]), a value of 1 shows that all the terms within the benchmark were also contained within the domain expert's ontology. As the recall value moves toward 0, the fewer the number of terms within the domain expert's ontology matched the benchmark ontology. The precision and recall values are shown in Table 4.2. Domain expert 1 (DE1) produced a mayfly ontology based on the same text sources as the benchmark. Domain expert 2 (DE2) produced a mayfly ontology based on his own text sources, and domain expert 3 (DE3) produced a starfruit ontology based on the same text sources as the benchmark.

One of the most important factors in determining the precision and recall values is the number of terms contained in each ontology. Given that there are nearly three times as many terms in DE2's ontology compared with the benchmark, the best precision we could ever expect would be 0.35 (if all the terms in the benchmark could also be found in the DE2 ontology, there would be a 100% recall). As such, the precision value of 0.125 is not ideal, but is also not as poor as one might initially think, although it still shows that even with the large number of terms included in DE2's ontology, only a small proportion actually matched with the benchmark's terms. A similar, but reverse point is true for DE1, who has less than half the total number of terms in his ontology compared to the benchmark, the high precision shows that most of the terms within his ontology match with the benchmark. However, this still means that DE1 did not include a substantial proportion of terms in his ontology that are contained in the benchmark.

The variety in the quantity of terms in each ontology reflects the different interpretations of what the domain experts thought was necessary to include in the ontology, and this is likely to be found whenever different people are asked to represent the same domain. It is reassuring, however, that most of the main concepts were identified by all of the domain experts, suggesting that the overall aims of the ontology

TABLE 4.2
Precision and Recall Values for Each Domain Expert

	T_b	T_{de}	Precision	Recall
Domain expert 1	91	37	0.838	0.341
Domain expert 2	91	256	0.125	0.352
Domain expert 3	70	173	0.266	0.657

Note: T_b is the total number of terms in the benchmark ontology. T_{de} is the total number of terms in the domain expert's ontology.

had been understood correctly. DE2 and DE3 seem to have gone into too much detail when trying to develop the ontology by including many terms that are not wholly related to the definition of mayfly. It is therefore likely that the creation of a semantic network would reduce the number of terms within the ontology. The conciseness, and therefore small recall value, of DE1's ontology shows that he has not actually included all the terms that are important and therefore might have not interpreted the ontology development quite as well as could be hoped. However, all three gave some useful results and seemed to grasp the fundamentals of the semiformal ontology approach.

4.6 FUTURE DIRECTIONS

Research into ontologies has progressed beyond simply developing them on an ad hoc basis and has started to identify the potential for reusing existing ontologies instead. Previous papers regard ontology "reuse" as a way of using existing ontologies in their entirety in place of creating a new one [15], which in itself is a reasonable approach. However, this work is studying an approach perhaps more akin to "recycling," where a new ontology is definitely required (i.e., one does not currently exist for the particular domain), but there is potential to use concepts and relationships from ontologies in related domains. Future work will therefore establish whether recycling parts of existing ontologies for the development of new ontologies aids the creation and reusability of the resulting ontology. This work will form the next stage of the research described in this chapter. Figure 4.2 shows the different ontologies that will be developed and then analyzed to assess the recycling approach.

The Ordnance Survey (OS) has already developed an ontology describing hydrological topographic features [16], and this will be used as the source of concepts for an ontology of the Environment Agency's (EA) freshwater biological sampling protocol and English Nature's (EN) monitoring protocol. Another ontology for the EA will be produced independently. The three resulting ontologies will be assessed according to a number of criteria that will aim to establish whether there is any difference between the two EA sampling ontologies in terms of:

- Integration of the EA and EN data sets using their ontologies
- Integration of the EA data set with the OS data set
- Ease of development

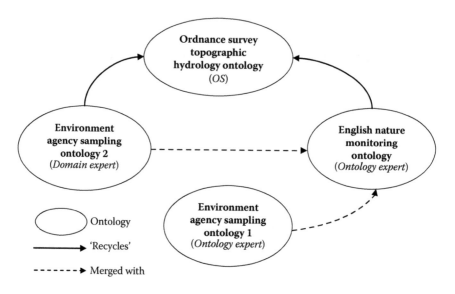

FIGURE 4.2   Schematic of ontology development for evaluating ontology recycling.

4.7 CONCLUSIONS

This chapter has described a methodology that enables domain experts to develop semiformal ontologies, thereby fully appreciating the important role that domain experts can and should have in the ontology development process. The methodology has been shown to produce successful results and the semiformal ontologies produced are structured in such a way that they can be translated into a fully formal ontology by an ontology expert as and when required. The next stage of introducing recycling to the ontology development process has great potential for enabling better cross-agency collaboration and will reduce the time required to produce ontologies, thereby encouraging the adoption of ontologies across domains where data integration is currently costly and time consuming.

ACKNOWLEDGMENTS

This work is funded by and carried out in collaboration with the Ordnance Survey of Great Britain.

REFERENCES

1. de Bruijn, J., *Using Ontologies: Enabling Knowledge Sharing and Reuse on the Semantic Web*, DERI Technical Report DERI-2003-10-29, 2003.
2. Fonseca, F. T., Egenhofer, M. J., Agouris, P., and Câmara, G., Using Ontologies for Integrated Geographic Information Systems, *Transactions in GIS*, 6, 231, 2002.
3. World Wide Web Consortium, Web Ontology Language (OWL), http://www.w3.org/2004/OWL/, 2007.

4. Fonseca, F. and Egenhofer, M., Ontology-Driven Geographic Information Systems, *7th ACM Symposium on Advances in Geographic Information Systems*, Kansas City, MO, 1999, 14.
5. Uschold, M. and Gruninger, M., Ontologies: Principles, Methods and Applications, *Knowledge Engineering Review*, 11, 93, 1996.
6. Gruber, T. R., Toward Principles for the Design of Ontologies Used for Knowledge Sharing, *International Journal of Human-Computer Studies*, 43, 907, 1993.
7. Blázquez, M., Fernandez, M., Garcia-Pinar, J. M., and Gomez-Perez, A., Building Ontologies at the Knowledge Level Using the Ontology Design Environment, in *Proceedings of the Knowledge Acquisition Workshop*, KAW98, 1998.
8. Uschold, M. and King, M., Towards a Methodology for Building Ontologies, Workshop on Basic Ontological Issues in Knowledge Sharing, held in conjunction with IJCAI-95, 1995.
9. Gómez-Pérez, A., Fernández-López, M., and de Vicente, A. J., Towards a Method to Conceptualize Domain Ontologies, *Working Notes of the Workshop on Ontological Engineering*, ECAI '96, 1996.
10. Corcho, O., Fernández-López, M., and Gómez-Pérez, A., Methodologies, Tools and Languages for Building Ontologies. Where Is Their Meeting Point? *Data & Knowledge Engineering*, 46, 41, 2003.
11. Arkive, www.arkive.org, 2007.
12. Fellbaum, C., *WordNet: An Electronic Lexical Database*, M.I.T. Press, London, 1998.
13. Brank, J., Grobelnik, M., and Mladeniæ, D., A Survey of Ontology Evaluation Techniques, presented at Conference on Data Mining and Data Warehouses (SiKDD 2005) Ljubljana, Slovenia, October 17, 2005.
14. Guarino, N., Toward a Formal Evaluation of Ontology Quality, *IEEE Intelligent Systems*, 19, 78, 2004.
15. Pinto, H. S. and Martins, J P., A Methodology for Ontology Integration, *Proceedings of the International Conference on Knowledge Capture*, 2001, 131.
16. Mizen, H., Hart G., and Dolbear C., A Two-Faced Approach to Developing a Topographic Ontology, *Proceedings of the GIS Research UK 14th Annual Conference GISRUK 2006*, University of Nottingham, 2006.

5 The Spatial Disaggregation of Great Britain and European Agricultural Land Use Statistics

Colin J. McClean

CONTENTS

OVERVIEW

Agricultural land use data are often collated on an annual basis by national and supranational agencies by sets of administrative polygons. These administrative regions are often inconvenient for the analysis of land use change in relation to other phenomena, such as biodiversity. This study introduces an approach that spatially disaggregates land use data to finer resolution output geographies. The approach first calculates models of land use activity at the original resolution, based on land cover map proportions. A simulation process follows whereby individual parcels of land use are allocated to the finer resolution geography on the basis of the models' scores. The method is applied to British agricultural census data and to European agricultural statistics. Results show very good fits for some land uses for both British and European data, despite the very simple models employed in the first stage of the approach. The spatial patterns obtained are seen to be stable after as few as 100 iterations of the simulation stage. The approach can be generalized to most input and output geographies.

5.1 INTRODUCTION

Spatially and temporally detailed land use data are an increasingly sought, yet generally lacking, base data set for environmental research [1]. Among the many sources of demand for such data is the requirement to understand how climate change is influencing global and local patterns of biodiversity. It is understood that human interactions with the distribution of biodiversity, in the form of land use change, need to be taken into account when modeling the current distribution of biodiversity and investigating future impacts on it [2–4]. Recent snapshots of land cover derived from remotely sensed data are proving highly valuable for teasing out the importance of climate versus land use change [5]. However, longer run time series are required to fully understand changes in biodiversity's spatial distribution over recent climate history. Such data are also required to determine how farmers adapt spatially to policy signals [6]. In the past, these signals have included development of international trade agreements and other external factors such as world commodity prices. Within Europe, reforms to the Common Agricultural Policy constitute one such source of policy signals. An understanding of how farmers adapt may prove invaluable in assessing how land use may change as a response to climate change.

Very good records of major agricultural land use in Great Britain (GB) have been collected since the 1860s, including areas of crops growth and livestock rearing [7]. Records were collected by an agricultural census, where all farmers gave information annually. Recently, the June Agricultural Census has become a very large survey, but the data remain of high quality with national coverage. For confidentiality reasons, individual records for farm businesses cannot be released. Results of the survey have been released using a number of different geographies over the years. At the highest spatial resolution, data have been released over time at the parish level, groups-of-parishes level, ward level, and now super output area level. All of these geographies have been developed with human activities in mind. As a result, these data cannot be integrated easily with data on biodiversity. Furthermore, changes in the spatial units over time do not allow straightforward comparison of agricultural activity across years. Other similar data sets exist for other areas of the world and suffer from similar limitations. For example, Eurostat releases statistics for agricultural land use at a number of different nomenclature of territorial units for statistics (NUTS) levels. In the case of the UK regions reported, these data come from the agricultural census [8].

Therefore, there is a need to spatially disaggregate data if they are to be compared through time using some common geography compatible with biodiversity-type data [9]. This need to change geographies has been appreciated by geographers for some time, particularly when dealing with data reported for differing administrative regions [10,11]. The modifiable areal unit problem (MAUP) has long been recognized as an important issue when analyzing spatially aggregated data reported for administrative units [11]. Put simply, the results of analyses of the same phenomenon using different spatial units or different sizes of spatial units can give substantially different results. The issue of resolution of spatial units is often referred to in nongeography-based disciplines as the ecological fallacy problem. Taking care to use data at scales and units appropriate for the study at hand is a major step toward mitigating against the MAUP. Tobler [10] introduced a form of pycnophylactic (or volume preserving) interpolation. This embraces the spatial autocorrelation still present in much aggregated spatial data; parts of neighboring spatial units close to one another are likely to be more related to one another than parts that are separated by greater distances. This form of interpolation can spatially disaggregate data from one set of aggregation boundaries, allowing reaggregation into different sets of boundaries. Such an approach has proved very useful when applied to aggregated count data such as population census data. Other work disaggregating population census data includes the kernel density approaches developed by Martin [12]. These methods all assume that sufficient spatial autocorrelation within data sets will allow distance decay effects, coupled with the variable being disaggregated, to guide interpolation.

A direct implementation of such approaches would not, however, work when disaggregating land use data, as these data consist of records of areas of multiple land uses, all of which might have some probability of being disaggregated onto the same unit of land. This is unlike population census data where only one set of counts is being disaggregated. However, previous work has disaggregated such data in the United Kingdom [13–15] and in the United States and Brazil [9,16]. All of these methods have used an additional spatial data set as a key to allow distribution of the spatially aggregated data to a finer resolution grid. In the UK studies the additional data have included some measure of land capability for agriculture. In the earliest work cited, an agricultural capability classification based on climate, soil, and terrain data was used. Agricultural land uses were allocated preferentially to the land deemed most capable of supporting that land use, in effect producing crop associations with land capability classes. Moxey et al. [14] recreated a similar agricultural capability classification from geographic information systems (GIS) data rather than from field survey. They then used an econometrics approach to link capability to land use. Given the land capability at the finer geographical resolution, the aggregated data are redistributed to land parcels that are most suitable.

Huby et al. [15] used a land cover map derived from satellite imagery as the land capability surface. Although land cover may change across time and the land cover map used is just one snapshot of land cover, it can be related to suitability across time by comparing land use data to it at the coarser geography of the land use for different time periods. Huby et al. used a computer intensive genetic algorithm to find the probabilities of each land use occurring on each land cover in a particular time period. The methods used in this chapter develop from that work, but the approach here demonstrates that a relatively simple and quick method can give results of sufficient quality for many purposes.

The work reported here attempts to disaggregate two different scales of land use data. Both are easily accessible via the Internet and other published sources: the GB agricultural census data from reports at the county level, and Eurostat regional agricultural statistics data at the NUTS 2 and 3 levels. The former are disaggregated via a 1 km grid resolution to a 5 × 5 km grid resolution, whereas the latter are disaggregated directly to a 10 × 10 km grid resolution. Data disaggregated by the methods are compared to land use data released for the same time period at higher resolution geographies to assess the success of the process.

5.2 METHODS

The spatial, aggregated land use data used in this study come from two sources. The first is for Great Britain, the data being from the annual June Agricultural Census/ Survey [7]. County- and regional-level spreadsheet data for 1950, 1980, and 2000 were downloaded for England. Similar summaries for Wales were obtained from Williams et al. [17]. Scottish county summaries were obtained as PDF files from the EDINA AgCensus project [18]. These data were amalgamated into one data set for 63 to 88 counties (depending on time period) including hectares grown of the following land use types: permanent grassland, temporary grassland, rough grazing, arable, vegetables, and oil seed. Data with which to compare results for the UK disaggregation came from Defra [7]. These were the agricultural census data at ward level for England in 2000.

The second source of data was the Eurostat Web site of the European Commission [8]. The general and regional data on agriculture were downloaded for areas of crops harvested and general land use. These give areas in thousands of hectares for 8 land use types and 19 crop types reported for a mixture of NUTS-level administrative regions. These data tables were amalgamated to give one table for 141 regions with the following nonoverlapping land use types: forest, grassland, permanent crop, durum wheat, soft wheat, barley, maize grain, maize fodder, potatoes, sugar, and rape seed. Regions where 15 or more years of records were included, resulting in a map that does not correspond to the extent of the European Commission at any one point in time. A second table was created to contain a more general classification of land use into four nonoverlapping types: forest, grassland, arable, and permanent crop. This second table was created to allow better comparison to GB land use types for model validation. Although data for multiple years have been passed through the disaggregation algorithm described, this chapter focuses on the results for just one year, 2000. This year provides a common time point between the decadal GB county data and the yearly European data. Digital administrative boundaries for the GB data were obtained from the UKBorders program Web site [19], whereas the equivalent boundaries for Europe were downloaded from the Eurostat Web site [8].

Similarly to the land use data, two sources of land cover data were used as keys in the disaggregation process for GB and Europe. For GB, the Centre for Ecology and Hydrology's Land Cover Map of Great Britain 2000 was used [20]. The 1 km^2 summaries of these data were processed to give proportions of each grid cell occupied by the following land covers: coniferous woodland, deciduous woodland, tilled land, marsh/rough grass, grass shrub heath, dwarf shrub heath, managed grass, heath/ moor grass, and urban and suburban. For Europe, the Corine land cover 2000, 250 m resolution raster data [21] were processed to give proportions of the following land cover types in 10 km^2 grid cells: urban, arable, orchard, pasture, other agricultural, forest, scrub, bare, inland wetland, and maritime wetland.

The algorithm used to disaggregate the land use data has two stages (Figure 5.1). For each individual land use recorded, a simple linear regression model of suitability is constructed. The dependent variable in each suitability model is the area of the land use found in each administrative region, and the independent variables are the proportions of land cover types found in the administrative regions. Predictions

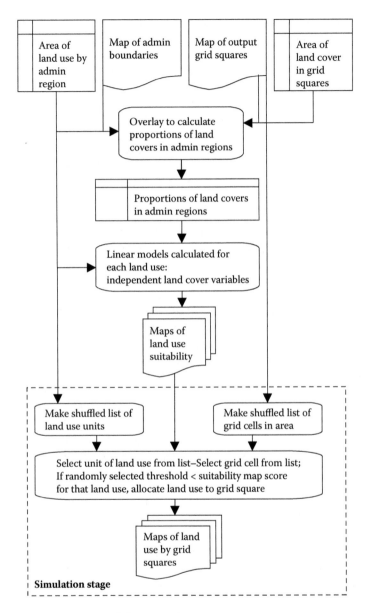

FIGURE 5.1   Flow diagram of disaggregation process.

of areas of each land use can then be made at a finer grid cell resolution using the models along with the proportions of land cover types found in the grid cells. This first stage was completed by using standard ordinary least squares regression functions in the R data analysis environment [22]. Given that these are national/regional models based on county-/NUTS-level data, it is not necessary that the estimates for each grid cell are particularly accurate. These surfaces are only used as a guide to the second stage of the disaggregation process. The second stage is a simulation process whereby randomly selected units of land use from the land use data for an administrative region are allocated to grid cells on the basis of the suitability scores estimated in the models from the first stage. This simulation stage is repeated a number of times to estimate a mean pattern of land use within grid cells.

In more detail, this program starts by reading in the areas of each of the output grid cells that overlap with input administrative boundaries. A constant check on how much land use has been allocated to grid cells that overlap two or more administrative areas can therefore be kept, thus avoiding the allocation of more land use to a cell than there is area within the cell. The suitability maps for the output grid cells for each land use, calculated in the first stage, are also read into the program. These values are standardized to an arbitrary 10-point scale for each land use.

The method proceeds by allocating land use to the output grid cells one administrative area at a time. A list of land use units of equal area (hectares when applied to the GB data, square kilometer [km²] units when applied to the European data) is created from the areas of each land use reported for the administrative unit. This list is then shuffled to ensure that random units of land use can be allocated at each step of the allocation process. A similarly shuffled random list of the grid cells that overlap the administrative unit is created. A unit of land use is then taken from the top of the list and becomes a potential piece of land use to allocate within the first cell in the grid cell list. The suitability of the grid cell for that land use is then considered. If it is above a randomly drawn number, the piece of land use is allocated to the grid cell, the area available in the grid cell is decremented, and the process moves on to the next piece of land use in the list. If the grid cell is deemed unsuitable, the next grid cell in the list is considered in the same way until a cell is found to be suitable for the land use being allocated. The algorithm continues until all pieces of land use in the list have been allocated. Everything is reshuffled, and the allocation procedure is repeated for a set number of iterations. Average amounts of land use in each grid cell are calculated at the end of the process by dividing through by the number of iterations.

When run on the GB data, the output grid was initially set at 1 km² resolution and final results are reported for a 5 × 5 km grid covering GB (after summing the values in constituent kilometer squares). The European data where disaggregated to a 10 × 10 km grid covering Europe.

To test how well the disaggregation procedure was performing, the results of the GB disaggregation for England were compared to the ward-level agricultural census data for 2000. The results for the European disaggregation for parts of England were compared back to the county-level agricultural census data. Unfortunately, even for the classification of land use into four types, only two land use types were directly comparable between GB and Europe. Only the results for arable and grassland were compared.

The algorithms were run using several different numbers of iterations, ranging from 1 to 1000, for the GB data, and results of the disaggregation for these different numbers of iterations are also considered.

5.3 RESULTS

Example results showing agricultural land use practices across GB are given in Figure 5.2, and examples of the European disaggregation are given in Figure 5.3.

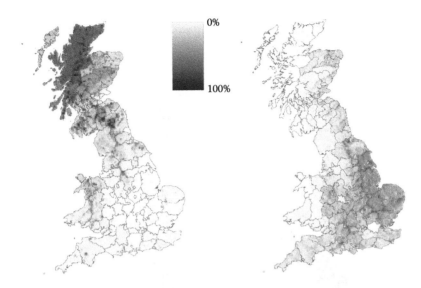

FIGURE 5.2 Maps showing 5 km × 5 km spatial disaggregations of county-level GB agricultural census data for rough grazing (left) and arable (right) land use types for 2000.

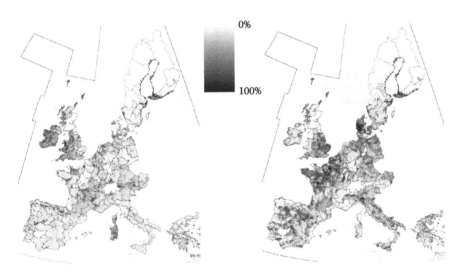

FIGURE 5.3 Maps showing 10 km × 10 km spatial disaggregations of Eurostat agricultural statistics data for rough grazing (left) and arable (right) land use types for 2000.

When the results are assessed in detail, some land uses are better predicted than others. Comparisons of the six land uses modeled for 2000 in GB with the equivalent land uses reported at the ward level are given in the form of scatter plots of observed versus predicted values in Figure 5.4. R^2 values associated with regressions

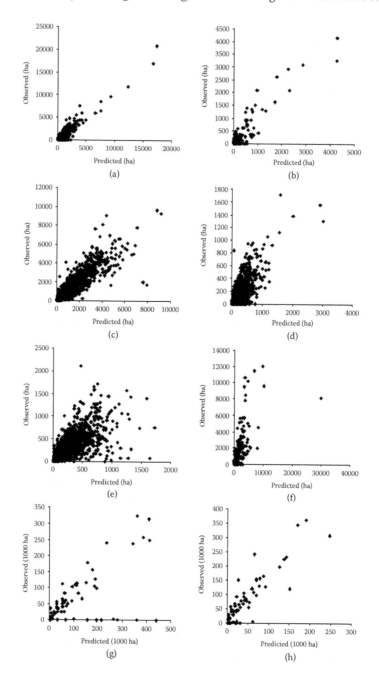

FIGURE 5.4 Scatter plots of predicted English land uses from disaggregations against observed land uses for ward-level English data: (a) arable, (b) vegetables, (c) permanent grazing, (d) oils, (e) temporary grass, (f) rough grazing; and European land uses from disaggregations against observed land uses from English county-level data, (g) grassland and (h) arable.

TABLE 5.1

Details of Spatial Disaggregation Results

Land Use	Validation R²	Stage 1 Model R²	Slope of Validation Regression	R² Iterations = 1
		Great Britain		
Arable	0.90	0.88	1.07	0.89
Vegetables	0.86	0.60	0.98	0.84
Permanent grass	0.82	0.74	0.99	0.76
Oils	0.65	0.74	0.60	0.63
Temporary grass	0.57	0.67	0.88	0.54
Rough grass	0.54	0.87	0.72	0.47
		Europe		
Grassland	0.89	0.53	1.32	—
Arable	0.80	0.28	0.60	—

of observed versus predicted values are given in Table 5.1 and range from 0.54 for rough grazing to 0.90 for arable.

In the case of the European disaggregations, Figure 5.4 shows scatter plots of the areas of grassland and arable land use in a number of English counties, observed in the agricultural census records, against the predicted values for the counties based on the disaggregation process. The observed records for grassland are obtained by summing the records for permanent grazing, temporary grazing, and rough grazing to attempt to make them comparable to Eurostat's grassland class. The R^2 value associated with a regression of observed versus predicted in the case of grassland is 0.89. In the case of arable, the R^2 value is 0.80.

The slopes of regression lines calculated are also reported in Table 5.1. Most GB land uses modeled tended to overpredict land uses for wards and coefficients are significantly different to 1 ($p < 0.05$), except for arable and permanent grazing. The European grassland type is generally underpredicted, and the European arable type is generally overpredicted.

If the regression models that form the basis of the first stage of the disaggregation process are considered, it can be seen that there is quite wide variation in their performance (Table 5.1). The GB models generally show a higher proportion of variation in the areas of land use explained by proportions of land cover than the European models. However, it can be seen that the R^2 values for these models do not correlate with the performance of the final land use models (Pearson's correlation coefficient, $r = -0.1$; p value, 0.857).

The number of iterations during the simulation stage of the algorithm has little effect on how well the predictions fit the observed ward-level data for the GB disaggregations. The lowest R^2 values are always for runs using only one iteration and these are also given in Table 5.1. The largest range of R^2 is 0.07 for rough grazing.

Spatial patterns do change with the number of iterations used in the simulation stage of the process. The stability of the spatial patterns increases with the number

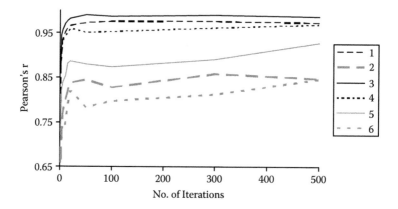

FIGURE 5.5   The correlation between maps produced using 1000 iteration simulations and each of the simulations using the smaller number of iterations shown on the x axis for 1) permanent grazing, 2) temporary grass, 3) rough grazing, 4) arable, 5) vegetables and 6) oils.

of iterations used. The correlation between the maps produced using 1000 iterations and each of the other numbers of iterations, for each land use, are plotted in Figure 5.5. From this, it appears that there is a tendency for the patterns to converge at less than 100 iterations.

There are no obvious spatial patterns in residuals; over- and underpredictions for individual wards tend to cancel out one another at the county level.

5.4 DISCUSSION AND CONCLUSIONS

Results are visually promising, showing realistic changes in agricultural land use practices across GB and Europe for the periods considered. However, maps of the European land uses demonstrate that there are many gaps in the underlying agricultural statistics from which they are created. NUTS regions are clearly visible where zero values in the base data for a land use have been reported instead of codes indicating that no records are available. For example, zero values are visible for central Scotland in both maps of arable and grassland (Figure 5.3). These gaps in data make the European data much less useful for studies of agricultural change across Europe, and less useful as a data set of relevance to studies of biodiversity distributions. The GB data are not without their own gaps, although the data for 2000 shown in Figure 5.2 only show one area of missing data, arable land use in the Outer Hebrides, a number of counties have missing data for the three time periods processed. Such limitations in data cannot be rectified by the approach taken here and can only be rectified by methods that predict land use activities from other variables. The first step of the approach taken here does, in effect, predict land use from land cover. However, there are many more elegant modeling approaches that can be taken if this is the goal [1,23].

The relatively simple disaggregation technique introduced here is seen to perform very well, in comparison with other approaches, for a number of land use types at both a larger, national (GB) scale and a smaller, regional scale (Europe). At the larger scale, R^2 values calculated seem very good for permanent pasture (>0.8), arable, and vegetable crops (>0.85). Moxey et al. [14], in a study limited to the Tyne river catchment in North East England using parish-level agricultural census data, estimated R^2 for permanent pasture at 0.71, while fits were better for cereals and other arable (0.87 and 0.81, respectively). A number of land uses in this study are seen to be predicted less well. Temporary and rough grazing results in Moxey et al. [14] were reported as having R^2 values of 0.70 and 0.92, respectively, although this study performed less well (temporary 0.57 and rough grazing 0.69, if one outlier removed). You and Wood [16] obtained correlation coefficients, rather than R^2 values, of 0.4–0.65 for a number of individual crop types in a study of Brazilian state-level statistics disaggregated to the municipality level using an entropy maximization approach. The oil seed land use type in this study would be the most comparable in terms of crop type resolution, giving a correlation coefficient of 0.80.

The GB land use types where the disaggregation process seems to be functioning best have near-to-one relationships of predicted to observed, though the vegetable class has a relationship where the gradient is significantly different from 1. The other three land use classes have a tendency in general to be overpredicted. This can be caused by high leverage points such as the very high prediction of rough grazing for one ward corresponding to Kielder Forest in Northumberland, England, a large forestry plantation in an area otherwise dominated by rough grazing. If this observation is removed, not only does the R^2 value rise to 0.67, but the gradient becomes 1.04. Another potential reason for overprediction is because the algorithm is less certain of where to distribute land uses such as temporary grassland, and so distributes these parcels of land use more evenly across the grid cells making up a county, thus overestimating the land use in many wards where the land use is quite rare. The highest percentage predicted cover for a 5 km × 5 km grid cell is 31% for temporary grassland, in comparison to 97% for rough grazing and 67% for arable.

Neither this study nor the previous studies cited have been able to consider the performance of the disaggregation at the output resolution, a fundamental problem of attempts to spatially disaggregate data. Despite the apparently good fits according to R^2 values, the individual finer resolution spatial units used for validation can be badly over- or underpredicted. The consequences of these errors need to be considered in terms of uses to which disaggregated surfaces might be put. The comparisons of GB disaggregations from county-level data to ward-level data was achieved by allocating land use to a 1 km resolution before summing across 100 m^2 cells making up the 1 km squares within wards. In effect, the analysis of results is being made on the 1 km^2 results, rather than the 5 km × 5 km output grid intended for the study and represented by the maps shown. Over half the wards used in the arable land use validation have areas less than 25 km^2, and three-quarters have areas less than two 5 km × 5 km cells. Errors at the 25 km^2 resolution are likely to be less severe, and areas reported within the counties are by definition conserved by the disaggregation process. Like previous studies, this study suffers from the lack of finer resolution land use data with which to compare results [14]. Detailed land cover survey data for some areas might be used in future studies, but there is always the problem of matching land cover survey classes to agricultural census land use classes, as well as the inherent limitations of field surveys [24].

In the case of the European scale disaggregations, it is reassuring that the trends in agreement are so strong. However, there are several reasons why there is not a one-to-one fit of predictions to observations. In fact, the county-level English agricultural census data are not true observations in terms of the European land use categories; they are themselves estimates based on grouping English land use categories into the equivalent European categories. There are no documents publicly available that explain the process by which English data have been reclassified into Eurostat categories. Indeed part of the suggested overpredictions of grassland and underpredictions of arable may be due the consideration of how to classify land that is recorded as temporary grassland in the United Kingdom. In this study, all of it has been grouped into the grassland type.

The merit of this two-stage disaggregation approach is shown by the fact that there is no clear relationship between the quality of the first-stage models and the quality of the final spatial disaggregations. The sometimes quite poor fit of county-/ NUTS regional-level models of land use to land cover proportions is to be expected given that these are the simplest of linear models, taking into account none of the pitfalls of ignoring collinearity in land cover variables and the likely spatial dependencies in both the supposedly dependent land use variables and the independent variables. They are also calculated from observations taken from a very wide range of environmental conditions and fits to the models will vary quite substantially across the spatial ranges of the data. However, all that is required of these models is to produce a reasonably reliable ranking of output cells within an administrative region. The exact magnitude of the differences between cells within an administrative region is not of great relevance, and, even more importantly, the correct variation in cell values between counties is of no relevance to the second simulation stage of the disaggregation process. The apparently rapid convergence of land use allocation patterns over quite a small number of iterations indicates that ranking of cells from the first stage is sufficient given the subsequent competition between land use parcels to be allocated to different cells within administrative region. This competition is guided by the makeup of land use proportions reported in the base data.

By using an administrative region-by-region approach, the inherent spatial dependencies in the finer resolution land cover data used as the land suitability key for the disaggregation, and those likely to be present in finer resolution land use data can be partially exploited. Particularly at the European scale, it is questionable whether the level of spatial autocorrelation at the spatial lags required to consider separations between NUTS 3 regions would be sufficient to guide a kernel-based approach or a pycnophylactic approach.

The use of a land cover map rather than a specially calculated land capability map, although not completely new [15], works well. A snapshot of land cover at one point in time captures land suitability across time. Agricultural capability maps may become outdated, just as land cover maps can. Advances in agricultural technology have occurred over time, leading to arable crops being grown on what might have been deemed most suitable as grazing land when capability maps were developed. However, both agricultural capability maps and land cover maps maintain the correct spatial patterns of suitability across time. It is just the land use activities that can occur on particular patches of land that change. By modeling the relationship between land use and land cover at each period for which data are to be disaggregated, the suitability of land for particular activities at that particular point in time is captured.

This study has investigated how this approach works on only two different sets of land use data. Many other areas of the world are covered by similar records and the European disaggregations indicate that the approach can be applied to relatively coarse geographies. It should be possible, for example, to disaggregate the United States Department of Agriculture records of agricultural activity at US county levels. Still coarser data, such as those from the UN Food and Agriculture Organization, might also be disaggregated below country level.

ACKNOWLEDGMENTS

Stuart Macdonald of Edina kindly provided older Scottish census records in PDF. Sarah Baulch, Lucy Bridgman, Lynsey Henderson, and Andrew Clark helped prepare initial land use spreadsheet data. I also thank two anonymous referees for very useful comments on an earlier draft of this manuscript.

REFERENCES

1. Aalder, I. H. and Aitkenhead, M. J., Agricultural census data and land use modelling, *Computers, Environment and Urban Systems*, 30, 799, 2006.
2. Parmesan, C. and Yohe, G., A globally coherent fingerprint of climate change impacts across natural systems, *Nature*, 421, 37, 2003.
3. Pearson, R. G. and Dawson, T. P., Predicting the impacts of climate change on the distribution of species: Are bioclimate envelope models useful? *Global Ecology and Biogeography*, 12, 361, 2003.
4. Thuiller, W., Araújo, M., and Lavorel, S., Do we need land-cover data to model species distributions in Europe? *Journal of Biogeography*, 31, 353, 2004.
5. Termansen, M., McClean, C. J., and Preston, C. D., The use of genetic algorithms and Bayesian classification to model species distributions, *Ecological Modelling*, 192, 410, 2006.
6. Schmit, C. and Rounsevell, M. D. A., Are agricultural land use patterns influenced by farmer imitation? *Agriculture, Ecosystems and Environment*, 115, 113, 2006.
7. Defra, The June Agricultural Census, http://www.defra.gov.uk/esg/work_htm/ publications/cs/farmstats_web/default.htm, 2007.
8. Eurostat, http://epp.eurostat.ec.europa.eu, 2007.
9. Howitt, R and Reynaud, A., Spatial disaggregation of agricultural production data using maximum entropy, *European Review of Agricultural Economics*, 30, 359, 2003.
10. Tobler, W. R., Smooth pycnophylactic interpolation for geographical regions, *Journal of the American Statistical Association*, 74, 519, 1979.
11. Openshaw, S., The modifiable areal unit problem, *Concepts and Techniques in Modern Geography*, 38, 41, 1984.
12. Martin, D., Mapping population data from zone centroid locations, *Transactions of the Institute of British Geographers NS*, 14, 90, 1989.
13. Hotson, J. M., *Land use and agricultural activity: An areal approach for harnessing the Agricultural Census of Scotland*, Working paper no. 11, Economic and Social Research Council Regional Research Laboratory Scotland, 1988.
14. Moxey, A., McClean, C., and Allanson, P., Distributing agricultural census cover data by areal interpolation, *Soil Use and Management*, 11, 21, 1995.

15. Huby M., Cinerby, S., Crowe, A. M., Gillings, S., McClean, C. J., Moran, D., Owen, A., and White, P. C. L., The association of natural, social and economic factors with bird species richness in rural England, *Journal of Agricultural Economics*, 57, 295, 2006.
16. You, L. and Wood, S., An entropy approach to spatial disaggregation of agricultural production, *Agricultural Systems*, 90, 329, 2006.
17. Williams, J., Digest of Welsh Historical Statistics: Agriculture, 1811-1975 computer file, AHDS History distributor, Colchester, SN: 4096, 2001.
18. Edina, Agcensus, http://edina.ac.uk/agcensus/, 2007.
19. UKBorders, UKBORDERSTM, http://edina.ac.uk/ukborders/, 2007.
20. CEH, CEH data holding land cover map, http://science.ceh.ac.uk/data/lcm/LCM2000.shtm, 2007.
21. European Environment Agency, Data Service, http://dataservice.eea.europa.eu/dataservice, 2007.
22. CRAN, The R Project for Statistical Computing, http://www.r-project.org/, 2007.
23. Rounsevell, M. D. A., Reginster, I., Araújo, M. B., Carter, T. R., Dendoncker, N., Ewert, F., House, J. I., et al., A coherent set of future land use change scenarios for Europe, *Agriculture, Ecosystems & Environment*, 114, 57, 2005.
24. Cherrill A. and McClean C., Between-observer variation in the application of a standard method of habitat mapping by environmental consultants in the UK, *Journal of Applied Ecology*, 36, 989, 1999.

6 Comparing Different Land Cover Data Sets for Agricultural Monitoring in Africa

Steffen Fritz, Linda See, Felix Rembold,
Michel Massart, Thierry Nègre,
and Craig von Hagen

CONTENTS

OVERVIEW

This chapter provides a comparison of four land cover data sets to determine which product is the most suitable for applications such as food security and monitoring of agricultural expansion. The land cover products compared are: the Global Land

Cover Map (GLC-2000), the MODIS land cover product (MOD12V1), the SAGE cropland database, and the AFRICOVER data set from the Food and Agricultural Organization (FAO). The data sets were first converted to the same resolution and the legend classes were then reconciled. A comparison was then made of the overall agricultural areas for the different data sets together with official national and subnational statistics. This was followed by a spatial comparison of the land cover maps; areas of high disagreement were identified between each land cover product and AFRICOVER. The results showed that uncertainties in the cropland distribution in African countries are very high. For example, MODIS had the tendency to underestimate cropland cover, whereas the GLC-2000 tended to overestimate cropland cover in those countries that are located at the northern transition zone of subtropical shrubland and semidesert areas. In this area, MODIS and SAGE showed a relatively similar cropland distribution. It was also demonstrated that even though overall cropland areas in administrative units (e.g., FAO national statistics) are not so far apart, the spatial distribution of these can vary and a high level of uncertainty exists when a spatial comparison is undertaken.

6.1 INTRODUCTION

In Africa there is a severe shortage of quantitative and qualitative information on vegetation cover and current land use at both national and regional levels. Remote sensing of land cover offers the potential to produce a rapid and up-to-date land use and land cover database for a variety of purposes. Accurate information within the agricultural domain is particularly important for crop monitoring for the purpose of food security. For example, the MARS-FOOD action of the European Commission Joint Research Centre was recently established to support the food aid and food security policies of the European Commission. The activities are aimed at improving existing methods in food early warning and crop monitoring, and at providing reliable and objective crop yield forecasts. Several techniques are used for extracting vegetation index temporal profiles from SPOT VEGETATION images, but in all cases land cover maps are needed to link the extracted indicators with the observed crops. This chapter will outline the comparison of four sources of land cover data to determine which product is the most suitable for agricultural monitoring and for the subsequent development of a crop mask, an important input to both food security and monitoring of agricultural expansion. Whereas in earlier papers a global comparison of land cover products was undertaken [1,2], this chapter focuses on the agricultural domain in Africa.

The land cover products used are the Global Land Cover Map (GLC-2000) [3], the MODIS land cover product (MOD12V1) [4], the SAGE cropland database [5], and the AFRICOVER data set from the Food and Agricultural Organization of the United Nations (FAO) [6]. Both the GLC-2000 and MODIS land cover products are at a resolution of 1 km^2, whereas AFRICOVER is available at a finer resolution as it is based on the visual interpretation of high resolution Landsat data, but with a minimum mappable unit of 75 to 200 ha, depending on the country. The first part of the methodology, therefore, involves the aggregation of the different land cover products to a compatible resolution since the products vary from very fine (AFRICOVER) to medium (MODIS, GLC-2000) to a coarser scale (SAGE) resolution. The second part of the methodology deals with the reconciliation of the legend categories taking into account the uncertainty in the definitions used in each land cover product. This is critical because the legend definitions that refer to the agricultural domain are not entirely compatible and three products use a range (e.g., agriculture 20% to 50%) rather than a single value. In the past, map comparisons have often taken the form of a Boolean or crisp approach, which is characterized by two main features: (1) legend categories from the different maps, which are often very different, are matched using a one-to-one mapping regardless of obvious incompatibilities; and (2) the resulting map comparison shows areas with 100% agreement or disagreement. The problem with this type of approach is that the resulting map comparison is problematic because the user is not certain whether the disagreement is real or just a function of semantic differences in the legend definitions. The approach used here allows for overlap between legend definitions to be taken into account and is based on previous work undertaken by See and Fritz [1,2].

Once the legend reconciliation is complete, the third part of the methodology involves a comparison with national statistics, where analysis is undertaken at both continental and national scales. The AFRICOVER data set is used as a reference data set against which the other three products are compared. Finally, a fuzzy logic approach to the comparison of all four land cover products is undertaken. The percentage disagreement between each of the three global land cover products—that is, the GLC-2000 for Africa, MODIS, and SAGE—is then calculated as outlined in previous studies (e.g., Fritz and See [1]). The methodology is then further extended, and the disagreement in terms of omission and commission is calculated using AFRICOVER as a reference data set. The results are discussed and recommendations are made regarding the suitability of the different land cover products for creating an agricultural mask for the AFRICOVER countries as well as those outside of the AFRICOVER area.

6.2 THE LAND COVER PRODUCTS AND FAO AGRICULTURAL STATISTICS

The first land cover map used in this study is the Global Land Cover 2000 (GLC-2000), which is a global product for the baseline year 2000, a reference year for environmental assessment. This data set was created in collaboration with partners around the world [3,7]. The GLC-2000 was developed using a bottom-up approach in which more than 30 research teams contributed to 19 regional windows, where the regional legends used the Land Cover Classification System as a common language to produce 22 global classes [3,8]. Together with the MODIS data set, it is currently the most recent global land cover product. The GLC-2000 for Africa was produced on a regional level with 27 different classes that are more refined and detailed, but they map directly onto the 22 global classes. Since the regional map was more detailed, it was used in this analysis. We refer to the regional African part of Global Land Cover 2000 in this chapter simply as GLC-2000.

The MODIS land cover product from Boston University (MOD12Q1 V004, 1 km) was created using the Moderate Resolution Imaging Spectoradiometer instrument on the NASA Terra Platform using data from the period mid-October 2000 to mid-October 2001. The MODIS land cover data set uses all 17 classes of the International Global Biosphere Project (IGBP) legend [9], and unlike the GLC-2000, a global classification approach has been used.

The Center for Sustainability and the Global Environment (SAGE) at the University of Wisconsin has developed a global cropland data set. It has been used in a number of recent research projects [10] even though it is based on relatively old AVHRR (Advanced Very High Resolution Radiometer) data [9]. SAGE uses a data fusion technique to integrate remotely sensed data derived from the Global Land Cover characteristics database [9] and administrative-unit level inventory data [11]. The SAGE data set is still used as the standard global data set for agricultural applications since it was calibrated with the use of national statistics [11]; in particular, it was refined to serve as a basis for crop-specific agricultural areas globally [12]. Even though it is still used in a number of modeling activities, the accuracy of this data set, especially in Africa, is questionable. The data set is based originally on AVHRR data, which have a number of reported problems such as poor sensor calibration, variable angular-induced pixel sizes [13], and, in particular, high georegistration errors [7]. We use the proposed methodology to determine the validity of this data set in the AFRICOVER countries and focus on the spatial disagreement with the AFRICOVER project.

AFRICOVER is a land cover data set that currently provides detailed, baseline agricultural land use information for 10 countries in Africa: Burundi, Democratic Republic of Congo, Egypt, Eritrea, Kenya, Rwanda, Somalia, Sudan, Tanzania, and Uganda [6]. Funded by the Italian government through the Italian International cooperation as part of an FAO project, it was originally conceived to help these countries in Africa set up a georeferenced database on land cover. Pending funding and donor support, the project will eventually be extended to the whole of Africa. Libya has recently been completed not under AFRICOVER, but under a parallel initiative of FAO using the same methodology. The project is currently starting operation in Dakar, Senegal, for the West African component, under the FAO-Global Land Cover Network umbrella. The AFRICOVER data set is based on visual interpretation of Landsat data. It is based on ground data and expert knowledge. Most images were acquired in the period between 1995 and 1999. To extract AFRICOVER agricultural statistics for a single country, the area of the country is first calculated. The AFRICOVER data are stored in polygons. Each polygon is then examined and the area of a given LCCS code is calculated. For single code polygons, the area is 100%. Polygons with more than one LCCS code contain mixed classes. In the case of two LCCS codes, the first class is assigned 60% and the second is given 40% of the polygon area. Similarly, for polygons with three codes, 40%, 30%, and 30% are used, respectively. In addition there are two special cases: (1) scattered clustered agriculture, which is assigned 35% for codes 2 and 3; and (2) scattered isolated agriculture, which is assigned 15% for codes 2 and 3. Some codes are then adjusted to make sure the total assigned is 100%. Torbick et al. [14] recently compared AFRICOVER to the GLC-2000. They found a 54% agreement overall for agriculture.

FAOSTAT [15] is an online, multilingual database currently containing over 3 million time-series records covering international statistics, including agriculture.

FAOSTAT is the official reference database for country statistics containing contributions from individual countries, which provide their national statistics online each year. Arable land in the FAO database is defined as land under temporary crops (double-cropped areas are counted only once), temporary meadows for mowing or pasture, land under market and kitchen gardens, and land temporarily fallow (less than five years). The abandoned land resulting from shifting cultivation is not included in this category. Data for arable land are not meant to indicate the amount of land that is potentially cultivable. This database is used and compared in those countries where permanent crops/tree crops play a negligible role, as permanent crops are excluded in some of the definitions of the other land cover products.

6.3 METHODOLOGY

The methodology can be divided into four main parts: (1) reconciliation of the four sets of legend classes; (2) aggregation of the AFRICOVER land cover map, GLC-2000, and MODIS to 5 minutes (approximately 10 km) to match the SAGE cropland database; (3) comparison of the overall agricultural areas for the different data sets together with official national statistics; and (4) spatial comparison of land cover maps for the data sets that record a negligible proportion of tree crops and direct spatial comparison with the AFRICOVER database. Hot spots or clusters of disagreement are identified between each land cover product and AFRICOVER.

6.3.1 RECONCILIATION OF LEGEND CLASSES

Table 6.1 contains the different definitions for agricultural land from the four land cover products. Even though it can be noted that the definitions are not identical, all the definitions have cropland in common. They slightly deviate when it concerns pasture, and only include pasture if it is intensive or cropland temporarily used for pasture. Since extensive pastures and grassland are excluded from all definitions we feel that the definitions agree sufficiently to allow for the comparison to be carried out. Moreover, since we only focus on countries where tree crops play a minor role, orchards, and tree plantations do not pose a problem in our comparative study. Generally researchers tend to use only one value for this type of comparison (e.g., see World Bank [16]), making certain assumptions, but in our approach we allow for the whole range to be considered and only record disagreement outside this range.

From this legend the lookup table linking the four land cover products is derived. Table 6.2 contains the lookup table for the GLC-2000 and AFRICOVER. Where there is any degree of overlap, complete agreement is assumed for simplicity and marked with an X. Similar tables showing agreement between each pair of land cover products were produced, but are not included in this chapter.

One of the problems encountered with allowing for any overlap to signify complete agreement is the presence of tree crops in the SAGE database. In contrast, the other global data sets do not have tree crops in their definition (cropland or cultivated and managed areas). The approach taken in Fritz and See [1] would result in the entire SAGE cropland database being mapped onto all the shrub and tree cover of

TABLE 6.1

Legend Definitions of Cropland from the Different Land Cover Products

Legend Type	Definition	Percent
MODIS	Croplands are lands covered with temporary crops followed by harvest and a bare soil period (e.g., single and multiple cropping systems). Note that perennial woody crops will be classified as the appropriate forest or shrub land cover type.	15%–60%
	Cropland/natural vegetation mosaics are lands with a mosaic of croplands, forests, shrublands, and grasslands in which no one component comprises more than 60% of the landscape.	60%–100%
GLC-2000 for Africa	Cultivated and managed: Areas with over 50% cultures or pastures. Regions of intensive cultivation and/or sown pasture fall in this class.	50%–100%
	Mosaic: Forest/cropland—The vegetation found here is formed by a complex of secondary regrowth, fallow, home gardens, food crops, and village plantations.	15%–50%
	Mosaic: Cropland/natural vegetation—At the southern end of the Sahelian belt, the croplands are mixed with natural vegetation and represent up to 30% of the cover.	15%–30%
	Irrigated agriculture—Agriculture depending on artificial water supply.	100%
SAGE	Arable land (including harvested cropland, crop failure, temporarily fallow or idle land, and cropland used temporarily for pasture) and land under permanent crops (such as cocoa, coffee, rubber, etc., including all tree crops except those grown for wood or timber). The harvested produce may be used for both human consumption and/or feed.	Calibrated in Africa with national statistics; no range
AFRICOVER	Pure continuous fields, herbaceous crops	80%–100%
	Mixed continuous fields, herbaceous crops	50%–80%
	Scattered clustered, herbaceous crops	20%–50%
	Scattered isolated, herbaceous crops	10%–20%
	Pure continuous fields, shrub or tree crops	80%–100%
	Mixed continuous fields, shrub or tree crops	50%–80%
	Scattered clustered, arable, shrub or tree crops	20%–50%
	Scattered isolated, arable, shrub or tree crops	10%–20%

other land cover maps. As a result, there would be hardly any disagreement recorded due to this large degree of overlap. Therefore, an alternative is to use only those countries that have a negligible amount of tree crops. A more reasonable comparison then becomes possible.

6.3.2 Aggregation of the Different Land Cover Products

The different land cover products were aggregated to the resolution of the SAGE database, that is, 10 km. This aggregation allows for a comparison of the different

TABLE 6.2
Example of Matrix between AFRICOVER and GLC-2000

AFRICOVER		Pure Continuous Fields, Herbaceous Crops	Mixed Continuous Fields, Herbaceous Crops	Scattered Clustered, Herbaceous Crops	Scattered Isolated, Herbaceous Crops
GLC-2000	Percent cover	80%–100%	50%–80%	20%–50%	10%–20%
Croplands	50%–100%	X	X		
Mosaic: Forest/ Cropland	15%–50%			X	X
Mosaic: Cropland/ Natural Vegetation	15%–30%			X	X
Irrigated Agriculture	100%	X			

land cover products at a grid level and allows for the identification of hot spots of disagreement. Furthermore, this aggregation will compensate for some geolocation problems [17].

6.3.2.1 AFRICOVER

To aggregate to 5 minutes (approximately 10 km), the values of the 10,000 individual pixels are added and then divided by 10,000 to arrive at an average percentage cropland. This is carried out twice, one for the minimum value of each class range and the other for the maximum value. This provides a percentage of cropland for each aggregated 100 km² grid and incorporates the uncertainty present in the definitions. In this way a minimum and maximum cropland fraction image is produced that records the percentage area of cropland on a scale from 0% to 100% found in each 100 km² grid square. The AFRICOVER fraction grid is shown in Figure 6.1. Since AFRICOVER differentiates between tree crops and herbaceous crops, the SAGE data set can be compared with the AFRICOVER data set if tree crops are included in the comparison.

6.3.2.2 Aggregation of MODIS and GLC-2000

In the same way that the AFRICOVER data set was aggregated, MODIS and GLC-2000 were aggregated to a 5 minute data set. The following classes with their definitions were aggregated: cultivated and managed (from the GLC-2000), cropland/ forest mosaic (from GLC-2000), cropland/natural mosaic (from GLC-2000), irrigated agriculture (from GLC-2000), cropland (from MODIS), and cropland/natural vegetation mosaic (from MODIS). A minimum and maximum cropland cover map was created for MODIS and GLC-2000 based on the minimum/maximum

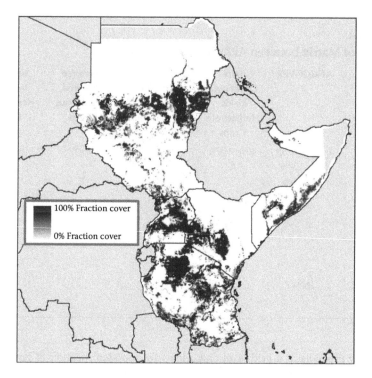

FIGURE 6.1   Example of AFRICOVER fraction image (AFRICOVER Max).

definitions of the different classes as set out in Table 6.1. For example, for the GLC-2000 class cultivated and managed, a minimum of 50% and a maximum of 100% was assigned.

6.3.3 COMPARISON OF AGRICULTURAL STATISTICS (FROM FAO) WITH LAND COVER PRODUCTS

In order to compare the different data sets with official FAO statistics, the minimum and maximum values for each land cover data set within the agricultural domain were taken and the overall cropland area for each data set was calculated on a country basis. Since the SAGE data set does not contain a range of values but a percentage value, and was calibrated with subnational statistics, the area of cropland for each country could be directly calculated. Official statistics for the overall national crop areas of AFRICOVER could also be obtained. Moreover, they do not use a range but rather a percentage value that is well informed by expert knowledge. In order to be able to compare the national data sets together with all the spatial data sets, only countries with a negligible proportion of shrub and tree crops, that is, less than 5% of arable area, were included in the analysis for the reasons explained earlier. It was then possible to directly compare all the data sets with the official national statistics obtained from FAO.

To identify the data set that coincides most closely with the official national (FAO) statistics, the root mean squared error [18] was calculated as follows:

$$\text{RMSE} = \sqrt{\frac{\sum_{i=1}^{n}(O_i - P_i)^2}{n}}$$

where O_i is the crop area as reported by FAO, P_i is the value derived from the land cover maps, and n is the number of countries. The RMSE has been used in the past as one of the standard measures for measuring accuracy in spatial analysis [19]. This type of error measurement penalizes larger errors more and can be useful when national statistics from different data sets are compared with official values.

6.3.4 CREATION OF PERCENTAGE DISAGREEMENT MAPS (BOOLEAN AND FUZZY)

The final step in the methodology is to create percentage disagreement maps between AFRICOVER and the other land cover products. Each pair of maps was compared on a pixel-by-pixel basis using the lookup table (Table 6.2). Where there is an X in Table 6.2, there is overlap in the definitions of the legend classes of a pair of given maps so the percentage disagreement is considered to be 0%. Although the overlap would appear to lend itself naturally to a fuzzy approach, this would be difficult to achieve in practice because the overlap is based on the nature of the definitions, which use different criteria between land cover maps. For this reason we have adopted a conservative approach and assume that the presence of any overlap equates to complete agreement. For all other pixel comparisons, the percentage difference is mapped onto a fuzzy set like that shown in Figure 6.2 to denote degrees of difference. For the Boolean approach, these same areas would have 100% disagreement. These Boolean and fuzzy differences are then mapped spatially. In order to illustrate how this fuzzy approach compares to a Boolean approach, we also show the difference between the two and calculate the commission and omission errors.

The three global land cover products (MODIS, GLC-2000, and SAGE) are first compared within Africa for those countries where tree crops play a negligible role, and maps of spatial disagreement highlighting hot spots of disagreement between MODIS and SAGE, MODIS and GLC-2000, and GLC-2000 and SAGE are produced. A more refined regional analysis is then undertaken for eastern African countries (Sudan, Uganda, Somalia, Kenya, and Tanzania) for which AFRICOVER is available. AFRICOVER is used as a reference against which the others are compared since it is based on a visual interpretation of high-resolution images and ground information. This also allows for the calculation of the area of omission and commission, where we borrow these terms from the well-known omission and commission error measures [20]. The area of omission is calculated as the overall area of disagreement in areas where AFRICOVER records a higher cropland fraction. This also includes the case where AFRICOVER records a certain cropland fraction and the other data set does not have any crops present. Conversely, the area of commission is calculated

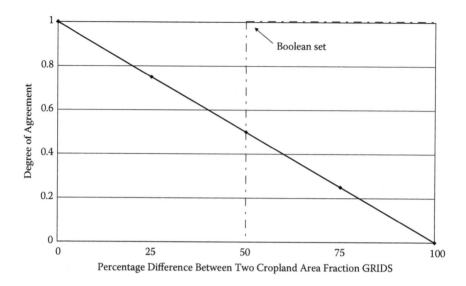

FIGURE 6.2 Fuzzy set showing degree of agreement for the percentage difference between cropland areas and the Boolean set (dashed line) with a threshold.

where AFRICOVER records a smaller cropland fraction or zero cropland fraction compared to the other data set. In this evaluation the minimum and maximum cropland fraction is taken into account. For example, if a given pixel in AFRICOVER records a minimum and maximum cropland fraction of 20% and 40%, respectively, while the other data set (e.g., GLC-2000) records 50% and 100% for the minimum and maximum values, then the percentage disagreement is 10%, which corresponds to a commission area of 10 km² (i.e., 10% of 100 km²). This analysis is performed on the same African countries as outlined above, that is, those with a negligible amount of tree crops to facilitate the comparison.

6.4 RESULTS

6.4.1 CONTINENTAL COMPARISON

The total crop areas for Africa derived by summing the FAO national statistics along with estimates from the different land cover products and the SAGE database are given in Figure 6.3. It is not surprising that the SAGE database gives total values that are close to the totals from the FAO statistics because this data set has been calibrated at a continental level using the best correlation (R^2). Any differences are due to the fact that SAGE was calibrated using older national statistics (from 1990). The GLC-2000 crop areas summed using the minimum value of the range also produce values at the continental level that are similar to the overall continental total derived from FAO. The minimum and maximum crop areas from MODIS both underestimate total crop area at a continental level.

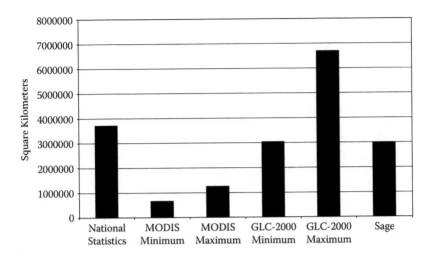

FIGURE 6.3   Total crop area from the FAO national statistics compared to the land cover products for Africa.

6.4.2 NATIONAL COMPARISON

Figure 6.4 provides the same comparison but at the national level for selected countries in Africa, representing a range of different ecosystems across the continent. It is clear from looking at the graph that there is a large amount of difference between the estimates, and the graph highlights the high amount of uncertainty when compared to national statistics. Moreover, certain patterns emerge. Whereas GLC-2000 tends to overestimate (both GLC-2000 minimum and GLC-2000 maximum), MODIS tends to underestimate (both for MODIS minimum and MODIS maximum). Similar patterns emerge when we compare the different data sets to subnational statistics (e.g., in Sudan we undertook such an exercise but it is not provided in this chapter), even though these patterns are less regular and more exceptions arise. Table 6.3 provides the RMSE for each of the countries and shows that the best fit is found using the GLC-2000 minimum figures followed by SAGE. The SAGE result is not surprising because this data set was calibrated using national statistics. The MODIS maximum and minimum both result in RMSE values that are slightly higher than the GLC-2000 minimum and SAGE; however, they are relatively close together. The GLC-2000 minimum and maximum, on the other hand, are further apart, with GLC-2000 maximum map producing the highest RMSE.

6.4.3 SPATIAL DISAGREEMENT FOR SELECTED AFRICAN COUNTRIES

Although comparing totals at a continental and national level provides an indication of how the different land cover products match official statistics, this exercise does not tell us anything about the accuracy of the spatial distribution of cropland. Figure 6.5a shows the spatial disagreement between the GLC-2000 and the MODIS land cover

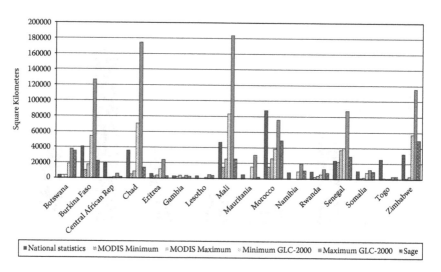

FIGURE 6.4 Area of arable land per country based on FAO national statistics.

TABLE 6.3

RMSE of Cropland Area Comparing Each Land Cover Product to National FAO Statistics

Land Cover Type	RMSE (km²)
GLC-2000 minimum	21064
GLC-2000 maximum	76802
SAGE	25109
MODIS minimum	27787
MODIS maximum	36504

---

products. There are areas of high disagreement in Mali, northeastern Burkina Faso, western Chad, and western Sudan. Figure 6.5b shows the map when comparing GLC-2000 with the SAGE data set. There are similar patterns to those seen in Figure 6.5a, with less severe disagreement in Burkina Faso and more in Zambia. Figure 6.5c shows the spatial disagreement when comparing MODIS and SAGE, which shows that there are few areas of severe disagreement. However, in Zambia, there are similar patterns to that seen with the GLC-2000 and SAGE comparison.

When comparing the patterns shown on the spatial disagreement maps with national statistics, there is an indication that the GLC-2000 in countries such as Burkina Faso, Chad, Senegal, Zimbabwe, and Mali (Figure 6.4) records cropland in areas where there is either very little or no cropland, since even the minimum cropland value for those countries is far higher than the official national statistics. On the other hand, in the country of Zambia, both SAGE and GLC-2000 are closer to the national statistics and MODIS appears to miss large areas of cropland, possibly in those areas where both the GLC-2000 and SAGE record cropland. However, final

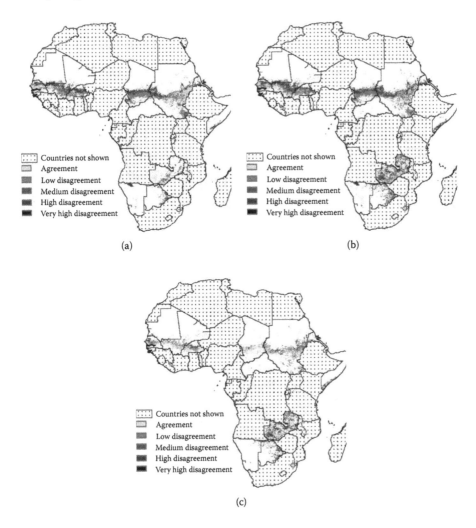

FIGURE 6.5   (a) Disagreement in the agricultural domain between GLC-2000 and MODIS land cover maps; (b) disagreement in the agricultural domain between GLC-2000 and SAGE; and (c) disagreement in the agricultural domain between MODIS and SAGE.

conclusions on these countries cannot be drawn as official national FAO statistics can also be prone to error and have to be treated with care. Clearly these are just general trends and the true spatial distribution of cropland in these areas has to be examined in more detail. We therefore present results of a more detailed analysis for those countries that are covered by AFRICOVER and have a high proportion of cropland, namely, Sudan, Tanzania, Uganda, Kenya, and Somalia.

6.4.4 SPATIAL DISAGREEMENT OF OMISSION AND COMMISSION ERRORS

For the AFRICOVER countries, a direct spatial comparison is performed and is shown in Figure 6.6a, Figure 6.6b, and Figure 6.6c. Figure 6.6a shows the area of spatial disagreement of omission and commission for AFRICOVER and GLC-2000,

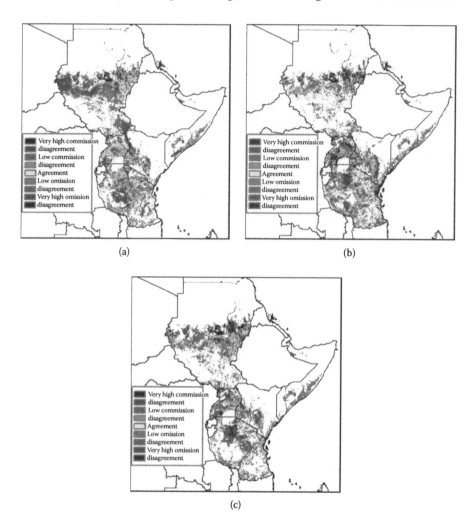

FIGURE 6.6 (a) Areas of omission and commission comparing AFRICOVER with the GLC-2000 in the agricultural domain; (b) areas of omission and commission comparing AFRICOVER with the SAGE data set in the agricultural domain; and (c) areas of omission and commission comparing AFRICOVER with MODIS in the agricultural domain.

Figure 6.6b is the comparison for AFRICOVER and the SAGE data set, and Figure 6.6c is the comparison with MODIS. In Sudan, we observe that all three land cover products (Figures 6.6a to 6.6c) miss a high proportion of cropland in the northern part, whereas the GLC-2000 shows large areas of commission in the central west. A high proportion of these areas of commission lie in Western Darfur where the provincial statistics also record three times the cropland area indicated by the GLC-2000 minimum. For Uganda, the GLC-2000 shows little omission and commission errors occurring in the central and southwestern part, whereas both MODIS, and in particular SAGE, record larger areas of omission in the central northern part. SAGE shows high areas of commission in the southwest around Lake Victoria. All data sets show high areas of omission in Tanzania, whereas MODIS shows hardly any in terms of areas of commission, except a very small proportion in the east. In Kenya, a similar pattern can be observed with hardly any commission errors for MODIS. In terms of omission, however, all three data sets miss cropland areas in the same places. The same pattern is also observed in Somalia.

6.4.5 Total Area of Omission and Commission by Country (Boolean and Fuzzy Approaches)

Figure 6.7a and Figure 6.7b provide the total area of omission for a Boolean compared with a fuzzy approach for MODIS, GLC-2000, and SAGE using AFRICOVER as a reference data set. Figure 6.8a and Figure 6.8b show the same comparisons for the commission error. In general it can be noted that the errors of commission and omission are lower for the Boolean approach. This is because the percentage disagreement is mostly below or equal to the 50% threshold for which agreement is recorded. Even though on an ordinal scale the Boolean method performs the same as the fuzzy approach, the proportional differences between the two methods for the different land cover types are significantly different (e.g., for Kenya the commission errors from the Boolean and fuzzy methods are very different). The analysis reveals that for Sudan, Uganda, and Tanzania, the GLC-2000 has the lowest omission error, whereas the areas of omission are more similar for Kenya and Somalia. In terms of commission error, MODIS has the lowest errors for all five countries (see Figure 6.8b), but on the other hand, consistently the highest omission.

6.5 DISCUSSION AND CONCLUSIONS

Good baseline information on agriculture is an important input to monitoring agricultural expansion and food security. A methodology for the comparison of land cover products was applied to the comparison of four recent land cover products. The overall crop area was validated with national statistics. AFRICOVER was used as a reference data set as a result of its good performance when compared with subnational statistics and since it was produced at a high resolution with visual interpretation and detailed ground information.

In order to have information on the disagreement of the data sets where AFRICOVER was not available, a spatial comparison of the global data sets MODIS versus SAGE, GLC-2000 versus MODIS, and GLC-2000 versus SAGE was carried out. Subsequently, a spatial comparison between the AFRICOVER database and the lower resolution land cover maps, namely, MODIS, GLC-2000, and the SAGE database, was undertaken. Agreement maps that use a methodology for the percentage aggregation of the land cover products was used to make an overall disagreement analysis at the resolution of the SAGE database for a Boolean as well as a fuzzy approach. Whereas the fuzzy approach more precisely shows the disagreement in terms of omission and commission error, the Boolean approach still captures the main discrepancies between the maps. By focusing only on those countries with a negligible proportion of tree crops and by using tree crops in AFRICOVER with the SAGE data set, we were able to compare the different data sets and thereby minimize the problems related to the incompatibility of the legends.

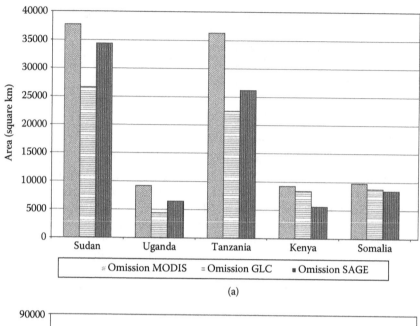

(a)

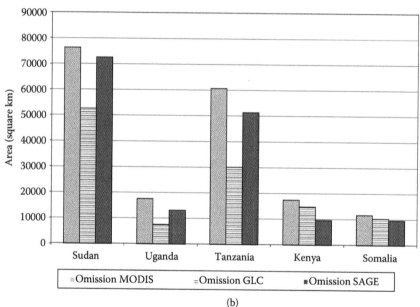

(b)

FIGURE 6.7 (a) Area of omission between AFRICOVER and different land cover products (Boolean approach); and (b) area of omission between AFRICOVER and different land cover products (fuzzy approach).

It has been shown that uncertainties in the cropland distribution in African countries are very high. High spatial disagreement between AFRICOVER and the other three data sets indicates that they have a number of limitations for certain applications within Africa. These preliminary results must be examined carefully as there

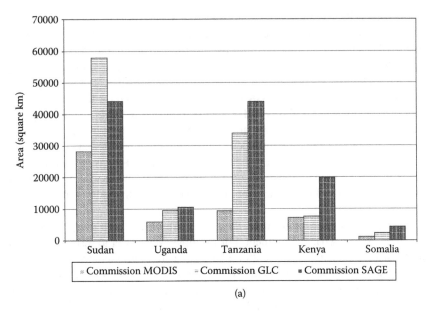

(a)

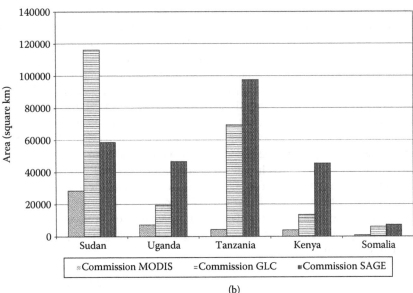

(b)

FIGURE 6.8 (a) Area of commission between AFRICOVER and different land cover products (Boolean approach) and (b) Area of commission between AFRICOVER and different land cover products (fuzzy approach).

is a certain degree of interannual variation of cropland area because the data sets have been produced at different times. Even though the comparison of products that were developed using different methodologies, and the use of different spatial scales and different definitions of cropland area is not straightforward, a number of patterns emerge when the described methodology is applied. We can conclude that

MODIS generally has the tendency to underestimate cropland cover, whereas the GLC-2000 tends to overestimate cropland cover in those countries that are located at the northern transition zone of subtropical shrubland and semidesert areas. In this area MODIS and SAGE show a relatively similar cropland distribution. Even though the SAGE database has been calibrated with national statistics, it does not perform better than the other two data sets overall, and has highlighted the fact that the SAGE data shows regional weaknesses and should be replaced in certain regions by more recent data sets such as GLC-2000 and MODIS, or ideally by a hybrid product that combines the best of the three products, depending upon region and country. It has also been demonstrated that even though overall cropland areas in administrative units (e.g., FAO national statistics) are not so far apart, the spatial distribution of these can vary and a high uncertainty exists when a comparison is undertaken at grid level.

The question remains as to what the causes are of this high disagreement over large areas on the African continent. For example, Mayaux et al. [21] clearly states that the distinction between agricultural land and natural grassland is extremely difficult if not impossible at the spatial resolution of 1 km, which is the resolution at which current global land cover products are produced. The reason given is that natural vegetation and natural grassland have very similar temporal profiles due to the very high proportion of rainfed agriculture in Africa. The different land cover products therefore used different criteria to differentiate between natural vegetation and grassland. As a result, some are more inclusive (overestimation, classifying grassland as cropland) while others are more exclusive (classifying cropland as grassland).

This work has highlighted the need for additional development of an uncertainty layer. This will allow those areas where there is a high range (e.g., 50%–100% cropland cover) to be quantified, and disagreement can be considered in combination with this uncertainty layer. Further work will focus on the selection of auxiliary information to help decide which map is better for those areas where AFRICOVER is not available, as well as the development of a hybrid map based on AFRICOVER, MODIS, and GLC-2000, which will allow the production of a data set at a 1 km resolution, especially with current initiatives to develop more precise national maps and the further enlargement of the AFRICOVER project. Furthermore, this research can be extended with respect to further statistical analysis of the spatial distribution of errors using some common techniques such as cluster analysis or geographically weighted regression [22].

The most conservative approach was followed in this study, allowing ranges of the legend definitions to be considered in the methodology and to focus on those countries where a direct comparison due to a low percentage of tree crops was possible. Nevertheless, we still find large areas of disagreement. The official national and subnational statistics are clearly not error free, but do, however, support some of the spatial patterns observed. The analysis allows us to focus on problematic zones and to identify those data sets that need to be examined in more detail in the countries where large discrepancies with national statistics, together with large areas of spatial disagreement with other data sets, were identified.

Furthermore, there is the issue of purpose. In the situation where these maps are used for land cover change projection, then both commission and omission have more or less the same weight, whereas for the purpose of crop monitoring, it can be advantageous to give a higher importance to omission errors. For this latter purpose the GLC-2000 may be more suitable. Even though it has higher commission errors than the other land cover products, it could be more suitable since it generally has lower omission errors. For all kinds of applications that use agricultural extent as an important input parameter, visualization of the disagreement helps us to gain a better understanding of the spatial distribution of the error. It also allows us to both obtain a better understanding of the other factors with which the disagreement coincides (e.g., patterns of a digital elevation model), and to better determine which data set might be more useful for a particular application.

This chapter has presented a methodology for the comparison of different land cover maps with different legend definitions in the agricultural domain. The differences between the GLC-2000, MODIS, and SAGE land cover products have been compared with FAO statistics as well as the higher resolution AFRICOVER data, and the disagreement has been visualized spatially. This visualization may help to identify the causes of the disagreement and highlight locations of distinct spatial patterns. This work can help to make researchers involved in land cover mapping aware of the current discrepancies and help to focus mapping efforts in those areas where disagreement is highest.

ACKNOWLEDGMENTS

This research was undertaken at the Joint Research Centre (Ispra, Italy) of the European Commission Italy and supported by the EC project GEO-BENE (www. geo-bene.eu), led by the International Institute for Applied Systems Analysis (IIASA). The authors would like to thank the FAO, which made its AFRICOVER full-resolution data set available. A more application-oriented and country-specific version of this work has been submitted to the *International Journal of Remote Sensing*.

REFERENCES

1. Fritz, S. and See, L., Comparison of land cover maps using fuzzy agreement, *International Journal of Geographic Information Science*, 19(7), 787, 2005.
2. See, L. M. and Fritz, S., A user-defined fuzzy logic approach to comparing global land cover products, 14th European Colloquium on Theoretical and Quantitative Geography, Lisbon, Portugal, 2005.
3. Fritz, S., Bartolomé, E., Belward, A., Hartley, A., Stibig, H. J., Eva, H., Mayaux, P., et al., *Harmonisation, mosaicing and production of the Global Land Cover 2000 database* (beta version), Office for Official Publications of the European Communities, Luxembourg, EUR 20849 EN, 2003.
4. Friedl, M. A., McIver, D. K., Hodges, J. C. F., Zhang, X. Y., Muchoney, D., Strahler, A. H., Woodcock, C. E., et al., Global land cover mapping from MODIS: Algorithms and early results, *Remote Sensing of Environment*, 83(1-2), 287, 2002.
5. Leff, B., Ramankutty, N., and Foley, J. A., Geographic distribution of major crops across the world, *Global Biogeochemical Cycles*, 18(1), 2004.
6. FAO, Land cover and land use: The FAO AFRICOVER Programme, 1998, Available at: http://www.fao.org/waicent/faoinfo/sustdev/EIdirect/EIre0053.htm, last updated June 1998.

7. Bartholomé, E. and Belward, A. S., GLC2000: A new approach to global land cover mapping from Earth Observation data, *International Journal of Remote Sensing*, 26(9), 1959, 2005.

8. Di Gregorio, A. and Jansen, L., *Land Cover Classification System: Classification Concepts and User Manual*, Food and Agriculture Organization of the United Nations, Rome, 2000.

9. Loveland, T. R., Reed, B. C., Brown, J. F., Ohlen, D. O., Zhu, Z., Yang, L., and Merchant, J. W., Development of a global land cover characteristics database and IGBP DISCover from 1-km AVHRR Data, *International Journal of Remote Sensing*, 21(6-7), 1303, 2000.

10. Foley, J. A., DeFries, R., Asner, G. P., Barford, C., Bonan, G., Carpenter, S. R., Chapin, F. S., Global consequences of land use, *Science*, 309(5734), 570, 2005.

11. Ramankutty, N. and Foley, J., Characterizing patterns of global land use: An analysis of global croplands data, *Global Biogeochemical Cycles*, 12(4), 667, 1998.

12. Ramankutty, N. and Foley, J., Geographic distribution of major crops across the world, *Global Biogeochemical Cycles*, 18, GB1009, 2004.

13. Miura, T., Huete, A. R., and Yoshioka, H., Evaluation of sensor calibration uncertainties on vegetation indices for MODIS, *IEEE Transactions on Geoscience and Remote Sensing*, 38(3), 1399, 2000.

14. Torbrick, N., Qi, J., Ge, J., Olsen, J., and Lusch, D., An assessment of Africover and GLC2000 using general agreement and videography techniques, *Proceedings of the Geoscience and Remote Sensing Symposium (IGARSS'05)*, IEEE International, 5005, 2005.

15. FAOSTAT, *Agricultural data, Food and Agriculture Organization of the United Nations*, 2005, available at: http://faostat.fao.org/faostat/collections?version=ext&hasbulk=0&subset=agriculture.

16. World Bank, *Degree of disagreement among cropland inventories in African, climate change and African Agriculture*, Policy Note No. 22, August 2006, CEEPA, available at: http://www.ceepa.co.za/docs/POLICY%20NOTE%2022.pdf#search=%22comparison%20of%20cropland%20area%22.

17. Klein, C., Dees, M., and Peltz, D. R., *Sampling Aspects in the TREES Project: Global Inventory of Tropical Forests*, University of Freiburg, Freiburg, Germany, 1993.

18. Gaile, G. L. and Willmott, C. J., *Spatial Statistics and Models*, D. Reidel, Dordrecht, Holland, 1984.

19. Siska, P. P. and Hung, I.-K., Assessment of kriging accuracy in the GIS environment, The 21st Annual ESRI International User Conference, San Diego, CA, 2001.

20. Congalton, R. G., A review of assessing the accuracy of classifications of remotely sensed data, *Remote Sensing of Environment*, 37, 35, 1991.

21. Mayaux, P., Bartholomé, E., Massart, M., Van Cutsem, C., Cabral, A., Nonguierma, A., Diallo, O., et al., *A land-cover map of Africa*, Carte de l'occupation du sol de l'Afrique, Publications of the European Communities, EUR 20665 EN, Office for Official Publications of the European Communities, Luxembourg, 2003.

22. Fotheringham, S., Brunsdon, C., and Charlton, M., *Geographically Weighted Regression: The Analysis of Spatially Varying Relationships*, Wiley, Chichester, 2002.

7 Using GIS to Identify Wildland Areas in the North Pennines

Stuart Blair, Linda See, Steve Carver,
and Peter Samson

CONTENTS

OVERVIEW

This chapter compares different approaches to defining and mapping wildland areas in the North Pennines to establish minimal-intervention areas where a rewilding programs might be introduced. These approaches were implemented in a geographic information system (GIS), and include visibility analysis, multicriteria evaluation, and the use of fuzzy logic. Overall, many of the approaches identified a very small percentage of wildland for the Allendale catchment. However, the fuzzy set and composite approaches provided the most appropriate means by which to identify continuous areas of high quality wildland that incorporates individual perceptions of wildland and attitudes toward anthropogenic features. Although the analyses were limited to the Allendale catchment, these techniques, when combined and placed in the context of the Wilderness Continuum Concept, were found to identify a number of high quality wildland areas potentially suitable for the location of rewilding programs, and could be easily extended to the whole North Pennines area.

7.1 INTRODUCTION

The term *wilderness* has been used in a variety of contexts within contemporary literature, although each instance associates the term with that of engendering human emotions of freedom, solitude, relaxation, and assimilation with nature [1]. Yet, the concept of wilderness has proven to be inherently difficult to define; this difficulty arises as a result of the dynamic, intangible, and very personal nature of wilderness [2]. At present there is no single recommended method for the definition or management of wilderness areas. However, the first statutory definition of wilderness came in 1964 with the United States Wilderness Act [3]:

> Wilderness, in contrast with those areas where man and his own work dominate the landscape, is hereby recognized as an area where the earth and its community of life are untrammelled by man, where man himself is a visitor who does not remain.

The US Forest Service (USFS) then devised the Roadless Area Review and Evaluation (RARE I) framework to undertake a wilderness inventory, where land was evaluated according to a series of quantitative and qualitative measures. The adopted indices were inherently subjective and difficult to measure due to a lack of guidelines. This led to the deployment of RARE II, which established more quantitative measures during the inventory [4].

In Australia an alternative approach to mapping wilderness was developed [5,6], which is based on the wilderness continuum concept (WCC). Unlike the approach adopted by RARE I and RARE II, Lesslie and Taylor's [5,6] wilderness inventory focused on the remoteness and primitiveness of wilderness areas, where remoteness is defined as being a function of proximity to settled land and accessibility from settled people. Lesslie and Taylor [6] developed four wilderness quality indicators: remoteness from settlement, remoteness from access, aesthetic primitiveness, and biophysical primitiveness. When placed in a managerial context, the method allows for the identification of attributes that prove beneficial and detrimental to wilderness quality.

More recently, Carver [7] proposed using a multicriteria evaluation (MCE) or weighted overlay approach to implement the WCC, since MCE is not restricted by specifying rigid thresholds, as adopted in most previous studies. Carver sought to identify the wilderness continuum within Britain, with raster data sets being given precedence due to their capacity to fully handle the continuity required by the WCC mapping problem. Using the wilderness indicators proposed by Lesslie and Taylor [6], the following data sets were derived: (1) remoteness from population; (2) remoteness from access; (3) apparent naturalness; and (4) biophysical naturalness. A weighted linear summation model was then applied. The main advantages facilitated by the use of the MCE approach of Carver are: the ability to simulate the importance that different individuals and pressure groups place upon specific indicators of wilderness quality; and composite maps can be constructed highlighting those areas that satisfy all the specified wilderness criteria.

Wilderness has also been defined using individual perceptions. Although the environments in which wilderness might be found have an objective ecological reality, and one usually excluding anthropogenic modification, what makes that reality explicitly wilderness rests truly with an individual, and her/his personal cognition, emotions, values, and experiences [8]. The problem lies in determining how to map individuals' perceptions of wildland qualities. In a similar manner to that used by Kliskey and Kearsley [8], Fritz et al. [2] composed an Internet questionnaire in which individuals were asked to evaluate the spatial and visual impact of anthropogenic features as described previously upon perceptions of wildland quality in the Cairngorms. Using the results from their online questionnaire, Fritz et al. constructed a series of fuzzy sets for visible and nonvisible anthropogenic features. Having performed the visibility analysis for each of the factors, the Euclidean distance was then calculated for each factor in order to acquire a data set for those areas where features were not visible but still impacted wildland quality. These layers were then converted to fuzzy perceptual map layers and combined using fuzzy logic operators.

It is clear from a brief overview of the literature that there is no one accepted method for defining and mapping wilderness or wildland. Rather there is a range of approaches. This chapter, therefore, compares a number of these methods for identifying the current extent of wildland areas. The study area is the North Pennines and is described in more detail in Section 7.2. Different approaches to defining and mapping wilderness, including visibility analysis, weighted overlay, and the use of fuzzy logic, are then described in Section 7.3. In Section 7.4 the results are compared visually and in terms of the amount and continuity of high quality wildland area suggested by each method, with final conclusions and recommendations following in Section 7.5.

7.2 STUDY AREA

The study area for comparing the different approaches to defining wilderness is the North Pennines, which was designated as an Area of Outstanding Natural Beauty (AONB) in 1988 [9]. This region is the second largest AONB in England and Wales, covering approximately 2000 km^2. In 2003 it became the first area in Britain to be awarded European Geopark certification by the United Nations Educational Scientific and Cultural Organisation [10]. With a resident population of only 12,000 people, the North Pennines AONB has been described as "one of the country's last expanses of wilderness, a high wild landscape of undulating heather-moorland and blanket peat" [11]. This notion of wilderness precipitates throughout many tourist brochures and Web sites, encouraging visitors to the area. Holdgate [12], however, remarks that there is no true wilderness remaining in England and that even the wildest areas have been modified by land management. Carver [13] and Fritz et al. [2] suggest that the terms *wildland* or *secondary wilderness* (i.e., areas recovering from former extensive human exploitation) are better representations of a landscape dramatically altered by a long history of settlement and exploitation. Moreover, the sense of naturalness and remoteness within many wildland areas is becoming increasingly threatened by potentially intrusive developments such as telecom masts and wind turbines [9]. Therefore, organizations such as the North Pennines AONB

Partnership want to identify and cartographically catalog the country's remaining wilderness expanses.

A five-year plan for land management has been formulated by the North Pennines AONB Partnership to address a number of issues, including the result of changes to the Common Agricultural Policy (CAP) and the abolition of CAP subsidies by 2013, as well as the increasing numbers of tourists wanting a wilderness or wild-land experience [9]. This investigation focuses upon Objective 14 of its management plan to "establish, monitor and review a major new 'minimal-intervention' trail site within the AONB" [9]. The minimal-intervention sites are intended for the purpose of introducing rewilding programs (i.e., areas where native vegetation/woodland species may be reestablished) into the North Pennines AONB, thereby continuing the aim of conserving and enhancing the AONB's natural integrity. Rewilding programs have already been undertaken and the impacts studied in both the United States and Italy [14,15]. The study compares the different approaches for one catchment, the Allendale catchment (Figure 7.1), but could easily be extended to the entire North Pennines.

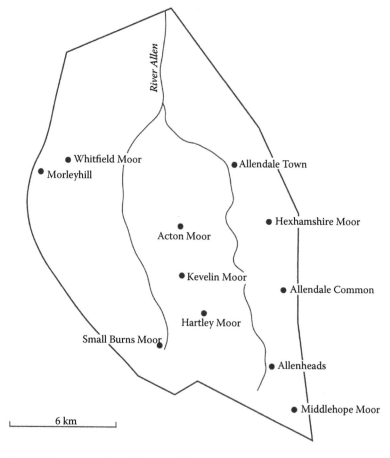

FIGURE 7.1   Features in the Allendale catchment.

7.3 METHODOLOGY

The literature clearly reveals that there is no single accepted method for defining and mapping wilderness. For this reason, different approaches are implemented and compared. These methods are drawn heavily from the work of Fritz [16], Carver and Fritz [17], and Carver [13]. Visibility analyses, weighted overlay, and fuzzy logic have all been applied by these authors and were chosen in this study to illustrate a range of potential methods. The difference lies in the area chosen for study, thereby assessing the transferability of these methods. It also includes a composite method that takes into account a combination of physical and perceived apparent naturalness with remoteness from anthropogenic features. Finally, it uses questionnaire data from the Cairngorms and transfers this to the North Pennines to determine whether reasonable wildland maps can be produced using this perceptual data.

As previously mentioned, the study is focused on the Allendale catchment, which is located within the northern half of the North Pennines AONB. Prior to any analysis, an 18 km buffer was included around this catchment to assess the visual impact of anthropogenic features occurring outside the catchment boundaries on wildland quality within the catchment [18,19] and to avoid any edge effects had this buffer not been included. The choice of an 18 km distance was taken directly from the Sinclair Thomas matrix (used by the UK Parliament and Campaign for the Protection of Rural Wales (CPRW) in assessing the environmental impacts posed by windfarms); 18 km was determined as the distance at which significantly intrusive features have little or no impact [20].

The determination of which anthropogenic features to include in this study was based on work by Lesslie and Taylor [6]. As previously mentioned, they identified four factors that together can be used to define wilderness areas: (1) remoteness from settlement; (2) remoteness from access; (3) aesthetic primitiveness, also called apparent naturalness; and (4) biophysical primitiveness or naturalness. These factors are considered through seven input layers: (1) built-up areas, which are defined as areas in which a number of houses are situated and in which there is permanent human population; (2) roads and tracks, both surfaced and unsurfaced vehicular access routes; (3) disused railway lines, which are routes that were once used for the transportation of mining products though have since become derelict or abandoned; (4) isolated buildings, such as huts, farmhouses, and derelict structures; (5) plantation woodlands; (6) reservoirs; and (7) quarries.

All of the above features were obtained from the topographic layer of the Ordnance Survey MasterMap with the exception of plantation woodlands and reservoirs. Reservoirs were digitized from existing paper maps; the procedure for producing a layer of plantation woodlands was more complicated. The National Inventory of Woodland Trees (NIWT), which was available from the Forestry Commission, was used as the starting point. Layers of known tree species locations available from English Nature were then subtracted from the NIWT layer, leaving plantation woodlands. These factors also match those used by Fritz [16] for the Cairngorms in Scotland, which therefore allows the use of his questionnaire of wildland perception to be applied in this area. The different approaches will now be described in the sections that follow.

7.3.1 Binary Visibility

In the first approach, a viewshed analysis was undertaken on each of the seven anthropogenic feature maps to determine the areas or grid cells that can see these features with the frequency of the occurrence. The frequency maps were then reclassed so that grid cells where features are visible were reclassed to 1 and not visible to 0. These seven binary feature maps were then multiplied together showing areas of invisibility to all features and areas where are one or more features are visible. This resulted in a map that produced some degree of anthropogenic feature visibility from virtually every viewpoint.

7.3.2 Cumulative Visibility

As a result of the limitations of a binary visibility approach, the original seven frequency maps were summed and then reclassified into categories of low, medium, and high wildland quality. The high wildland quality category was then further subdivided into three classes: moderately high, high, and very high, thereby considering the WCC as specified by Lesslie and Taylor [6]. The resulting map is shown in Figure 7.2a.

7.3.3 Perceptions of Wildland Quality

The online questionnaire developed by Fritz [16] was used to determine peoples' perceptions of wildland quality. The questionnaire asked respondents to evaluate the perceived impact of a variety of anthropogenic features upon wildland quality when situated at near, medium, and far distances from the observer location. The answers were then translated into a numerical value, ranging from 6 (very strong impact) to 1 (no impact), providing a means by which to quantify the perceived impact to wildland quality posed by specific anthropogenic features. Respondents were also asked to state the length of the walk in minutes that is required to be in low, medium, and high quality wildland. Using different aspects of the questionnaire, five approaches were then implemented, as described next.

7.3.3.1 Simple Access Model

A simple buffering approach was first undertaken, where Euclidean distance from road features was calculated as a proxy for remoteness. This definition of remoteness assumes equal ease of travel in all directions, irrespective of land cover and topography [21]. The distance in each pixel was then divided by 1.389 m/s, which equates to a constant 5 km/hr walking speed. This output was in turn reclassified to delineate areas requiring 0–33 minutes, 33–89 minutes, and >89 minutes to access from points of mechanized access (with these numbers reflecting the average values of the walking times required to be in low, medium, and high quality wildland areas as indicated by the questionnaire respondents).

7.3.3.2 Anisotropic Access Model

The next approach, referred to here as anisotropic buffering, is similar to the simple access model but walking times were corrected for land use and slope. This model

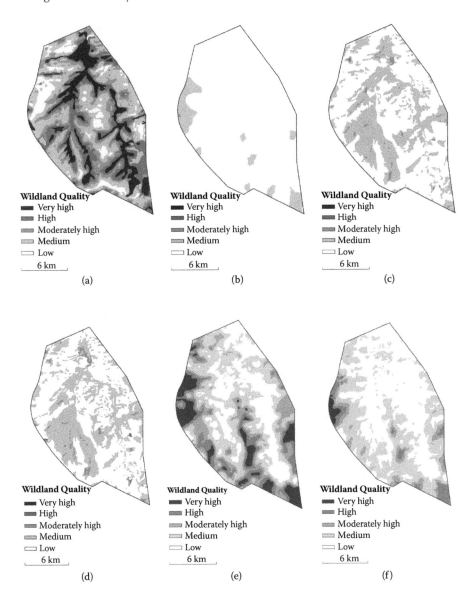

FIGURE 7.2 The output maps of wildland areas for the Allendale Catchment derived from the (a) cumulative visibility (b) anisotropic access model (c) rule-based approach (d) weighted overlay and (e) fuzzy set approach and the (f) combined approach.

extends the concept of remoteness to incorporate Langmuir's correction using the vertical relative moving angle (VRMA) field. The VRMA identifies the slope between selected source and target cells, and may be converted into a vertical factor to indicate relative walking times using a correction factor (see Carver and Wrightham [21] for more details and for the full set of correction factors). Once the correction had been applied, the resulting map was then reclassified into areas of low, medium, and high wildland quality based on a constant 5 km/hr walking speed.

7.3.3.3 Rule-Based Approach

In the rule-based approach the answers from the questionnaire of Fritz [16] were used to derive a set of rules, which was then applied to the seven factor maps to produce a map of wildland quality. Using an average of the respondents' answers regarding distances for near, medium, and far, the impact values assigned to the seven anthropogenic features were used to identify wildland areas using the if–then rules; an example of rules for visibility of a road feature is:

> **IF** you are near (0–695 m) to a visible road **THEN** this has an impact value of 5 on wildland quality.
>
> **IF** you are a medium distance (695–3394 m) to a visible road **THEN** this has an impact value of 5 on wildland quality.
>
> **IF** you are far (>3394 m) from a visible road **THEN** this has an impact value of 4 on wildland quality.

The average distance values were derived from questionnaire responses, defining near as 0–695 m, medium as 695–3394 m, and far as greater than or equal to 3394 m. The distances were not predefined in the questionnaire; respondents were asked to indicate what they felt were near, medium, and far distances. These rules were then applied on a pixel-by-pixel basis. This process was repeated for each of the seven factors and the maps were averaged to produce a composite map that was then reclassed into low, medium, and high wildland quality, where high was further subdivided into high, moderately high, and very high according to the WCC.

7.3.3.4 Weighted Overlay

A weighted overlay was utilized where the weights were determined by the mean wildland impact value given by respondents of the questionnaire. Answers to the questionnaire were coded from 0 to 5 to denote no impact to a very strong impact and were used to produce a mean wildland impact. The seven factors were standardized in order to enable a meaningful comparison. Table 7.1 shows the weights assigned to each factor. These weights were then multiplied against their corresponding standardized linear-distance factor maps and summed. As with the rule-based approach, the map was reclassified into low, medium, and finer categories of high wildland quality.

7.3.3.5 Fuzzy Set Approach

An additional approach undertaken to assess the extent and quality of wildland within the study area incorporates fuzzy set theory into a weighted overlay procedure [22]. Distance was first transformed into fuzzy membership functions using information from the questionnaire and then processed using the rules. The left-hand, mid, and right-hand values of triangular membership functions were identified using the minimum, mean, and maximum values specified by respondents of the questionnaire. These results were then used to construct the overlapping input fuzzy

TABLE 7.1
Weights Adopted in the Weighted
Overlay Analysis

Criterion	Weighting Value
Buildings	0.40
Quarries	0.20
Reservoirs	0.12
Roads	0.12
Plantation woodland	0.10
Railway lines (disused)	0.05
Structures	0.01
Total	1.00

sets, with three fuzzy sets (representing near, medium, and far distances) being created for each of the seven anthropogenic factors as described previously. Once fuzzy membership functions for near, medium, and far distances had been constructed, the if–then rules were used to process the data. This was repeated for each of the seven factors. The resulting maps were then averaged, which, as described by Jiang and Eastman [22], is a perfect fuzzy operator for representing attitudes at a middle point, and proves highly desirable from a managerial context in representing the majority of perceptions. Classification into degrees of wildland quality was the final step.

7.3.3.6 Combined Approach

Finally, after looking at the results from applying the aforementioned methods and considering the advantages and disadvantages of each one, a final combined method was adopted in which the cumulative visibility, anisotropic buffering, and fuzzy set wildland maps were averaged. The choice of these three approaches provides a combination of physical and perceived apparent naturalness with remoteness from anthropocentric features, all of which are important factors affecting wildland quality. Averaging is only one method of combination. It would, of course, be reasonable to weight these layers given expert knowledge, which could be easily modified in future analyses.

7.4 RESULTS AND DISCUSSION

Figures 7.2a to 7.2f and Table 7.2 show the resulting maps and percentage of high quality wildland areas respectively for the Allendale catchment as generated by application of the different approaches. When analyzed using a binary visibility approach, a total of 5.25 ha of wildland is identifiable, equating to 0.02% of the total Allendale catchment. The result is due to visually detrimental features such as roads-and-tracks and built-up areas affecting the aesthetic naturalness. The largest area of wildland identified is an area of ~1 ha in size, found along the Knight's Cleugh tributary in Hexhamshire Moor, a Special Site of Scientific Interest (SSSI). The results of the cumulative viewshed analysis (Figure 7.2a) evaluated using the WCC [6] appear

TABLE 7.2

Extent of High Quality Wildland Areas
Identified within the Allendale Catchment

Method	% High Wildland Quality
Binary visibility	0.02
Cumulative visibility	36.85
Simple access model	0.00
Anisotropic access model	0.36
Rule-based approach	2.94
Weighted overlay	1.67
Fuzzy set approach	37.79
Combined approach	10.33

to provide a more appropriate means by which to assess aesthetic primitiveness and naturalness. The results show that 8961 ha or 37% of the study area is at the top end of the WCC, with areas of higher wildland quality typically being found within the East and West Allen valleys. However, these are also the locations in which settlements such as Allendale Town and Allenheads have evolved. Regions of higher altitude (Whitfield Moor and Acton Moor) have lower aesthetic primitiveness/wildland quality as a consequence of the increased spatial area (and therefore anthropogenic features) visible from higher altitudes, compared with locations situated within the incised valleys. Baban and Parry [23] and CPRW [20] cite that the effects of atmospheric refraction and earth-curvature may increase individuals' perceptions of solitude [24]. As noted by Fisher [25], given these factors, viewshed analyses typically overestimate the number of visible features.

The next set of approaches considered the physical remoteness of regions within the Allendale catchment, derived as the relative time taken to access areas (by foot) from points of mechanized access [17]. Table 7.2 shows that there were no areas of high quality wildland identified for the simple access model and thus no map is provided. Due to the density of road-and-track-networks within this area, the highest quality of wildland identified by this simple buffering process is medium (i.e., 33–89 minutes-walk), with the area identified being of relatively small spatial extent (i.e., 71.75 ha). Although this technique presents a means by which to identify increasingly remote regions, such an approach provides only a rudimentary means by which to assess this indicator of wildland quality [17], neglecting crucial geographical factors influencing pedestrian off-road access such as topographic variables (e.g., slope, altitude), land cover (e.g., vegetation type/height), meteorological conditions, and barrier features (e.g., water bodies).

To facilitate the improved simulation of remoteness, the model developed by Carver and Fritz [17] was applied to the Allendale catchment with the results for the anisotropic access model presented in Figure 7.2b. Upon applying the correction factor for topography, a ~7% increase is cited to those areas classified as medium quality wildland, with this approach also classifying 13.25 ha situated upon Morleyhill Fell as high quality wildland. Those areas requiring increased access times are those situated at higher elevations, with large areas of open and dense moorland vegetation, where the effects of rapid increments in elevation and slope angle reduce walking speeds and accessibility (i.e., Morleyhill/Mainsrigg Fell, Kevelin Moor, Hartley Moor, the Combs/Allendale Common, and Middlehope Moor/Wolfcleugh Common). Such areas also show a reduced number of built-up areas (due to harsher weather conditions), with many of the highlighted areas being SSSIs, which may explain the limited number of access features and their increased physical remoteness. Given the improved accuracy with which physical remoteness from points of linear access may be interpolated, this approach appears to provide a more appropriate assessment of wildland quality than the previous approach. Owing to the incorporation of a number of physical factors affecting pedestrian off-road access, it is hypothesized that these results, when placed in a spatial-decision-support-system (SDSS), may identify those areas most likely to be sought out by recreationalists seeking increasingly solitary wildland experiences [3,24].

In addition to physical remoteness, perceived remoteness (i.e., the presence/absence of human features within the landscape) also holds significant value in determining perceptions of wildland quality [3,8,26]. The remaining approaches therefore concentrated upon identifying locations most suitable for the minimal-intervention sites according to individuals' perceptions of wildland and what attributes contribute to providing a high quality wildland experience. The first approach undertaken was that of processing the if–then rules generated by answers to Fritz's [16] wilderness questionnaire. Using this approach, ~4% of the study area (Figure 7.2c) is found to tend toward the higher end of the WCC, equating to 769.75 ha. This technique provides a simple and transparent means by which to process the impacts of an individual's attitude to various anthropogenic features. Furthermore, this approach allows the correlation between distance and the assignment of impact values to be considered using terminology commonly associated with the cognitive manner in which humans typically gauge/describe distance (i.e., in terms of near, medium, and far objects).

Figure 7.2d shows the results of the weighted overlay analysis. Upon reclassifying results to indicate high, medium, and low quality wildlands, ~98% of the study area is classified as low to medium quality. The total area identified as being of high wildland quality is 404.75 ha, approximately 365 ha less than that identified by directly processing the if–then rules. Those areas identified by the weighted overlay analysis as tending toward the higher end of the WCC are typically situated in areas surrounded by terrains of high slope angles (e.g., Smallburns Moor) or in areas following the numerous gills created by the East/West River Allen's tributaries (e.g., Knight's Cleugh within Hexhamshire Moor). Given the physical geographical attributes of such areas and the criteria used throughout the weighted overlay analysis, it is not surprising that such areas were identified as being of potentially higher wildland quality due to their increased relative distance from anthropogenic features. Lesslie et al. [27] also note that, unlike techniques based upon the simple addition of wildland quality indicators, such as those used to construct the cumulative frequency visibility map (Figure 7.2a), weighted overlay-analyses do not assume that the factors contribute uniformly to overall wildland quality (see Habron [3,24] and Henderson

[26]) or that a unit of measurement for one indicator has equivalence with another. Indeed, when placed in the context of a SDSS, the individual factor maps contributing to the overall wildland quality may be used independently to identify those features with the most impact [16,27]. Given the uncertain future facing many upland farmlands in the area, such approaches can be seen to facilitate the simulation of potential land use change by removing certain human features, and thus determine the types of changes that will result in improved wildland quality [16]. Such approaches also offer a relatively transparent means of identifying areas potentially most suitable for rewilding while accommodating updates/additional data sets.

The next approach to identifying perceived wildland within the Allendale catchment saw Fritz's [16] questionnaire results being translated into the spatial domain using fuzzy logic. The results of this analysis are shown in Figure 7.2e. Regions within the near visible proximity of highly built-up areas and road/track features are denoted as low to medium quality wildland, whereas those situated at higher elevations or in far locations are classified as tending toward the very high end of the WCC. Comparing these results with those previously discussed, the fuzzy set approach produces results that diverge quite dramatically from those previously described. It also resulted in a significant increase in the percentage of area identified to be in the higher end of the WCC.

The final approach was the composite approach, which combined results from different approaches, recommending 10% of the catchment for identification of minimal intervention sites, as shown in Figure 7.2f. This final wildland map contains areas that are considered both physically and perceptually remote while also retaining increased levels of aesthetic naturalness from surrounding anthropogenic features. The locations identified as being potentially most suitable are Middlehope Moor and Morleyhill/Mainsrigg Fell within Whitfield Moor.

7.5 CONCLUSION

For local areas of relatively high wildland quality, land managers increasingly require detailed information about the location of wildland resources to address a wide range of policy, planning, and management issues. Inventories such as the one conducted within this study are becoming increasingly necessary, especially as the wildness and remoteness of the United Kingdom's more natural landscapes are threatened by potentially intrusive developments such as installation of mobile telephone masts and wind farms [9]. This study reflects recent efforts to provide a critical comparison of different GIS techniques adopted in wildland inventories. Overall the majority of approaches identified a very small percentage of wildland, which renders identification of potential sites for rewilding a difficult task. They do, however, support the view that subsequent to the industrial revolution and agricultural intensification witnessed within the Allendale catchment, as throughout England, no geographically sizeable areas of true wildland remain [7,12,16,24]. However, the fuzzy set and composite approaches provide one method of identifying a greater amount of contiguous high quality wildland, which represents a starting point for considering the location of rewilding sites. The methods also allow individual perceptions of wildland and attitudes toward anthropogenic features to be incorporated.

It would be possible for wildland managers in this area to carry out questionnaires on perceptions of wildland that might feed into future analyses specific to this area. At the same time they might consider addressing the additional impact of noise, which has been shown to negatively affect the wilderness experience of visitors to national parks in North America, as summarized by Mace et al. [28], but which was beyond the scope of this study.

REFERENCES

1. Ewert, A. and Hollenhorst, S., Adventure recreation and its implications for wilderness, *International Journal of Wilderness*, 3(2), 21, 1997.
2. Fritz, S., Carver, S., and See, L., New GIS Approaches to Wild Land Mapping in Europe, USDA Forest Service Proceedings RMRS-P-15, 2, 120, 2000.
3. Habron, D. Visual perception of wild land in Scotland, *Landscape and Urban Planning*, 42, 45, 1998.
4. Hendee, J., Stankey, G., and Lucas, R., *Wilderness Management*, Fulcrum Publishing, North American Press, Colorado, 1990.
5. Lesslie, R. G. and Taylor, S. G., Wilderness in South Australia, Centre for Environmental Studies Occasional Paper 1, Centre for Environmental Studies, University of Adelaide, 1983.
6. Lesslie, R. G. and Taylor, S. G., The wilderness continuum concept and its implications for Australian wilderness preservation policy, *Biological Conservation*, 32, 309, 1985.
7. Carver, S., Mapping the wilderness continuum using raster GIS, in *Raster Imagery in Geographic Information Systems*, Morain, S. and Lopez-Baros, S., Eds., OnWord Press, New Mexico, 283, 1996.
8. Kliskey, A. and Kearsley, G., Mapping multiple perceptions of wilderness in southern New Zealand, *Applied Geography*, 13, 203, 1993.
9. NPOAONBMP, The North Pennines Management Plan 2004-2009, Available at: http://www.northpennines.org.uk/index.cfm?articleid=4369, 2004.
10. English Heritage, Heritage Counts 2004 in the North East, Available at: http://www.english-heritage.org.uk/heritagecounts/2004_pdfs/NORTHEAST.pdf, 2004.
11. BBC, Seven Natural Wonders, http://www.bbc.co.uk/england/sevenwonders/info/, 2005.
12. Holdgate, M., Benefits Beyond Boundaries; Work in the UK's Protected Areas, Council for National Parks, London, UK, 2003.
13. Carver, S., Wilderness Britain? Social and Environmental Perspectives on Recreation and Conservation, UK, ESRC Funded Seminar Series, http://www.geog.leeds.ac.uk/people/s.carver/wildbrit_web/newsletters.html, 1999.
14. Noss, R. F., A checklist for wildlands network designs, *Conservation Biology*, 17, 1270, 2003.
15. Hochtl, F., Lehringer, S., and Konold, W., "Wilderness": What it means when it becomes a reality—a case study from the southwestern Alps, *Landscape and Urban Planning*, 70, 85, 2005.
16. Fritz, S., Mapping and Modelling of Wild Land Areas In Europe and Great Britain: A Multi-Scale Approach, Unpublished PhD Thesis, School of Geography University of Leeds, UK, 2001.
17. Carver, S. and Fritz, S., Mapping remote areas using GIS, in *Landscape Character: Perspective on Management & Change, Scottish Natural Heritage*, Usher, M., Ed., The Stationary Office, London, 1999.
18. Fisher, P., First experiments in viewshed uncertainty: Simulating fuzzy viewsheds, *Photogrammetric Engineering and Remote Sensing*, 58, 345, 1992.

19. Fisher, P., Farrelly, C., Maddocks, A., and Ruggles, C., Spatial analysis of visible areas from the bronze age cairns of Mull, *Journal of Archaeological Science*, 24, 581, 1997.
20. CPRW Report, Memorandum by the Campaign for the Protection of Rural Wales (CPRW), UK, The United Kingdom Parliament, available at: http://www.publications.parliament.uk/pa/ld199900/ldselect/ldeucom/18/18a04.htm, 1999.
21. Carver, S. and Wrightham, M., Assessment of Historic Trends in the Extent of Wild Land in Scotland: A Pilot Study, Scottish Natural Heritage Commissioned Report No. 012, 2003.
22. Jiang, H. and Eastman, R., Application of fuzzy measures in multi-criteria evaluation in GIS, *International Journal of Geographical Information Science*, 14, 173, 2000.
23. Baban, S. M. J. and Parry, T., Developing and applying a GIS-assisted approach to locating wind farms in the UK, *Renewable Energy*, 24, 59, 2001.
24. Habron, D., Defining the characteristic landscape attributes of wild land in Scotland, in *Landscape Character: Perspective on Management and Change, Scottish Natural Heritage* Usher M., Ed., The Stationary Office, London, UK, 1999.
25. Fisher, P. F., Algorithm and implementation uncertainty in viewshed analysis, *International Journal of Geographical Information Systems*, 7(4), 331, 1993.
26. Henderson, N., Wilderness and the nature conservation ideal: Britain, Canada and the United States, *Ambio*, 21, 394, 1992.
27. Lesslie, R., Mackey, B., and Preece, K., A computer based method of wilderness evaluation, *Environmental Conservation*, 15, 225, 1988.
28. Mace, B. L., Bell, P. A., Loomis, R. J., Visibility and natural quiet in national parks and wilderness areas: Psychological considerations, *Environment and Behaviour*, 36, 5, 2004.

8 Representations of Environmental Data in Web-Based GIS

Peter Mooney and Adam C. Winstanley

CONTENTS

OVERVIEW

The Geographic Information Systems (GIS) community is using the vast potential of the Internet to disseminate geospatial information. Web-based GIS software and services are key components in the distribution of geospatial data. Web-based GIS provide government departments, local authorities, and environmental agencies with unprecedented opportunities to offer online access to their environmental information and related services for citizens. Web-based GIS offers access to information services 24 hours a day, 7 days a week, 365 days a year. In order for Web-GIS to be successful in delivering environmental information, the representation of the input data sets and output delivery formats/structures must be suitable to both the Internet delivery medium and the intended audience. In the majority of cases this will involve conversion and remodeling of existing data resources. This chapter discusses representations of environmental data for delivery and dissemination using Web-based GIS in order to serve a variety of stakeholders: policy makers, scientists, media, and the general public. We summarize the major issues for delivering complex geospatial data about the environment using this medium. Prioritization of metadata collection and geospatial data interoperability is a crucial factor in delivering effective Web-GIS tools. The INSPIRE Directive will greatly increase the number of available data sources and the use of Web-based GIS for environmental information provision in the future will be discussed.

8.1 INTRODUCTION

Environmental issues are now topics of conversation for the general public. Coffee-break conversations often include global environmental issues such as climate change or changes in weather patterns, and more localized issues such as air quality, water quality, and waste disposal. Previously, environmental data access and distribution was confined to policy makers, analysts, and the environmental science community. Traditionally, GIS was a technology that ran only on large computer systems, eventually migrating to desktops, and, more recently, the software has been increasingly available across the Internet. This has provided many opportunities to provide access to data previously unavailable to the public [1]. The Internet-enabled general public is now an important stakeholder for governments and environmental agencies across the world. The public's awareness of environmental issues coupled with almost universal access to the Internet means that there is a great opportunity for Web-based GIS to provide the services that stakeholders require. There is also an increasing recognition that government agencies holding environmental information collected at public expense must make these data assets accessible. This accessibility includes easier and quicker access to this environmental information for little or no charge.

This chapter discusses the use of Web-based GIS from the perspective of users who will use Web-based GIS for either information gathering purposes (i.e., accessing local authority information) or to download geospatial data. In this chapter we predominantly focus on users of Web-based GIS. This user audience comprises GIS specialists, decision makers, and the general public. If the user group has no formal GIS skills we refer to them as nonspecialist users. The core discussion topic in the chapter is how Web-based GIS systems are developed from a software perspective, managed by organizations, and used to deliver visualization and geospatial data access services for users. This involves a discussion of the most popular types of representations for environmental data being delivered using Web-based GIS and the role metadata has to play as a representation of data and software services. Web-GIS presents a unique opportunity in data and information provision to make more data available to a wider range of people.

The chapter is organized as follows. Section 8.2 outlines an overview of Web-based GIS by discussing the key components in delivering GIS to the Internet. This is followed by a brief overview of some of the most well known and widely deployed Web-GIS server and client software technologies. Section 8.3 discusses the core focus of the chapter by looking at how best to represent environmental information to allow it be used by a wide variety of Web-based GIS applications and tools. Section 8.4 looks at metadata and its importance to Web-based GIS and to access to geospatial data in general. The chapter closes with a summary of the main conclusions and some ideas for further work and discussion.

8.2 AN OVERVIEW OF WEB-BASED GIS

Web mapping is loosely defined as the process of designing, generating, and delivering maps on the Internet. Web GIS brings GIS functionality to Web mapping. In GIS vernacular, Web mapping and Web-based GIS are often used synonymously despite the fact that they do not necessarily refer to equivalent technologies. During the mid-1990s, several innovations made it possible to develop Web-GIS solutions: support for vector graphics in Internet browsers, the birth of Javascript and JAVA, and ESRI's entry into Web mapping, among other technological developments. Such software development was very complex and was only carried out by large organizations with large software development budgets. For other users, delivering map-based information on the Internet was facilitated in many cases by generating the necessary mapping as JPEG or GIF images and then implementing a mix of HTML and Javascript to allow users to click on features in these maps, query and retrieve other map output, or access data download. This situation has greatly changed, and users of Web-based GIS perform many of the operations they can perform on their desktop GIS. Examples of such operations include: pan and zoom, feature (point, line, polygon)-based querying, addition or removal of layers, path distance calculations, area estimations, and buffer zone queries. In theory, any spatial analysis or spatial statistics functionality available in standard desktop GIS can be implemented within Web-based GIS. However, there is one very important consideration. Desktop GIS benefit from the processing power of the local computer, have immediate access to disk storage, and leverage high-end graphics visualization. In Web-based GIS, all operations must be performed in real time. Factors such as bandwidth capacity, network latency, and Internet browser type must be taken into consideration. In addition to this, Web users expect near to instantaneous responses from information systems [2]. For this reason only lightweight queries (any query to which the user is delivered a response in close to real time) are usually implemented. Large-scale queries are performed offline.

8.2.1 WEB-BASED GIS TECHNOLOGIES

Web-GIS applications can be classified according to whether they are server or client systems. These shall be discussed in more detail in the next sections. Before this discussion we will outline the two key architectures upon which Web-based GIS are developed. The first is the Open Geospatial Consortium (OGC) specifications and the second is AJAX (Asynchronous Javascript and XML).

The OGC is an international voluntary consensus standards organization. The OGC specifications are a collection of specifications or standards developed to assist in achieving interoperable geospatial technology at both the software service level and the geospatial data specification level. The most widely implemented OGC specifications are as follows: Web Map Services (WMS) provide specifications on how the delivery and rendering of maps as images should be implemented in software;

Web Feature Services (WFS) specify the standards for the exchange of raw geographic feature data over the Internet; and Web Coverage Services (WCS), which are similar to WFS and are used for the exchange of geographic coverage data such as segmented curves, grids, and digital terrain models. The OGC specifications have greatly influenced the direction of recent Web-based GIS developments. WMS and WFS make it much easier to publish, visualize, and exchange any geospatial data over the Internet. WMS essentially creates maps in popular image formats (PNG [Portable Network Graphics] or JPEG formats) of the requested geographical area. The generated image is rendered within the Web browser. Users do not have to copy large geospatial data sets to local systems in order to visualize the output in their desktop GIS. WFS allows users to access and download subsets of larger geospatial data sets directly into their Web-based or desktop GIS.

AJAX is primarily an Internet application development technology that supports the development of Web pages that permit interactivity without the need for Web page refreshing. This is particularly useful when map-based interfaces are embedded into Web-based applications. If the application is driven by AJAX, the client user can query and interact with the map display without the need for Web page refreshing. Many readers will be already familiar with this technology from using Web sites featuring Google Maps or Microsoft Virtual Earth mapping.

If organizations do not wish to manage Web-based GIS server or client software directly from their own systems, Google Maps and Virtual Earth, for example, allow developers to access their API (application programming interface) to develop customized location-based services and Web mapping services. This means that such users do not have to manage Web-based GIS services on their own systems and can take advantage of the large cartographical resources of Google and Microsoft. Scharl [3] remarks that this approach by Google and Microsoft has exposed the deficiencies of traditional GIS by tapping into what GIS end users really want: simplicity, accessibility, immediacy, responsiveness, and low-cost development of Web mapping services.

The use of AJAX technology has shown organizations and users that access to geospatial information and the building of Web mapping services can be easy and intuitive. Consequently, general Internet users with little or no GIS, cartography, or geospatial data handling expertise are equipped with a very powerful set of mapping tools and services. Such is the simplicity of using the Google Maps or Microsoft Virtual Earth APIs that in just a few minutes users can integrate mapping into their Web sites or blogs by simply copying and pasting embeddable code. The very simple representation of geospatial information using KML (Keyhole Markup Language; see Section 8.3.3) has empowered a vast community outside of GIS. Scharl [3] remarks that this opens a new age of cartography, that people are now "spurred by space photography, mobile phones, and new ways of annotating Web content [such] that the ancient art of cartography is now on the cutting edge."

8.2.2 KEY WEB-BASED GIS SERVER SOFTWARE TOOLS

Geospatial data is usually stored on a server machine to make it accessible via the Internet. In order to provide spatial query functionality, cartographic visualizations, and spatial data exchange, special GIS software must be running on these server machines. This software is referred to as GIS server software. This section gives a brief overview of four of the most popular GIS server software tools.

ESRI ArcIMS (Arc Internet Map Server) [4] is a Web map server developed by ESRI. Web-mapping applications developed on ArcIMS offer tools that users of other ESRI products will be familiar with—zooming, toggling between layers, identify and query, find and measure, buffering, select by feature, and printing of the cartographic output. ArcIMS can be customized and extended using programming environments such as JAVA, ASP, and .NET. It is particularly well suited to organizations and individuals who already use other ESRI products (ArcSDE and Geodatabases) to mange their geospatial data resources.

MapGuide Open Source [5] is an open source version of Autodesk's MapGuide Enterprise. MapGuide is a Web GIS server allowing developers to create interactive Web mapping applications using AJAX. It offers access to most vector and raster data formats including SDF vector file formats and CAD-based data access. MapGuide offers quality cartographic output, uniform access to raster and vector data formats, application development in several languages such as PHP, .NET, and JAVA, and access to mapping and feature data from other publicly available WMS and WFS on the Internet.

MapServer [6] is also an open source development environment. It delivers high-quality cartographic rendering of *spatial* data (maps and images), as well as access to *spatial* data (vector and raster). It also features a fully featured development environment allowing developers to create applications to interact with MapServer using tools such as PHP, Ruby-on-Rails, C#, and JAVA. Many organizations use MapServer as the access point to their geospatial data repositories to take advantage of the excellent WMS, WFS, and WCS implementation. Consequently, MapServer is often used in conjunction with a dedicated Web-based GIS mapping client, where these clients render the output from MapServer in client browsers and applications.

GeoServer [7] is a JAVA-driven geospatial data server providing organizations with WMS, WFS, and WCS implementations. In addition to vector and raster support it offers mature and stable support for a large variety of geospatial databases such as PostGIS, Shapefiles, ArcSDE, and Oracle. As it is a JAVA servlet-based application it can be run within any JAVA servlet container environment. A large deal of the functionality offered by GeoServer—reprojection, access to geospatial databases, and raster manipulation—is handled transparently by GeoServer itself. It uses the extensive JAVA GIS toolkit called GeoTools to accomplish this.

8.2.3 WEB-BASED GIS CLIENT SOFTWARE TOOLS

There are a large number of software tools available that offer easy-to-use client-side Web-mapping application development and access to geospatial data services. This section provides an overview of three of the most commonly used tools. Web-based GIS clients normally connect to publicly accessible WMS, WFS, and WCS offered by data provider organizations running Web-based GIS server technologies such as those mentioned in Section 8.2.2. WMS are also offered by Google (through Google Earth and Google Maps) and Microsoft (through Windows Live and Virtual Earth).

ESRI ArcIMS features a rich set of Web-publishing capabilities allowing organizations to develop scalable and interactive interfaces to their geospatial data resources. The key advantage of using ArcIMS as the development environment is that the overall cost of ownership is controlled as ArcIMS also controls the Web-GIS server functionality and management. GeoMedia WebMap by Intergraph is a long established Web-based map visualization and analysis software tool. It can access and view a large number of geospatial data formats in their native format without conversion or translation and also has extensive Scalable Vector Graphics (SGV) support, which is a World Wide Web Consortium (W3C) graphics format specification. The Free and Open Source OpenLayers project is a pure Javascript library for displaying map data in Web browsers with no server-side dependencies. The OpenLayers Javascript API provides developers with full control over OpenLayers-powered maps from within Javascript on a Web page. OpenLayers offers no server capabilities but can connect to a wide variety of WMS, WFS, and WCS.

Figure 8.1 is an example of a Web-based GIS client accessing remote GIS services and rendering a map within the Web-browser. The example uses OpenLayers API to access an OGC WMS at University of Leeds, UK, and a KML file which is at the Environmental Protection Agency, Ireland. The Javascript API hides all complex details of accessing WMS or other data resources and renders the map in the browser quickly and easily. The rendered map is shown in Figure 8.2. Readers familiar with development of Web-GIS client interfaces in Google Maps and/or Microsoft Virtual Earth will notice the similarities between these two Javascript APIs and the OpenLayers Javascript API.

8.3 REPRESENTING ENVIRONMENTAL DATA FOR WEB-BASED GIS

The tools discussed in Section 8.2 offer remarkable flexibility in terms of the wide range of geographical data formats and representations they can both import for analysis and

```
<script type="text/javascript">
var map, layer;
function init(){
        map = new OpenLayers.Map( 'map' );
        var    layer = new  OpenLayers.Layer.WMS( "Global  MODIS  Blue  Marble",
http://iceds.ge.ucl.ac.uk/cgi-bin/icedswms?", {layers: 'bluemarble_1'} );
        map.addLayer(layer);
        map.addLayer(new        OpenLayers.Layer.GML("NOAA        Bouy        Locations",
"http://erc.epa.ie/kml/bouys.kml", {transparent: 'true', format: OpenLayers.Format.KML}));
        map.addControl( new OpenLayers.Control.LayerSwitcher() );
        map.setCenter(new OpenLayers.LonLat(-92.6835, 32.395), 5);
}
</script>
```

FIGURE 8.1 A simple example of using the OpenLayers Javascript API to access a WMS data source and a KML file.

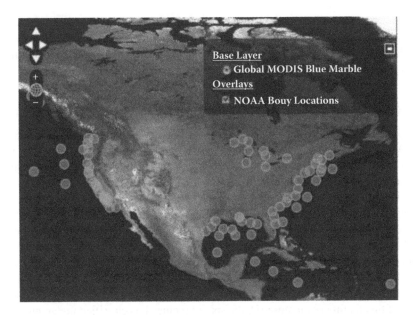

FIGURE 8.2   The cartographical output from the Javascript function in Figure 8.1. The screenshot features the pan and zoom, and layer switcher functionality offered by the Web-based GIS client.

visualization, and explore. The underlying principle for all of these tools is that the geospatial data is represented in standardized formats with several prominent standard formats emerging, for instance, ESRI Shapefiles, MapInfo TAB format, and ArcSDE. Choosing the correct format for storing environmental data is crucial in preventing duplication of work and waste of resources at a later stage associated with converting to other formats or fixing problems in the current format. Mooney and Winstanley [8] comment that large amounts of geospatial data from the environmental domain is stored and manipulated in software storage formats that are unsuitable for the purpose. Incorrect choice of format often makes analysis of the data set difficult while at the same time diminishing the opportunities for data reuse by other scientists. An example of this is using MS Excel to store time series, location-based, environmental monitoring data. Performing advanced location-based statistical queries on these data sets is difficult because MS Excel does not have the range of tools that can natively perform such analysis. The next sections outline the key features of three methods of representing environmental data that allow better integration with Web-based GIS systems.

8.3.1 PostGIS: A Spatially Enabled Database

A spatial database is a database that is optimized to store and query data related to objects in geographic space—points, lines, and polygons. PostGIS [9] is a spatial language extension to the well-known PostgreSQL database server. It implements the OGC Simple Feature Specifications for SQL standard, which defines the operations required to insert, query, manipulate, and manage spatial data objects. ESRI

Shapefiles can be loaded directly into PostGIS using the internal PostGIS Shape Loader tool. Flat files of coordinates and attributes may also be loaded directly into the database, with PostGIS performing the appropriate coordinate system conversions if necessary.

8.3.2 Geographic Markup Language (GML)

GML was developed by the Open GIS Consortium and is a spatially enabled dialect of XML. GML provides both a vendor neutral and a software implementation neutral format that is optimally suited for distribution over a network. Plain GML files as well as compressed GML files may be streamed so that a user does not have to wait until an entire file is completely downloaded before opening; this greatly enhances usability in a networked environment. GML 2 (and now GML 3, recently approved by the OGC) can enable the linking of features in one GML file to those in another GML file through use of hyperlinks.

In Figure 8.3 a small sample of GML is shown from the Ordnance Survey UK's OS MasterMap (Basingstoke Free Sample Dataset). The feature described is a line feature that is obstructing other features. It is visually easy to understand the attributes of this feature (its feature ID, the spatial themes it is classified under, etc.). The GML structure (with the assistance of the appropriate GML Schemas) parsing and extracting from GML data files is reasonably straightforward. As GML is an OGC

```
<osgb:TopographicLine fid='osgb1000000334391475'>
<osgb:featureCode>10046</osgb:featureCode><osgb:version>1</osgb:version>
<osgb:versionDate>2001-11-07</osgb:versionDate>
<osgb:theme>Land</osgb:theme><osgb:theme>Rail</osgb:theme>
<osgb:accuracyOfPosition>1.0m</osgb:accuracyOfPosition>
<osgb:changeHistory>
<osgb:changeDate>1970-01-01</osgb:changeDate>
<osgb:reasonForChange>New</osgb:reasonForChange>
</osgb:changeHistory>
<osgb:descriptiveGroup>General Feature</osgb:descriptiveGroup>
<osgb:physicalLevel>50</osgb:physicalLevel>
<osgb:physicalPresence>Obstructing</osgb:physicalPresence>
<osgb:polyline>
<gml:LineString srsName='osgb:BNG'>
<gml:coordinates>278235.950,187050.750 278231.100,187064.450 278227.250,187075.950
</gml:coordinates>
</gml:LineString></osgb:polyline>
</osgb:TopographicLine>
```

FIGURE 8.3 An example of a feature represented in GML from the Ordance Survey UK OS MasterMap data set.

standard, it is one of the default import and export representations in many GIS software tools. All of the tools mentioned in Section 8.2 import GML natively and can display the contents of the data set without conversion to an intermediate structure or format.

Figure 8.4 shows a segment of GML for a point-based data set. The data set represents the location of ground-based air quality monitoring stations under the control of the Irish Environmental Protection Agency. The easting and northing of each station is represented in addition to other parameters such as the type of aerosols monitored at that station. The spatial characteristics of the station are represented as standard within a geographical layer. The URL of a data processing service is provided. If users query this location from a Web browser-based map display, they are provided with the link to the data access service. This service provides visualizations of the previous year's air quality monitoring data and access to the raw data in several different formats.

8.3.3 KEYHOLE MARKUP LANGUAGE (KML)

The key building block behind the types of applications we see on Google Maps and Google Earth is the KML language. KML is an XML variant. The entire schema is available from the Google Web site. KML allows users to represent their geospatial information in such as way that it can be overlayed onto the base maps within Google Earth or Google Maps. The word *mashup* is the now accepted technical term for when a geographic layer is created in this manner. Geographical features such as points, lines, and polygons are expressed in KML as ordered lists of (latitude, longitude) pairs (expressed in decimal degrees). Primarily for visualization purposes, the popular uptake of mashups and Google Earth and Google Maps has

```
<ogr:AirStations fid = "1002">
<ogr:geometryProperty>
        <gml:Point>
        <gml:coordinates>253870.2,206530.8</gml:coordinates>
        </gmlPoint>
<ogr:Site>Emo Court (Co. Laois)</ogr:Site>
<ogr:Type>Fixed Station</ogr:Type>
<ogr:LocationType>Rural</ogr:LocationType>
<ogr:Operator>EPA</ogr:Operator>
<ogr:Paramters>Ozone,SO2</ogr:Parameters>
<ogr:DLink>http://erc.epa.ie/air/query.jsp?location=EmoCourt</ogr:DLink>
</ogr:geometryProperty>
</ogr:AirStations>
```

FIGURE 8.4 An example of a GML representation of a point feature. The coordinates of the point feature are specified, accompanied by some attribute information. The <org:Dlink> attribute allows users to be redirected to a related service where more detailed data can be obtained corresponding to this feature.

seen KML develop into something of a de facto standard for exchange of data sets with reasonably noncomplex sets of geographical features (points, lines, polygons). KML has recently been submitted to the Open Geospatial Consortium with a view to standardizing its usage for geobrowser applications. Many of the tools mentioned in Section 8.2 can import KML data sets and display these data sets as base layers or overlays. Export to KML is also available in many cases. Due to the simplicity in representation of geographic features with KML, it is relatively straightforward to write converter software that extracts output from a geospatial data set and outputs the entire data set (or selected subset) into a KML representation. With a short investment of time in learning how to create KML, nonspecialist GIS users can begin creating mashups using GPS data they have collected themselves. As GPS technology becomes more ubiquitous, these type of data streams flow from a diverse and unconstrained set of sources: athletes tracking their training runs, birdwatchers following transects and sightings of birds, and hikers mapping out paths in woods and forestry.

Figure 8.5 illustrates how the software in Section 8.2 and the format, storage, and representation options of the software are integrated by organizations in building a spatial data infrastructure. At the back end, geospatial data is stored in a mixture of spatially enabled databases, relational databases, and file systems. Web-based GIS server software can then access these data sources using connections (such as JDBC for JAVA-based applications) or natively (such as ArcIMS with ESRI Shapefiles). A rich client interface can be offered using Web-based GIS client software. End users can retrieve geospatial data as images, feature data (in GML format) as KML, or be directed to other services such as in Figure 8.5.

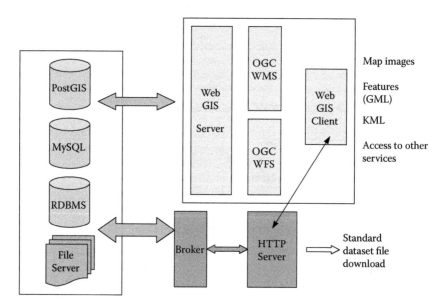

FIGURE 8.5 A schematic of the possible configuration of a spatial data infrastructure offering both GIS server and client functionality.

8.4 THE ROLE OF METADATA IN WEB-BASED GIS

The discussion in the previous sections focused on software issues related to access to geospatial data. Numerous studies have established that although the value of geospatial data is recognized by government, scientific communities, and society [10], the effective use of geospatial data in Web-based GIS is inhibited by poor knowledge of the existence of these data, poorly documented information about the data sets, and data inconsistencies. The loss of time and resources in searching for existing spatial data or establishing whether they may be used for a particular purpose is a key obstacle to the full exploitation of the available data [11]. Metadata is an important resource in its own right. It provides a high-level representation of the contents, physical properties, and geographical characteristics of the data set. In the case of some GIS tools metadata can be consumed as a data resource if the representation is a recognized metadata standard such as ISO 19115. Metadata has become a term conveniently ignored or avoided by those required to provide or manage data. Consequently, large collections of environmental and other geospatial data become "data tombs" [12], seldom visited or maintained, key data sets never emerging from "grey dusty archives" or "slowly rotting" [13] because users do not know of their existence or are denied access to them. Metadata is seen as boring [14]. Consequently, many researchers see little or no academic or workplace recognition for the task of providing or maintaining metadata [8]. As the volumes of geospatial data being generated increases, particularly in the area of environmental monitoring and environmental science, this situation must not be allowed to continue. Some authors remark that it is ironic that at a period where the volume of data generated in scientific research is at an all-time high, the practice of documenting and accessing these resources is at an all-time low [15].

The concept of metadata was introduced to provide orientation in a space of continuously growing data and information resources [16]. Metadata provides information about the data but does not include the data itself. In the majority of cases the metadata accompanying a geospatial data resource is represented within a tabular structure in a separate digital file. It may also be represented as rows within a database table (or set of related tables). Some earth science disciplines (climatology, weather forecasting) use machine-independent data formats that support the creation, access, and sharing of array-represented scientific data. Examples of these data formats include NetCDF (Network Common Data Form) and HDF (Hierarchical Data Format). These formats are self-describing, allowing a software application to interpret the structure, contents, and representation of the data set file without any outside information. The metadata is embedded within the file format. For Web GIS, metadata is usually stored in files with XML format or stored within tables of relational databases and converted to XML. Using style sheets, the XML representation of the metadata is presented in a clear, human-readable representation (usually in conjunction with HTML) in the Web GIS.

8.4.1 METADATA REPRESENTATIONS

Web-GIS metadata can actually be considered as a data resource. When search tools are provided through the Web-GIS interface, the search results are extracted from the metadata resources. The metadata then usually provides links to data visualization and data extraction services. Without high-level representation through metadata, the data sets themselves are essentially standalone digital objects. The metadata provides an explanation of what the data resources and map layers available in the Web GIS represent. Waller and Sharpe [2] remark that the importance of metadata cannot be overestimated: "It adds a whole new dimension, providing extra richness of contextual or descriptive information at the point of access." The perception among many scientists and creators/maintainers of geospatial data is that the creation and maintenance of geospatial metadata is a laborious and unnecessary data management task.

Many GIS software packages (for example, ESRI ArcCatalog) will automatically create metadata in a standardized format such as ISO 19115, as required by the INSPIRE Directive. Many of the fields are filled in automatically. These fields include information about the parties responsible for the creation, analysis, and maintenance of the data set resources; file types, sizes, file content, and other computer representations; and in the case of GIS formats (such as Shape Files), the geographical extent, scale, and projections used. These information fields are vital in metadata provision. However, it is often the case that fields requiring manual input from the scientific expert (such as fields related to the quality or preprocessing of the data set) are often left blank or only partially filled in. As a result, it is very difficult for a third party to make an assessment of the fitness for use (quality) and the fitness for purpose (the problem they are trying to solve). This usually causes such third parties to download the partial or entire data set into their GIS or analysis software in order for them to make these preliminary assessments. Metadata is also a core component of OGC services. Figure 8.6 shows a subset of the metadata returned from a GetCapabilities request from a WMS. In this example, the information contained in the metadata can be used automatically by the Web-based GIS client to perform the necessary coordinate transformations.

```
- <Layer queryable="0" opaque="0" noSubsets="0">
  <Title>GDR_E</Title>
<SRS>EPSG:4326</SRS><SRS>EPSG:4269</SRS><SRS>EPSG:4267</SRS>
  <LatLonBoundingBox minx="-150" miny="40" maxx="-47" maxy="90" />
  <BoundingBox SRS="EPSG:4269" minx="-123.6486" miny="48.8696555569"
  maxx="-123.0846040092" maxy="49.1189004763" />
- <Layer queryable="0" opaque="0" noSubsets="0">
  <Name>AtlanticDEM</Name>
  <Title>Atlantic Canada Digital Elevation Model</Title>
  <SRS>EPSG:4326</SRS>
<LatLonBoundingBox     minx="-72.000849"     miny="39.996615"     maxx="-48.002030"
maxy="51.997853920" />
  </Layer>
```

FIGURE 8.6 Subset of the XML from a GetCapabilities() call to the WMS at National Resources Canada. The XML shows the base layer GDR_E with the layer AtlanticDEM as a sub-layer. The SRS denotes the coordinate system transformations allowed on these layers.

8.5 CONCLUSIONS

There are many advantages to providing GIS services on the Internet and World Wide Web, and many mapping and visualization services are being made available using Web-based GIS systems. Stachowicz [17] summarizes the most important advantages for users accessing these services as (a) that there are no software downloads required because usually an up-to-date Web browser is all that is required; (b) users are offered one interface to potentially many separate services, for example, geospatial data integrated from an environmental organization, human health research, and land-user planning; and (c) users can access these services in a 24/7/365 manner. The National Science Board [18] argues that much of the scientific data being collected today are "born digital"; there is no analog or paper counterpart. Additional scientific data are being converted to digital representations and disassociated from their analog representations. For these data resources, the public cannot merely make an appointment with a local authority office and physically visit to browse the files. Web-based GIS allows authorities to provide "always on" services, where appropriate users can browse map-based representations of these data sets and possibly download the data to their own computer for further analysis.

8.5.1 METADATA AND IMPROVING ACCESS TO GEOSPATIAL DATA

As the volume of geospatial data about the environment continues to grow, so too does the need to properly document these data resources with metadata. Without metadata these data resources may lie dormant and undiscovered on the Internet. This invisibility may give rise to duplication of effort in creating, accessing, and managing these data resources. The INSPIRE Directive states that geospatial data must be managed as close to the source as possible. In relation to this, the representation of key knowledge about these data resources in metadata must be performed initially as close to the original data creator source or scientific expert group. Without this vital specialist knowledge about the data, resources may be left in "dusty archives and grey literature" [19] or lost in laboratory or field notebooks. The management overhead of convincing busy scientists and analytical staff that they need to maintain and manage metadata for their data resources is considerable. However, the exposure of geospatial data through Web services for consumption by a wide variety of users, who may be using Web GIS to access these services, is greatly hindered if the users do not have access to accurate and complete metadata.

An important feature of Web-based GIS is that their usage can indirectly assist with digital resource curation and ensuring long-term access to geospatial data resources. Many scientific funding organizations highlight urgent needs to invest in data curation and data recovery: "Very substantial amounts of data have already been lost and even greater losses are imminent as the ability to recover data stored on obsolete technologies declines exponentially" [20]. Web-GIS systems are encouraged to access the underlying geospatial data in its original format (or some converted standard representation) and to present results of Web queries in open and accessible data format representations. Metadata is used to document the existence of these data sets and the services available to access them. Reichman and Uhlir [21] state that "big science or mega science—NASA, NOAA, ESA," often openly share their data and results in public repositories. However "small science," independent investigator-driven research remains dominant in most scientific fields. Traditionally, data from such studies have been extremely heterogeneous and unstandardized with few individuals making their data sets available through public repositories or even openly sharing them. In many mashup creations using Google Maps, for example, non-GIS skilled developers have implemented methods called "screen scraping." This involves extracting data from digital documents or maps in a nonautomated manner or without the use of a formal conversion schema from the old model to the new model. Transformation errors commonly occur and consequently the accuracy of the newly created data set may be difficult to verify. The data or information being scraped is usually neither documented nor structured for convenient parsing.

8.5.2 Closing Remarks

Some authors remain skeptical regarding the actual impact of Web GIS on the public. Kingston [1] remarks that "there is still little evidence available as to how much the public are using such systems." Kingston goes on to state that despite the vast amounts of money invested in e-government, a reluctance by local and national government to divulge data on access and usage leads one to suspect "that current usage of Web-GIS for e-Government is relatively low compared with more traditional methods" [1]. Mooney and Winstanley [22] emphasize the importance of using Web-log usage patterns taken from server machines running Web-based geospatial services, and analyzing this information to assist in delivering better quality of service to the end user, as well as to target areas of specific weakness. Combining spatial and nonspatial data presents unique challenges to data management and access over the Web. Few Web-based GIS environments have tools "out-of-the-box" for importing data and validating that data against a metadata profile as it is loaded into a database. Presently, end users must manually check if the coordinate system specified in the metadata is equivalent to the coordinate system in the corresponding data set.

We feel that there will not be a sudden demise of desktop GIS in favor of Web-based GIS in the foreseeable future. While the cartographical visualizations and interactivity of Web-based GIS are continually improving, desktop GIS will remain a core component of GIS. This is due in no small part to the ability of desktop GIS to run large-scale GIS queries involving very large data sets and complex spatial algorithms and database queries. As discussed previously, Web-based GIS remains restricted in this regard due mainly to bandwidth considerations and network latency. Desktop GIS will become more Internet integrated. Desktop GIS users will no longer download several large spatial data sets directly to their local hard-disk drive or network drive in order to perform some GIS tasks. Instead these users will use WFS to download the precise subsets of the larger data sets, or WMS to retrieve map layers. It is at this point that metadata has one of its most crucial roles to play in informing potential users that spatial data exists and providing information to allow these users to make judgment on the data set's fitness for purpose and fitness for usage.

Web-based GIS allow organizations to expand their geospatial data holdings without any interruption to the end user or data consumer. Legacy data sets can be converted to the agreed spatial data representations and made available for access from the Web-based GIS. With some intermediate work on updating certain aspects such as style sheets and color schemes to accommodate these additional data sets, these newly available data resources are quickly available to the organization's stakeholders. Web-based GIS must be highly scalable. Successful applications take advantage of networks with high bandwidth while working efficiently to avoid problems with slow networks and low bandwidth. The storage and representation of geospatial data is important in this regard. Very large environmental data sets should be stored in such a manner that queries can be more efficient, with smaller data download size and smaller data transfer requirements for OGC services. The main goal, at least for the foreseeable future, will be harmonization through interoperability in a service-based architecture rather then full-blown harmonization of the underlying data models. Representation issues will be prevalent in both cases but much more manageable and achievable by using harmonization through interoperability services.

REFERENCES

1. Kingston, R., Public participation in local policy decision-making: The role of web-based mapping, *The Cartographic Journal*, 44(2), 138, 2007.
2. Waller, M. and Sharpe, R., Mind the Gap: Assessing Digital Preservation Needs in the UK, Report Prepared for Digital Preservation Coalition, Available at: http://www.dpconline.org, accessed September 2007.
3. Scharl, A., Towards the geospatial web: Media platforms for managing geotagged knowledge repositories, in *The Geospatial Web—How Geo-Browsers, Social Software and the Web 2.0 are Shaping the Network Society*, Scharl, A. and Tochtermann, K., Eds., Springer, London, 2007, 3.
4. ESRI, The ArcIMS Internet Map Server, http://www.esri.com/software/arcgis/arcims/index.html, accessed March 2008.
5. MapGuide, The MapGuide Open Source Platform, http://mapguide.osgeo.org/, accessed March 2008.
6. MapServer, MapServer Open Source Development Environment for Web-Based GIS, University of Minnesota, USA, http://mapserver.gis.umn.edu/, accessed March 2008.
7. GeoServer, A JAVA Driven GIS Data Server, http://geoserver.org/display/GEOS/GeoServer+Home, accessed March 2008.
8. Mooney, P. and Winstanley, A. C., Improving Environmental Research Data Management, *Enviro Info 2007*, Warsaw Poland, September 2007.
9. PostGIS, Spatial Database Extension for PostgreSQL Databases, http://postgis.refractions.net/, accessed March 2008.
10. Nebert, D., Ed., *Developing Spatial Data Infrastructure: The SDI Cookbook*, 2004, Available at: http://www.gsdi.org/docs2004/Cookbook/cookbookV2.0.pdf.
11. INSPIRE, Directive 2007/2/EC of the European Parliament and of the Council of 14 March 2007 establishing an Infrastructure for Spatial Information in the European Community (INSPIRE) Official Journal of the European Union L108, Volume 50, 25 April 2007. Available at: http://eur-lex.europa.eu/JOHtml.do?uri=OJ:L:2007:108:SOM:EN:HTML, accessed March 2008.
12. Han, J. and Kamber, M., *Data Mining: Concepts and Techniques* (The Morgan Kaufmann Series in Data Management Systems), Morgan Kaufmann, San Francisco, 2006.

13. European Territorial Management Information Infrastructure (EteMII) White Paper on Reference Data, Metadata and Interoperability through Standards, http://www.ec-gis.org/etemii/, accessed March 2008.

14. Comber, A., Ahlqvist, O., Fisher, P., Harvey, F., Gahegan, M., and Wadsworth, R., Can metadata ever be interesting?: The case for expanded metadata, in Billen, R., Drummond, J., Forrest, D. and Joao, E., Eds., *Proceedings of the GIS Research UK 13th Annual Conference*, 6–8 April 2005, University of Glasgow, 53–57.

15. Bulterman, D. C. A., Is it time for a moratorium on metadata? *IEEE Multimedia*, 11(4), 10, 2004.

16. Pillmann, W., Geiger, W., and Voigt, K., Survey of environmental informatics in Europe, *Journal of Environmental Modelling and Software*, 21, 1519, 2006.

17. Stachowicz, S., Geographical Data Sharing—Advantages of Web Based Technology to Local Government, Proceedings of 10th EC GI & GIS Workshop ESDI State of the Art, Warsaw, Poland, June 2004.

18. NSB, Long Lived Digital Data Collections: Enabling Research and Education in the 21st Century, Report of the National Science Board, The National Science Foundation, Arlington, Virginia, May 2005, available at: http://www.nsf.gov/pubs/2005/nsb0540/, accessed September 2007.

19. Wilson, J. R., Data and Information Management Strategy and Plan of the Global Ocean Observing System (GOOS), Intergovernmental Oceanographic Commission, Paris, France, GOOS Report No. 13 (IOC/INF-1168), 2002.

20. Leach, P., Strong, D., and Wood, G., Canadian National Consultation on Access to Scientific Research Data (NCASRD), Web-Based Report, available at: http://ncasrd-cnadrs.scitech.gc.ca/tdfdesignreport_e.shtml, accessed Sept 2007.

21. Reichman, J. H. and Uhlir, P. F. A contractually reconstructed research commons for scientific data in a highly protectionist intellectual property environment. Law and contemporary problems, 66, 315, 2003.

22. Mooney, P. and Winstanley, A. C., Evaluating Interfaces to Publicly Available Environmental Information, Human Computer Interaction 2007: Human Interface Part I, Beijing, China, July 2007.

9 Developing and Applying a Participative Web-Based GIS for Integration of Public Perceptions into Strategic Environmental Assessment

Ainhoa Gonzalez, Alan Gilmer,
Ronan Foley, John Sweeney, and John Fry

CONTENTS

OVERVIEW

The intrinsic spatial nature of development plans poses specific requirements on the analytical tools applied to support Strategic Environmental Assessment (SEA) processes. Geographic information systems (GIS), with their mapping and analytical potential, can assist and enhance the various stages of SEA. A method has been developed to apply GIS as a support tool to assist SEA of land use plans in the Republic of Ireland. This chapter describes one phase in the development and testing of the method during the preparation of County Development Plans, a participatory Internet-based GIS tool

117

developed to communicate and gather information in a spatially specific format. The aim of the web site was to promote and expand the use of GIS in public participation and, thus, allow for the incorporation of spatially specific public perceptions in SEA. The results revealed that the integration of public perceptions into the assessment through GIS stimulates debate and provides an overall scientific and social view of the relative environmental significance and vulnerability of the different areas. However, current issues in relation to availability and quality of spatial data constrained the applicability of GIS. Furthermore, complexity of the technology, data disclosure issues, and statutory consultation requirements restricted its implementation and use, affecting the adequacy and the level of public opinion gathered through the Web site.

9.1 INTRODUCTION

European Directive 2001/42/EC [1], also known as the Strategic Environmental Assessment (SEA) Directive, requires an assessment of the potential effects of certain plans and programs (e.g., for land use or waste management) on the environment. The SEA process requires a number of steps to be undertaken (Figure 9.1) during the preparation of the plan or program to anticipate, assess, and mitigate any environmental issues associated with the implementation of the plan/program's objectives and actions. All European Union (EU) member states, except Luxembourg, have transposed the SEA Directive into national legislation and have implemented it, particularly in land use planning [2]. The strong spatial and temporal dimensions of land use plans pose specific requirements in relation to the analytical tools applied to support SEA processes. The intrinsic spatial nature of land use plans solicits their presentation in graphic format. Similarly, temporal variation can often be represented in visual form by spatially illustrating changes over time. Furthermore, it is estimated that up to 85% of government data—used to support policy, plan, and program making—have spatial components [3,4] and can therefore be mapped using geographic information systems (GIS). The graphic display and analytical potential of GIS can significantly contribute to the SEA of development plans by facilitating and enhancing the various stages of the process.

SEA processes and the integration of environmental concerns into planning can be positively influenced by public participation [5,6]. The SEA Directive and the related Århus Directive 2003/35/EC [7] make mandatory provisions for public participation in the assessment of potential environmental effects of certain plans and programs in the EU. It is argued that involving the affected public and interest groups enhances the level of legitimacy, transparency, and confidence in the decision-making process [6,8]. Methods such as submission of written comments, public hearings, workshops, and interviews, as well as more modern forms of consultation such as Internet-based forums, are acceptable forms of participation in the EU [7]. Selection of appropriate public participation techniques is necessary to ensure that citizens are given enough time and scope to participate in an effective manner while avoiding undesirable time delays in the decision-making process [8]. Although public participation methods have been widely explored, systems for influential inclusion of public concerns and interests in environmental assessment have seldom been defined [9].

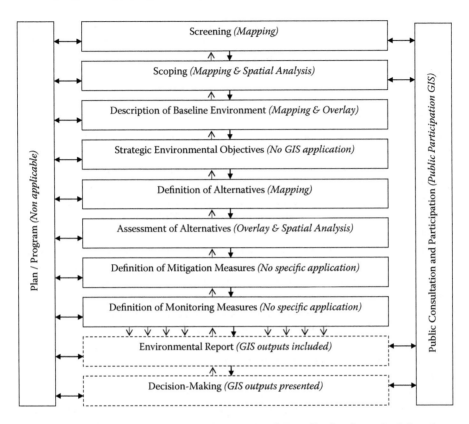

FIGURE 9.1 SEA stages, their correlation, and the GIS application for each of the stages. Note that the iterative nature of the process is illustrated by the continuous communication with the planning process. However, the participative aspect of SEA is illustrated referring only to those SEA stages where the directive requires public and stakeholder involvement. The feedback between processes indicated by the two-headed arrows represents the continuous reappraisal and adjustments required in the process.

Contemporary European planning practice shows an increasing trend toward electronic-based or e-planning (e.g., G-Plan, the Internet-based planning system used by Irish local authorities), as well as toward Internet-assisted information and consultation (e.g., e-tax and e-voting). In addition, the application of technology and computer-based models is common practice in some phases of environmental assessment [10,11]. GIS constitute a useful tool for conveying and presenting information by overlying geographically referenced data, thus facilitating the assessment of the location, extent, and spatial interaction of environmental factors.

Unfortunately, GIS packages tend to require skilled knowledge of the system to operate them, as applications normally have a technological rather than usability focus [12,13]. However, recent developments are leading to more user-friendly software interfaces, and usability barriers are being reduced, as indicated by a number of studies where GIS has been successfully used in participatory processes to facilitate spatial comprehension, enhance transparency, and stimulate debate [5,14–16]. In light of this, a GIS-based Web site has been developed for public participation in

SEA and incorporated into two Irish land use planning SEAs. This tool provides the means of viewing and gathering data in a spatially specific format and, consequently, facilitates the integration of public perceptions into the environmental assessment of development plans.

9.2 METHODOLOGY

This chapter describes one phase in the development and testing of GISEA—a methodological application of GIS to SEA. The resultant GIS-based methodology, suitably adapted to the requirements of each SEA stage (Figure 9.1), is being tested by incorporation within real SEA case studies of land use plans in Ireland to assess its usefulness from an environmental planning perspective. This chapter presents the results derived from the case studies of Mayo and Kilkenny county development plans (CDP).

Since personal evaluations of importance can vary widely, a participatory approach to SEA was considered necessary to define a valuing scale that was legitimate and fair to all involved in the assessment process. Therefore, as part of the methodology, a participatory Internet-based GIS tool was developed (hereafter referred to as the GISEA Web site). The aim was to both promote and expand the use of GIS in public participation and, thus, allow for the incorporation of spatially specific public perceptions in SEA. Environmental criteria and their value of significance were determined at the scoping stage based on scientific fact and expert knowledge. Subsequently, the GISEA Web site gathered public perceptions in relation to the importance (weight) of previously defined aspects and other environmental issues, as well as on proposed alternative actions. The objective of this approach was to ensure that articulation of values from most affected parties, including the public, were incorporated into the computerized GISEA methodology for a holistic assessment.

The ArcGIS family of products was chosen as the platform for developing the method since it provided the versatility and tools needed to achieve the research objectives. The ArcIMS interface (i.e., the server GIS used for developing the public participation Web site) was edited to develop a user-friendly and easy to understand system that would not require specific GIS skills and could be manipulated with basic Web-browser knowledge. Therefore, the viewframe and tools available in ArcIMS were adapted to the requirements of the research. This included an enhanced browser, improved user interaction, and incorporation of a database, as well as display of tools, and questionnaires specific to the chosen case studies. This was achieved by programming and editing the scripts on the ArcIMS files in several computer languages, including PHP, JAVA, HTML, SQL, and Visual Basic.

The GISEA Web site follows a number of steps that guide the user through the public consultation process (Figure 9.2), with an introductory Web page describing the purpose of the site. Users are asked to select three environmental criteria of concern. These selected criteria are essential for validating the significance of environmental factors. The GIS-based Web pages subsequently display the relevant environmental information, and users can view and interact with these spatial data. Personal perceptions and comments on the displayed environmental information and the proposed alternatives can be submitted via questionnaires, which are gathered

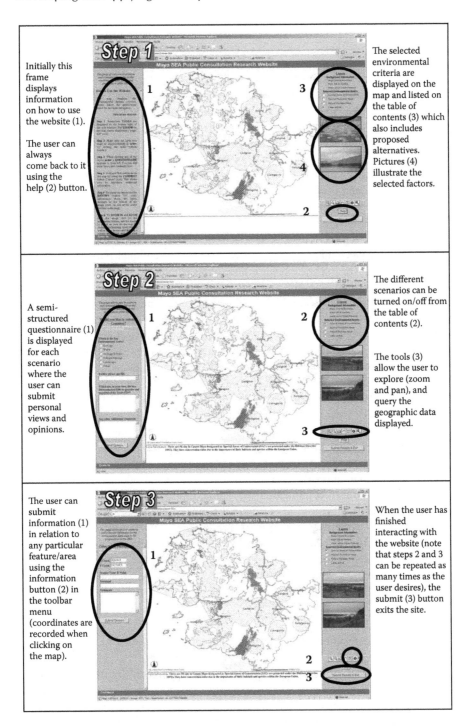

FIGURE 9.2   Details of the GISEA ArcIMS Web page.

on a database for future analysis. In addition, and to avoid limiting the submission of comments to the previously established environmental factors, a supplementary tool is provided to allow comments to be recorded on any particular location or feature on the map (by recording the X and Y coordinates). Once the user finishes exploring the information and submitting opinions, the browser continues to a final Web page where users are asked to comment on the usefulness of the site.

The tool was pilot tested and subsequently amended to incorporate changes and improve its user interface. It was then made available during the development of two real-life SEAs to evaluate its applicability in the Irish planning system. These have allowed preliminary conclusions to be drawn in relation to the limitations, opportunities, barriers, and benefits derived from the availability of a GIS tool for public participation. It must be noted that the Web site was intended to complement rather than replace existing practices and techniques and traditional public participation methods by ensuring that stakeholders have timely access to information and are provided with a mechanism to have a say outside conventional participatory processes. Therefore, the Web site results were to be compared and incorporated with other participatory outcomes for completion and consolidation.

The overall objective was to validate the chosen environmental criteria, to gather opinions in a spatially specific format, and to incorporate these into the environmental assessment of the proposed alternatives. The qualitative comments and opinions submitted could also be evaluated and summarized in the SEA's environmental report. Perceptions in relation to environmental criteria of concern were used in the form of weighted values for assessing the relevance and consequent vulnerability of the environmental resources in the region. Multicriteria analysis was applied, and existing GIS tools used to automatically detect the degree of overlap of thematic layers (i.e., environmental data) and to determine areas of potential vulnerability (i.e., the higher the number of overlapping key environmental factors, the greater the vulnerability). This was done by converting feature spatial data to raster format and reclassifying them to allow the GIS to undertake automated calculations.

The weighted linear combination algorithm proposed by Chrisman [17] was adapted by subtracting the division factor that averages the output value. This adaptation was made to avoid neglecting potential cumulative effects as it was considered that the vulnerability of each area was directly related to the number of environmental criteria that overlapped at one location (i.e., pixel). The following equation was applied to combine the number of environmental factors, and their significance and weight:

$$V_n = \Sigma W_j V_j$$

where

V_n refers to the resultant vulnerability value for the area/pixel that relates to the total number (n) of criteria that overlap in the area.

W_j refers to the significance or sensitivity value for each criterion (j) according to scientific opinion. To standardize categorizations it was established that highly sensitive environmental factors (e.g., surface waters designated as being at risk (1a) under the Water Framework Directive or landscapes classified as highly sensitive in the CDP) equated to 10, and sensitive factors (e.g., surface waters designated as being potentially at risk (1b) under the Water Framework Directive or landscapes classified as sensitive in the

CDP) equated to 5. A value of 0 was given to the cells that had no occurrence of environmental constraint.

V_j refers to public weighting and includes the subjective judgments from stakeholders and the general public on the perceived vulnerability of each criterion (j) considered. The weighting values (V_j) are used as a "strengthening" factor. Those aspects of concern (i.e., the three criteria selected the highest number of times) were perceived as more important and, thereby, given a weight of 1.5 that increased their significance. The criteria perceived as neutral (i.e., unselected criteria or criteria selected the fewest times) still had scientific significance and were therefore given a weight of 1.

The computer model undertook the weighted overlay process and reevaluated the data. The results provided a thematic map reflecting the composite vulnerability of each area according to both scientific opinion and public perception. The results were also computed and provided in quantitative form to complement and further facilitate the understanding of key environmental aspects within the study area, as well as the succeeding assessment of alternatives.

9.3 RESULTS

9.3.1 Assessing the Usability of the Web Site through Pilot Tests

Pilot studies were carried out to assess the usability and overall user-friendliness of the GISEA public participation tool. These pilot tests targeted 61 undergraduate and postgraduate students with basic or no GIS knowledge. Results revealed that total lack of GIS skills could limit the understanding of the displayed maps and affected performance; 75% of the students with some GIS knowledge were able to complete all the steps indicated in the Web site, whereas this value was only 39% for individuals with no GIS skills. The majority (66%) of users that completed the process found the Web site easy to use and navigate. The graphics were perceived as a good way of communicating environmental information. However, 30% indicated that the absence of background Ordnance Survey Ireland (OSI) maps and a readily available legend (i.e., an alternative to having to select the legend menu) were major drawbacks. Other observations highlighted the necessity to improve the guidance on how to use the Web site, to reduce the amount of information displayed, and to enhance the browser structure. The Web site was consequently amended to improve the overall interface and incorporate those suggestions derived from the pilot studies. New instruments to facilitate its use were included (such as an animated demonstration on how to use the Web site and interactive questionnaires), together with representative photographs for the different areas and environmental considerations and, where possible, OSI base maps.

9.3.2 Applying the Web Site in Practical Case Studies

The Web site was launched as part of the SEA of two CDPs in the Republic of Ireland. The GISEA Web site was adapted to the requirements of each case study, providing the flexibility necessary to reflect and incorporate both the regulatory requirements and the planning information needs. The planning teams involved in both SEAs perceived it as a complementary participative instrument and supported public access to the Web site by providing GIS data and including a link on the organizations' official Web sites. However, the authorities responsible for the SEAs already had a formal method for gathering public submissions, derived from the statutory planning procedures. This formal requirement constrained the effective application of the tool in the case studies, and a number of other factors also limited the usability of GIS during the participative stages of SEA.

The SEA process in County Mayo envisaged an experimental public consultation program using the GIS-based Web site to validate the environmental objectives. Unfortunately, the work program for revising the existing CDP and the consequent SEA process were delayed, thus affecting the public consultation stage. In addition, issues of political will and confidentiality, and fears over early disclosure of information affected both the timely provision of the Web site and the disclosure of certain layers of information, such as OSI base maps and considered alternatives. This, in turn, affected the evaluation of proposed alternatives by the general public. The GISEA Web site, therefore, only displayed the environmental data used during the SEA process. Furthermore, despite initial enthusiasm, the forward planning team questioned the usability of a GIS-based participatory Web site, indicating that GIS-based interfaces are complex tools that only technically skilled personnel would be able to use. Although the GISEA Web site was made publicly available within the County Council's official Web site on May 4, 2007, access was gained via a series of links in additional Web pages and the GISEA Web site link was addressed as a research study rather than an additional public consultation tool. All of the above aspects had implications on the usability of the tool. A limited number of hits were registered (a single hit from Mayo, four from Dublin, six from the rest of the country, two from Germany, and one from London). Moreover, no comments were submitted to the GISEA Web site during the public consultation period (April 10 to June 21, 2007). During the consultation period, the County Council received 56 written submissions and 22 online submissions.

A modified version of the research tool (which included OSI maps and a specific questionnaire for each of the proposed alternatives displayed in the Web site) was subsequently launched during the public consultation stage of the Kilkenny CDP revision. The forward planning team involved in that SEA process actively supported the publication of the GISEA Web site through appropriate license agreements and provision of all relevant data. It was anticipated that the tool would be launched at the initial stages of the SEA process to facilitate all consultation procedures and promote the transparency of the decision-making process. However, a number of practical considerations affected its implementation. As with the Mayo CDP, the statutory information and submission channels limited the effectiveness and applicability of the tool. Although no limitations were imposed on disclosing data, delays in the definition of proposed alternatives affected its early incorporation in the process. Similarly, changes and delays in the scheduled work program affected the timely incorporation and, thus, the availability of the tool. Although access to the GISEA Web site required fewer intermediary Web pages, the official link also addressed the GISEA Web site as a research study rather than an additional public consultation tool. The Web site was made available on the August 15, 2007. No comments were submitted to the GISEA Web site during the public consultation period

(August 10 to October 19, 2007). There were a limited number of hits registered, none of them apparently from Kilkenny (thirty-four from Dublin; two from Cork; four from the rest of the country; and one from each of Australia, United States, and Spain). During the consultation period, the County Council received 208 written and 46 online submissions.

9.3.3 INTEGRATING PUBLIC PERCEPTION INTO THE ENVIRONMENTAL ASSESSMENT

An initial test was undertaken utilizing a simple version of the methodology to assess the level of acceptance by the SEA and the forward planning teams regarding representing environmental vulnerabilities in a composite map (Figure 9.3). This version avoided the more complex computation model designed as part of the full methodology. Feedback indicated that conveying results in overlay format facilitated the combined assessment of multiple factors, enhancing the identification of key nodes of environmental sensitivity. Subsequently, with public perceptions still being gathered, the full methodological approach was applied as a pilot study in which assessment was undertaken by ascribing weighting values to each environmental criterion. A distinction was made between high and moderate sensitivity factors (see Section 9.2), as some of the environmental variables considered already incorporated a sensitivity classification (Table 9.1).

The software computed those environmental sensitivities that co-occur in each pixel cell to obtain a total sensitivity value (i.e., environmental vulnerability to development) for each particular area. The pixel cell size adopted for the pilot test

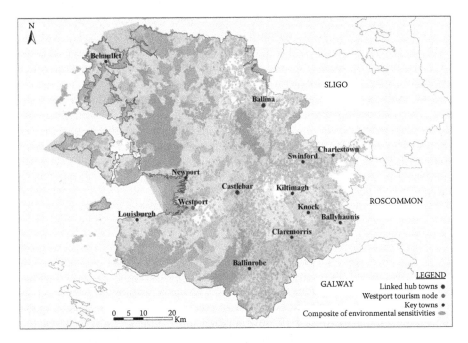

FIGURE 9.3   Results of the overlay indicating composite environmental vulnerabilities in County Mayo. The darker the shaded area, the higher the environmental sensitivity.

TABLE 9.1

Ascribed Relative Vulnerability of the Key Environmental Aspects Considered in the Assessment

Environmental Criteria	Sensitivity Value (W_j)
Designated natural heritage areas	10
Special areas of conservation (Natura 2000 sites)	10
Special protection areas (Natura 2000 sites)	10
River basins at significant risk	10
Lakes at significant risk	10
Coastal waters at significant risk	10
Designated national monuments	10
Sensitive landscape protection policy areas	10
River basins probably at significant risk	5
Lakes probably at significant risk	5
Coastal waters probably at significant risk	5
Ground waters probably waters at significant risk	5
Total	100

calculations was 30 m × 30 m; this size can be adjusted to provide a higher level of detail for larger scale, or geographically smaller area, assessments. Figure 9.4 illustrates the total vulnerability for that area, assuming the vulnerability values indicated in Table 9.1 and incorporating an equal-weighted value for each cell (i.e., assuming all criteria have the same relevance according to public opinion). For example, the total sum for three moderately sensitive factors ($3 \times W_j = 5$) and two highly sensitive factors ($4 \times W_j = 10$) co-occurring at a given location with an equal weighted value ($V_j = 1$) applied to each, would score 55 and thus render that particular area extremely vulnerable in environmental terms (Table 9.2).

The resulting map (Figure 9.4) provides a graphic representation of the location, interrelationship, and extent of areas vulnerable to impact, classified according to the various levels of vulnerability (Table 9.2). It also allows quantitative analysis by calculating the number of pixels under each environmental vulnerability category. Table 9.3 illustrates the type and extent of environmentally vulnerable areas in the county (e.g., 5.5% of the county is highly vulnerable in environmental terms).

Proposed scenarios can be evaluated against the vulnerability map, rapidly identifying those areas of proposed urban expansion or economic development that conflict with areas of significant environmental vulnerability. Representation of codified results (by color and with spatially definite variables) allows fast identification of potential incompatibilities and viable alternatives, informing the decision-making process in a concrete and transparent manner. The breakdown of the results in percentages (relating to perceived possible environmental impacts of implementing the plan) also contributes to a more effective comparison of alternatives, as well as to the definition of spatial indicators that can facilitate the monitoring and auditing phases of SEA.

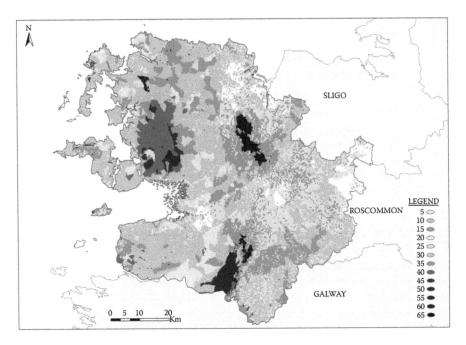

FIGURE 9.4 Results of weighted overlay indicating areas of varying degrees of environmental vulnerability in County Mayo.

TABLE 9.2
Vulnerability Classes According to Weighted Overlay Scores

Vulnerability of the Area	Weighted Overlay Score
Low vulnerability	5–20
Moderate vulnerability	20–30
Vulnerable	30–40
High vulnerability	40–50
Extreme vulnerability	50–60
Acute vulnerability	>60

9.3.4 ASSESSING PERCEPTIONS IN RELATION TO THE GIS-BASED SEA METHODOLOGY

Planners and technicians involved in the case studies were interviewed to gain further insight into the potential benefits and limitations of applying GIS to SEA. Summarizing the survey findings and maintaining the focus on the public participation aspect of the methodology, it can be argued that the responses were largely positive: spatial data and GIS were considered to provide clearer and spatially specific information that improved understanding of environmental and planning

TABLE 9.3

Quantification of Environmentally Vulnerable Areas in County Mayo

Environmental Vulnerability	Area (km²)	Percentage (%) of Total County Area
Low vulnerability areas	3552.16	60.5
Moderate vulnerability areas	937.92	16.0
Vulnerable areas	855.84	14.5
High vulnerability areas	325.28	5.5
Extreme vulnerability areas	180.8	3.5

issues, facilitated plan making, and better informed decision making. The majority of respondents indicated that the main benefit of GIS derived from its potential to overlay information in a spatially specific manner. Graphic representation and the quantitative computation of results were perceived as enhancing the comprehensiveness and transparency of the SEA process.

Interviewees generally perceived that, if used properly (i.e., ensuring data quality and avoiding complex analysis and intricate representations), maps can promote, debate, and assist public participation and consultation processes. However, several respondents noted that in reality the public does not commonly engage in forward planning processes and, moreover, the lay public may have educational impediments for reading and understanding maps, a barrier that could be exacerbated when using GIS-based interfaces.

9.4 DISCUSSION: ASSESSING THE APPLICABILITY OF THE TOOL

GIS has been recognized as a useful tool for assisting environmental decision making [18–20], and the methodology for employing GIS to assist the various SEA stages revealed a number of strengths and weaknesses (Table 9.4). The case studies highlighted that GIS has the potential for improving the transparency of the information available to the public, and the spatial analysis of combined quantitative and qualitative data. Similarly, the availability of a Web-based participatory tool can facilitate public consultation processes by providing an alternative and complementary way of informing the public and allowing them to remotely submit views and comments. However, it is still considered an expensive solution that requires high levels of spatial understanding and technological skill to use (Kingston, personal communication). Moreover, Kingston et al. [21] suggest that the levels of participation are directly related to the geographical scale, with the greater participation occurring at more localized scales. Several of the interviewed practitioners confirmed this observation by highlighting the limited participation levels of the general public in forward planning.

Notwithstanding the findings of an international questionnaire indicating that Internet-based GIS can facilitate participative processes [22], it can be argued that there is a somewhat limited scope for GIS during the consultation procedures of SEA.

TABLE 9.4

Key Strengths and Weaknesses of the Method Observed during the Case Studies

Strengths	Weaknesses
• Enhanced transparency of both SEA and planning processes	• The reliability of results depends largely on the availability and quality of baseline information/GIS data
• Spatially specific assessment of issues and alternatives	• The method relies on GIS knowledge/expertise
• Improved information delivery and easier interpretation of results by planners and decision makers	• Existing formal procedures for public participation can affect the effectiveness of participatory GIS
• Speed of applicability derived from the availability of a systematic methodology	• Fear of early disclosure can affect the use of GIS and divulging of outcomes
• Controlled subjectivity of the assessment (as a result of the inclusion of public perception values)	• There is a tendency to interpret overall results in a quantitative manner (and not all environmental aspects or planning decisions are quantifiable)
• Facilitated comparison among alternatives and case studies	• Comparison among different studies/alternatives requires availing this method

The apparent division between computer-skilled and "traditional" citizens [23,24], the complexity of the system, and variable access to the technology [14,25] can affect its applicability. In line with international opinion, complexity of the technology and issues associated with data disclosure and statutory consultation requirements restricted the implementation and use of GIS during the case studies. In addition, the majority of received submissions were provided in written form, despite the availability of e-mail submission options on the County Council Web sites. Therefore, it can be argued that computer literacy or reservations in relation to technology are a basic barrier to e-participation. This issue is aggravated when using additional and more complex technologies such as GIS.

However, despite the constrained use of GIS during the consultation process of the case studies, the planners involved considered GIS as information media to benefit the spatial understanding of both environmental aspects and planning processes. This agrees with published findings that data analysis through GIS produces a synergistic effect, enhancing collaboration and understanding, as well as improving both the quality and accuracy of results [20,26].

The majority of environmental GIS applications rely on mapping and simple overlay operations to examine where resources or vulnerabilities co-occur [11,27], but this general approach does not give consideration to the relative importance and vulnerability of the different environmental factors. Significant attempts have been made to incorporate qualifiers that stress the relative significance of environmental considerations. Such approaches commonly translate public perceptions and scientific opinion into weighted values. However, there is still a significant gap between experimental and practical application of participatory GIS and very few real-life case studies have been published (examples include Kingston et al. [14], Jordan and Shrestha [28], and Weiner and Harris [29]). Current real-life environmental studies largely rely on basic GIS operations [27]. Taking into consideration the work undertaken by a number of researchers (e.g., Kingston et al. [14] and Antunes et al. [30]), this application introduced weighting values derived from both expert opinion and public participation for each relevant data set. This approach provided a new dimension to the existing SEA methodologies by incorporating an innovative approach to the strategic assessment of land use plans. However, a number of fairly predictable limitations were observed, such as the need for GIS expertise, and data availability and accuracy issues, similar to those noted by Joao and Fonseca [11] and Vanderhaegen and Muro [27]. Moreover, reservations with regard to the usability of the tool and willingness to share and disclose spatial information varied between the case studies. Despite the perceived potential of the tool to assist and enhance participative processes in SEA, the aforementioned factors are considered to significantly affect the uptake of participatory GIS in the context of the Irish planning system.

9.5 CONCLUSION

Spatial data and GIS have the potential to facilitate and improve methodological aspects of environmental assessment (e.g., Joao [10], Antunes et al. [30], Agrawal and Dikshit [18], Steadman et al. [19], and Semmens and Goodrich [31]). Similarly, e-planning has huge potential to improve public participatory processes [32]. The provision of a complementary and alternative participatory GIS tool via the Internet has the potential to promote public involvement and enhance the transparency of the process by means of explicit display of information that reaches more people [33,34]. Despite this, the Irish case studies exposed a number of technical issues (e.g., computer and GIS knowledge/skill requirements, as well as spatial literacy) and institutional problems (e.g., copyright, confidentiality, and regulatory requirements for formal consultation) that significantly influence the usability of GIS-based public consultation in SEA.

The integration of public perceptions through GIS adds a new dimension to existing SEA methods and fulfills the requirements of Article 17 of the SEA Directive, which establishes that opinions expressed by the public are to be taken into consideration [1]. Having taken this approach, the results provide a composite scientific and social view of the relative environmental significance and vulnerability of the different areas, providing a more holistic view of the potential issues. It was observed through the case studies that the spatial representation and analysis of environmental considerations allows further scrutiny and contributes to a better understanding of the environmental implications of a planning decision. The consequent graphic and quantitative representation of the results allows a rapid and effective identification of most viable development scenarios/alternatives. The case studies indicate that GIS maps help stimulate debate and perform as a support tool in SEA by providing the mappable aspects. The GISEA methodology moves toward a more comprehensive and better informed decision-making process.

It must be noted, however, that current issues in relation to availability and quality of spatial data significantly hamper the effective application of GIS techniques in all

SEA stages. Data confidentiality and licensing issues also limit the extent to which GIS can be used. Furthermore, GIS skill and knowledge requirements and the more strategic and nonspatial nature of certain planning policies and objectives at the SEA level restrict the applicability of GIS in a number of steps in the environmental assessment process (e.g., public participation, assessment of alternatives, and definition of mitigation measures). In all cases, results derived from the spatial assessment need to be carefully scrutinized for validity and complemented with other forms of scientific knowledge and data if they are to be accountable. Resolution of complex environmental and planning decisions goes beyond the use of spatial data and the application of a systematic technology. Significant developments (at the education and technology levels) are still required to improve the efficiency of GIS in public participation processes. Similarly, more practical applications of spatial inclusion of public perceptions are needed to assess the real contribution of the methodology to participative environmental planning. Further research in relation to both participative SEA processes and governance issues in current planning procedures could also help identify feasible methods for the effective incorporation of public perceptions into decision making.

REFERENCES

1. Commission of the European Communities (CEC), Directive 2001/42/EC on the assessment of the effects of certain plans and programmes on the environment, Luxemburg, Official Journal of the European Union, 197, 30, 21.07.2001.
2. Fischer, T. B., Environmental Assessment in Land Use Planning in European Union Member States. Environmental Policy Advisory Service and Environmental Management Project, GTZ–Deutsche Gesellschaft für Technische Zusammenarbeit, Task 1 of project 02.2164.8-001.00, 2006.
3. Chan, Y. and Easa, S., Looking ahead, in *Urban Planning and Development Applications of GIS*, Easa, S. and Chan, Y., Eds., American Society of Civil Engineers, Reston, VA, 2000.
4. Wicks, P., INSPIRE Directive, Unpublished paper presented at the Irish Organisation for Geographic Information (IRLOGI) Conference, Dublin, Ireland, 2006.
5. Al-Kodmany, K., GIS and the artist: Shaping the image of a neighbourhood in participatory environmental design, in *Community Participation and Geographic Information Systems*, Weiner, D., Harris, T. M., and Craig, W. J., Eds., Taylor & Francis, London, 2002.
6. Risse, N., Crowley, M., Vincke, P., and Waaub, J.-P., Implementing the European SEA directive: The member states' margin of discretion, *Environmental Impact Assessment Review*, 23, 453, 2003.
7. Commission of the European Communities (CEC), Directive 2003/35/EC—Providing for public participation in respect of the drawing up of certain plans and programmes relating to the environment and amending with regard to public participation and access to justice council Directives 85/337/EEC and 96/61/EC, Official Journal of the European Union, L 156, 25.6.2003, 2003.
8. von Seht, H., Requirements of a comprehensive strategic environmental assessment system, *Landscape and Urban Planning*, 45, 1, 1999.
9. Gonzalez, A., Gilmer, A., Foley, R., Sweeney, J. and Fry, J., Dynamics of a decision support system in strategic environmental assessment implementation, *Proceedings of the IAIA '05 Ethics and Quality, Annual Conference of International Association for Impact Assessment*, Boston, 2005.

10. Joao, E., Use of geographic information systems in impact assessment, in *Environmental Methods Review: Retooling Impact Assessment for the New Century*, Porter, A. and Fittipaldi, J., Eds., The Army Environmental Policy Institute, Atlanta, 1998, 154.

11. Joao, E. and Fonseca A., Current use of geographical information systems for environmental assessment: A discussion document, Research Papers in Environmental and Spatial Analysis (No. 36), Department of Geography, London School of Economics, 1996.

12. Jordan, G., *A Public Participation GIS for Community Forestry User Groups in Nepal: Putting People Before the Technology*, Department of Agriculture & Forestry, University of Central Lancashire Penrith, Carlisle, UK, 1998.

13. Sieber, R. E., Geographic information systems in the environmental movement, http://www.ncgia.ucsb.edu/varenius/ppgis/papers/sieber.pdf, 1998

14. Kingston, R., Carver, S., Evans, A., and Turton, I., Web-based public participation geographical information systems: An aid to local environmental decision-making, *Computers, Environment and Urban Systems*, 24(2), 109, 2000.

15. Bojórquez-Tapia, L., Diaz-Mondragón, S., and Ezcurra E., GIS-based approach for participatory decision making and land suitability assessment, *International Journal of Geographical Information Science*, 15(2), 129, 2001.

16. Wood, J., "How green is my valley?" Desktop geographic information systems as a community-based participatory mapping tool, *Area*, 37(2), 159, 2005.

17. Chrisman, N., *Exploring Geographic Information Systems*, John Wiley & Sons, Chichester, 1999.

18. Agrawal, M. L. and Dikshit, A. K., Significance of spatial data and GIS for environmental impact assessment of highway projects, *Indian Cartographer*, 4, 262, 2002.

19. Steadman, E. J., Mitchell, P., Highley, D. E., Harrison, D. J., Linley, K. A., McFarlane, M., and McEvoy, F., Strategic environmental assessment and future aggregate extraction in the East Midlands region, British Geological Survey Report CR/04/003N, 2004.

20. Bettes, L., Successfully integrate environmental and transportation planning, http://www.geoplace.com/uploads/OnlineExclusives/quantm.asp, 2005.

21. Kingston, R., Carver, S., Evans, A., and Turton, I., Virtual decision making in spatial planning: Web-based geographical information systems for public participation in environmental decision making, Proceedings of the International Conference on Public Participation and Information Technology, Lisbon, Portugal, 1999.

22. Gonzalez, A., Gilmer, A., Foley, R., Sweeney, J., and Fry, J., Technology-aided participative methods in environmental assessment: An international perspective. *Computers, Environment and Urban Systems*, 32, 303, 2008.

23. Furlong, S. R., Interest group participation in rule making: a decade of change, *Journal of Public Administration Research and Theory*, 15(3) 353, 2005.

24. Scott, D. and Oelofse, C., Social and environmental justice in South African cities: Including invisible stakeholders in environmental assessment procedures, *Journal of Environmental Planning and Management*, 48(3), 445, 2005.

25. Carver, S., Public participation using Web-based GIS, *Environment and Planning B*, 28(6), 803, 2001.

26. Andrienko, G., Andrienko, N., and Gitis, V., Interactive maps for visual exploration of grid and vector data, *ISPRS Journal of Photogrammetry and Remote Sensing*, 57(5-6), 380, 2003.

27. Vanderhaegen, M. and Muro, E., Contribution of a European spatial data infrastructure to the effectiveness of EIA, *Environmental Assessment Review*, 25(2), 123, 2005.

28. Jordan, G. and Shrestha, B., A participatory GIS for community forestry user groups in Nepal: Putting people before the technology, *Participatory learning and action notes 39, International Institute for Environment and Development*, 2000, available at: http://www.geog.ntu.edu.tw/course/gislucc/GIS_Paper%5Cforestry.pdf.

29. Weiner, D. and Harris, T. M., Community-integrated GIS for land reform in South Africa, *Urban and Information Systems Association Journal*, 15(APA II), 61, 2003.

30. Antunes, P., Santos, R., and Jordao, L., The application of geographical information systems to determine environmental impact significance, *Environmental Impact Assessment Review*, 21(6), 511, 2001.

31. Semmens, D. J. and Goodrich D. J., *Planning change: case studies illustrating the benefits of GIS and landuse data in environmental planning*, 2005, available at: http://www.tucson.ars.ag.gov/AGWA/docs/pubs/hypesd.pdf.

32. Kingston, R., *The role of participatory e-planning in the new English local planning system*, 2006, available at: www.ppgis.manchester.ac.uk/downloads/e-Planning_LDFs.pdf.

33. Ceccato, V. and Snickers, F., Adapting GIS technology to the needs of local planning, *Environment and Planning B: Planning and Design*, 27, 923, 2000.

34. Ball, J., Towards a methodology for mapping "regions of sustainability" using PPGIS, *Progress in Planning*, 58, 81, 2002.

Section 2

Modeling the Natural Environment

10 Keynote Paper
Challenges for
Environmental Modeling

Richard Aspinall

CONTENTS

OVERVIEW

This chapter explores some issues in modeling the environment based on coupling models with GIS and other spatial data technologies. The purpose is to identify opportunities for further development of environmental modeling with geographic information systems (GIS). The chapter argues that there is considerable opportunity for GIS to expand its capability to address a broader range of environmental modeling through a focus on three challenges: (1) representation of environment, (2) representation of process, and (3) representation of time. Representation of environment refers to descriptions of environment based on a variety of complex descriptors or models, including dynamic representations, rather than static landscape descriptors based on geometry of landforms and land cover types. Network representations of landscape structure and condition are also considered as part of the representation of environment. Representation of processes includes mechanisms for addressing dynamic processes producing system change through time and space. Representation of time discusses approaches that address the limitation of static data and data structures in GIS, specifically with respect to dynamic process models, and argues for improved conceptualization of time as well as methods for managing temporal concepts in analysis.

10.1 INTRODUCTION

This chapter explores some issues associated with modeling the natural environment, specifically in relation to approaches based on coupling models with geographic information systems (GIS) and other spatial data technologies. The purpose of this is to identify opportunities for further development of GIS-based environmental modeling. I take a broad and inclusive definition of GIS in this context and use the term to include a variety of spatial data management and analysis technologies, such as remote sensing as well as statistical and other analyses of spatial data.

The use of GIS for environmental modeling relies heavily on the tools available within, or readily linked to, whatever GIS is being used. It also depends on data sets that describe environment conditions and variability in GIS. Typically the environment that is considered in GIS-based models refers to the land surface, rather than atmosphere, subsurface, or ocean environments, although there are developments in representation and analysis in GIS that promote the 3-D capabilities of GIS [1–4] and that support models for atmospheric, ocean, and subsurface environments. GIS-based tools include kriging, various forms of regression, and other software; whereas data sets include imagery that describes land cover or digital elevation data that describe topography and a variety of topographically related variables [5].

There are many different types of model. Jeffers [6] provides a taxonomy of models used in environmental modeling, including matrix, stochastic, multivariate, optimization, topological, and dynamic. Some models are well known in different areas of environmental science, for example, the state–factor model of soil formation [7], neutral model in landscape ecology [8], and process models such as CENTURY [9–11]. Other modeling approaches representing some of the families of models described by Jeffers are also well known in GIS, including kriging, statistical and spatial–statistical models, agent-based models, and cellular automota. Other analytical approaches exploit terrain geometry to estimate topographically related variables or estimates of enviromental condition [12]. These range from relatively simple derivative variables such as slope gradient and aspect [13], or flow accumulation [14], to more complex properties such as soil moisture accumulation [15] or solar radiation input [16].

In general, much GIS-based environmental modeling, in which modeling functions are integral to the GIS software, concentrates on the spatial elements of data—the location or geometry—rather than on process and dynamics. This approach has some merits. Many topographic derivatives are process related, and analysis of elevation data can be used to map variables that have a clear relation to processes (for example, those that are influenced by gravity and potential/kinetic energy, or by the geometry of the sun–earth surface relationships, as for radiation input). These variables may also provide valuable input to process models.

I do not elaborate further on the large number of studies that have been carried out with these modeling tools and data sets, and refer the reader to the current journal literature as well as reference and synthesis texts for GIS and environmental modeling [17–19] for examples. Instead, in this chapter I argue that there is considerable opportunity for GIS to expand its capability to address a broader range of environmental modeling through a focus on three challenges: (1) representation of environment, (2) representation of process, and (3) representation of time.

10.2 REPRESENTATION OF ENVIRONMENT

Environmental models are typically based on a representation of landscape that is achieved with topography (via a digital elevation model [DEM]) and land cover (e.g., from satellite sensor imagery). These representations serve well for many environmental processes, particularly in hydrology, landscape ecology, and in relation to species distributions, soils, and landforms. For example, predictive models for plants [20,21], vegetation [22–24], and animals [25–27] provide an example of environmental modeling that is often based on representation of environment with elevation data and land cover. These models use a variety of independent environmental variables to model distribution of an organism of interest based on sample locations of presence and absence. Independent predictor variables are chosen for their potential process-based association with the occurrence of the organism being modeled; variables chosen may have direct or indirect effects on distribution, or may represent a resource variable that influences distribution or growth [28]. The most frequently used range of variables includes topographically derived variables and land cover. Models are based on generalized linear models or generalized additive models, although many other modeling approaches are also used [23]. Output is a representation of environment measured as an estimate of the probability of occurrence of the species or group of interest being found at each location. In addition, the output field of probabilities from predictive models provides a new representation of environment: the probability that a location is suitable for the species of interest. Other models, for example, of dispersal, can be built on this representation (see Section 10.3).

In other cases, environmental models need not be related to environmental descriptions based on elevation or land cover. In these cases, representation of environment based on other descriptions and measurements may be used. For example, Brown and colleagues [29] present a representation of environment for elk and wolves in Yellowstone National Park that is based on the ideal free distribution (IDF), a model that combines process relationships for feeding energetics and inter- and intraspecific interaction. The IDF model was originally developed to describe the optimal distribution of mobile consumers in relation to their food resources by measuring the relative utility or fitness of discrete habitat patches for feeding [30]. Farnsworth and Beecham [31] showed that the IDF is a limiting case of a general model based on diffusion relationships. In their model, Brown et al. implemented the IDF distribution as a raster/field representation by making the functional relationships in the IDF model relate to spatial neighborhoods of grid cells rather than patches, as in the original formulation. The model can be run iteratively to allow intraspecific interaction to develop over time, and patterns of distribution change. The output from the analysis is a field representation of environments for elk and wolves based on predator–prey–grazing interactions. Currently, no data are collected that allow this model of environment or representation of wolf and elk distribution to be tested; both the representation of environment and the model output can be considered as spatial and process hypotheses in need of testing. The model of environment in this case reflects a dynamic inter- and intraspecific social landscape rather than a static landscape based on the geometry of landforms and land cover types.

Network representations of landscape and environmental structure and condition have also been developed in GIS for use in modeling; this explicitly bases environmental modeling on a network data structure, as opposed to object and field structures. For example, Urban and Keitt [32] used networks to represent environmental relations for different animal species. Network representations support landscape ecological models of metapopulations and dispersal, not only facilitating description of connectivity of landscapes for a particular species, but also allowing investigation of possible movements between patches. As for the previous example (Brown et al.), few field data are collected to allow this form of representation of environmental condition to be tested, and the approach is currently more appropriately used for hypothesis generation.

10.3 REPRESENTATION OF PROCESS

Process representation in a GIS-compatible format provides a second challenge for GIS to expand its capabilities for environmental modeling. Two categories of process are of interest: (1) dynamic processes producing system change through time and space, and (2) spatial processes.

10.3.1 PROCESS DYNAMICS

Process-based understanding is commonly sought in environmental sciences. Process models in geomorphology, biogeography, ecology, hydrology, and climatology, as well as in other areas of environmental science, represent processes with a variety of mathematical forms, including simultaneous equations describing chemical mass balance for watershed scale chemical transport [33]. These forms of analyses are neither described simply, nor coupled easily, with spatial data structures. One approach has been to focus on coupling GIS with other software that implements dynamic models of processes, for example, STELLA [34,35]. Another has been to redefine the modeling problem. Raper and Livingstone [36] provide an example of a GIS-centered approach to process geomorphology that is based on a reconceptualization of space–time, processes, and landforms. They present this using a case study from coastal geomorphology, and their approach is implemented in an object-oriented GIS. Other approaches to couple GIS with process models focus on identification of processes within models [37]. More recently, agent-based models have also been linked with GIS to model change in environmental systems, modeling dynamic processes in human, rather than physical environmental, systems and using the results to assess impact of change on environmental systems [38,39].

10.3.2 SPATIAL PROCESSES

More progress has been made in coupling spatial process models to GIS. A variety of approaches are suited to modeling spatial processes, and their linkage or implementation in GIS is more straightforward than for dynamic process models since the data structures are more suitable for these approaches. This does not, however, mean that GIS data structures are always ideal in terms of computational efficiency.

Examples of methods that model spatial processes include spatial statistical methods (e.g., geographically weighted regression [40,41], forms of regression models that include attention to spatial structures and dependencies [24,42,43], and geostatistical methods such as kriging [44]. Kriging has found application in many areas of environmental science (see, for example, Syed et al. [45], Jerrett et al. [46], Chappell [47], Cesaroni et al. [48], and many others for examples from hydrology, climate, geomorphology, and biogeography).

Cellular automata (CA) have also been used for modeling environmental phenomena for a number of years [49,50], and their use has grown recently. A cellular automaton combines a focus on spatial relationships with rules that describe processes, and an explicit temporal dimension that introduces dynamics to both the model and its representation. CA are used successfully in models of spatial diffusion processes, such as urban growth [51,52], effects of zoning [53], and fire spread [54]. The CA approach to process modeling has also expanded to disciplines beyond geography and environmental science, for example, by the use in civil engineering in attempts to couple systems dynamic process models with GIS for modeling water resources [55].

10.4 REPRESENTATION OF TIME

The third challenge for GIS-based environmental modeling is to improve the representation and management of time in GIS. Dynamic models, as described earlier in relation to representation of process, aim to address the question of how environmental systems operate through understanding process. GIS and GIS databases are, however, relatively static and are not able to address dynamics, not least because temporal concepts are not well implemented in GIS [56].

Several approaches have been followed to address this limitation of GIS. First, simple GIS-based differencing of data, usually with accompanying statistical analysis, is routinely used to explore and model land cover change from a time series of snapshots of land cover data [57–60]. The shorter the time interval between snapshots, the more closely the time series represents a record of continuous change over time. Second, raster GIS have been developed with programming tools that support analysis and coding of environmental system dynamics [61–64]. Although these systems manage environmental data to be dynamic through the dynamics coded in the computational tools, the time steps of the computation are not always clearly related to time beyond the program cycles. Third, considerable effort has been directed toward implementing space–time prisms [65–68], although these are mostly directed at transportation and human mobility rather than environmental process models. Fourth, cellular automata, described earlier in relation to implementation of spatial processes within GIS, are also used to represent temporal change and dynamics within GIS [49,54,57,69,70], notably because they provide explicit handling of time [70]. Although the time step of the CA is represented in the CA program code, the relationship of this program time to process time or scaling—which translates the program time to world time related to the phenomenon being modeled—is not usually explicit. A fifth approach is to reconceptualize environmental modeling as processes operating in space–time (rather than in space and time), and implement this conceptualization in GIS, possibly using object-oriented approaches [36]. This approach presents time as a dimension of the database structure, and when implemented in an object-oriented GIS, provides a record of the evolution of geographic objects. The management of time in a model or analysis is, however, most likely to be external to the computational implementation. Sixth, more recent computational developments provide event- and action-oriented approaches [71] that offer potential for environmental modeling since they move from the evolution of geographic (or other) events toward event chronicles that can represent process.

Although each of these approaches incorporates time in analysis, they all provide only limited control over the manner and metrics by which process is related to time. Specifically, the conceptualization of time in some of the approaches is based on a programming clock in which one iteration of the program represents a single time step in the operation of a process. However, representing time using a program cycle or step does not make explicit the relation of the program time with process, clock, calendar, or other time. For environmental systems, processes operating at multiple scales of time (and space) may need to be included, and flexible tools for implementing dynamic models and their temporal and spatial domains will be required.

Time may be represented as a specific instant over which a process occurs, or a period over which a process continues; this should be made explicit in a model of a dynamic process, and should be capable of control in implementation of a dynamic model in a GIS. In addition to being a product of processes, environmental change and environmental system dynamics also reflect the past history of the environmental system, formalized as path dependence [72]. This may also be usefully developed and included in a representation of time for process modeling in GIS.

10.5 COMBINING REPRESENTATION OF ENVIRONMENT, PROCESS, AND TIME

Progress in each of the three challenges described here has potential to support a much wider range of environmental modeling with GIS and provide new insights into spatial, temporal, and location-specific change, as well as dynamics of environmental systems. There are several reasons why this would be a welcome and valuable addition to the capabilities of GIS. First, it would allow GIS and GIS-based models to be applied to significant questions concerning the basic science of environmental systems, including how they operate and how they change. Second, there are strong applied uses for dynamic models of environmental systems that are place- and time-specific. For example, current interest in ecosystem services, particularly regulating services, is dependent on clear measurement of location-specific environmental function. Description and analysis of ecosystem services developed using process models in GIS would have dual advantages of being place-based and based on a functional understanding of environmental processes. Similarly, there is growing interest in linking policy and practice to science with attention being paid to place-based planning, including spatial planning for sustainability and issues of vulnerability and resilience of environmental systems (and of communities to changes in environmental systems). Process models based on appropriate representations of environments, and with explicit understanding of space and time, will provide place- and time-based information of value to science–policy–practice links.

REFERENCES

1. Karssenberg, D. and De Jong, K., Dynamic environmental modeling in GIS: 2. Modeling error propagation, *International Journal of Geographical Information Science*, 19, 623, 2005.
2. Kaufmann, O. and Martin, T., 3D geological modeling from boreholes, cross-sections and geological maps, application over former natural gas storages in coal mines, *Computers & Geosciences*, 34, 278, 2008.
3. Matejicek, L., Engst, P., and Janour, Z., A GIS-based approach to spatio-temporal analysis of environmental pollution in urban areas: A case study of Prague's environment extended by LIDAR data, *Ecological Modeling*, 199, 261, 2006.
4. Pinho, J. L. S., Vieira, J. M. P., and do Carmo, J. S. A., Hydroinformatic environment for coastal waters hydrodynamics and water quality modeling, *Advances in Engineering Software*, 35, 205, 2004.
5. Wilson, J. P. and Gallant, J. C., Eds., *Terrain Analysis: Principles and Applications*, John Wiley & Sons, Chichester, 2000.
6. Jeffers, J. N. R., From free-hand curves to chaos: Computer modeling in ecology, in *Computer Modeling in the Environmental Sciences*, Farmer, D. G. and Rycroft, M. J., Eds., Clarendon Press, Oxford, 1991, 299.
7. Jenny, H., *Factors of Soil Formation: A System of Quantitative Pedology*, McGraw-Hill, New York, 1941.
8. O'Neill, R. V., Gardner, R. H., and Turner, M. G., A hierarchical neutral model for landscape analysis, *Landscape Ecology*, 7, 55, 1992.
9. Parton, W., Tappan, G., Ojima, D., and Tschakert, P., Ecological impact of historical and future land-use patterns in Senegal, *Journal of Arid Environments*, 59, 605, 2004.
10. Parton, W. J., Scurlock, J. M. O., Ojima, D. S., Gilmanov, T. G., Scholes, R. J., Schimel, D. S., Kirchner, T., et al., Observations and modeling of biomass and soil organic-matter dynamics for the grassland biome worldwide, *Global Biogeochemical Cycles*, 7, 785, 1993.
11. Paustian, K., Parton, W. J., and Persson, J., Modeling soil organic-matter in organic-amended and nitrogen-fertilized long-term plots, *Soil Science Society of America Journal*, 56, 476, 1992.
12. Hodgson, M. E. and Gaile, G. L., A cartographic modeling approach for surface orientation-related applications, *Photogrammetric Engineering and Remote Sensing*, 65, 85, 1999.
13. Jones, K. H., A comparison of algorithms used to compute hill slope as a property of the DEM, *Computers & Geosciences*, 24, 315, 1998.
14. White, D. and Fennessy, S., Modeling the suitability of wetland restoration potential at the watershed scale, *Ecological Engineering*, 24, 359, 2005.
15. Gritzner, M. L., Marcus, W. A., Aspinall, R., and Custer, S. G., Assessing landslide potential using GIS, soil wetness modeling and topographic attributes, Payette River, Idaho, *Geomorphology*, 37, 149, 2001.
16. Dubayah, R. and Rich, P. M., Topographic solar-radiation models for GIS, *International Journal of Geographical Information Systems*, 9, 405, 1995.
17. Clarke, K. C., Parks, B. O., and Crane, M. P., Eds., *Geographic Information Systems and Environmental Modeling*, Prentice Hall, New Jersey, 2002.
18. Goodchild, M. F., Parks, B. O., and Steyaert, L. T., Eds., *Environmental Modeling with GIS*, Oxford University Press, New York, 1993.
19. Goodchild, M. F., Steyaert, L. T., Parks, B. O., Johnston, C., Maidment, D., Crane, M., and Glendinning, S., Eds., *GIS and Environmental Modeling: Progress and Research Issues,* Wiley, New York, 1996.

20. Austin, M. P., Modeling the environmental niche of plants: Implications for plant community response to elevated Co2 levels, *Australian Journal of Botany*, 40, 615, 1992.
21. Rew, L. J., Maxwell, B. D., and Aspinall, R., Predicting the occurrence of nonindigenous species using environmental and remotely sensed data, *Weed Science*, 53, 236, 2005.
22. Franklin, J., Predictive vegetation mapping: Geographic modeling of biospatial patterns in relation to environmental gradients, *Progress in Physical Geography*, 19, 474, 1995.
23. Guisan, A. and Zimmermann, N. E., Predictive habitat distribution models in ecology, *Ecological Modeling*, 135, 147, 2000.
24. Miller, J., Franklin, J., and Aspinall, R., Incorporating spatial dependence in predictive vegetation models, *Ecological Modeling*, 202, 225, 2007.
25. Eyre, M. D., Rushton, S. P., Luff, M. L., and Telfer, M. G., Predicting the distribution of ground beetle species (Coleoptera, Carabidae) in Britain using land cover variables, *Journal of Environmental Management*, 72, 163, 2004.
26. Ferrier, S., Watson, G., Pearce, J., and Drielsma, M., Extended statistical approaches to modeling spatial pattern in biodiversity in northeast New South Wales. I. Species-level modeling, *Biodiversity and Conservation*, 11, 2275, 2002.
27. Austin, G. E., Thomas, C. J., Houston, D. C., and Thompson, D. B. A., Predicting the spatial distribution of buzzard Buteo buteo nesting areas using a geographical information system and remote sensing, *Journal of Applied Ecology*, 33, 1541, 1996.
28. Austin, M. P. and Smith, T. M., A new model for the continuum concept, *Vegetatio*, 83, 35, 1989.
29. Brown, D. G., Aspinall, R., and Bennett, D. A., Landscape models and explanation in landscape ecology: A space for generative landscape science? *Professional Geographer*, 58, 369, 2006.
30. Fretwell, S. D. and Lucas, J. H. J., On territorial behaviour and other factors influencing habitat distribution in birds, *Acta Biotheoretica*, 19, 16, 1970.
31. Farnsworth, K. D. and Beecham, J. A., Beyond the ideal free distribution: More general models of predator distribution, *Journal of Theoretical Biology*, 187, 389, 1997.
32. Urban, D. and Keitt, T., Landscape connectivity: A graph-theoretic perspective, *Ecology*, 82, 1205, 2001.
33. Luo, Y. Z., Gao, Q., and Yang, X. S., Dynamic modeling of chemical fate and transport in multimedia environments at watershed scale—I: Theoretical considerations and model implementation, *Journal of Environmental Management*, 83, 44, 2007.
34. Costanza, R. and Maxwell, T., Spatial ecosystem modeling using parallel processors, *Ecological Modeling*, 58,159, 1991.
35. Costanza, R. and Gottlieb, S., Modeling ecological and economic systems with STELLA: Part II, *Ecological Modeling*, 112, 81, 1998.
36. Raper, J. and Livingstone, D., Development of a geomorphological spatial model using object-oriented design, *International Journal of Geographical Information Systems*, 9, 359, 1995.
37. Bian, L. and Hu, S., Identifying components for interoperable process models using concept lattice and semantic reference system, *International Journal of Geographical Information Science*, 21, 1009, 2007.
38. Castella, J. C., Trung, T. N., and Boissau, S., Participatory simulation of land-use changes in the northern mountains of Vietnam: The combined use of an agent-based model, a role-playing game, and a geographic information system, *Ecology and Society*, 10, 27, 2005.
39. Evans, T. P. and Kelley, H., Multi-scale analysis of a household level agent-based model of landcover change, *Journal of Environmental Management*, 72, 57, 2004.
40. Zhang, L. J. and Shi, H. J., Local modeling of tree growth by geographically weighted regression, *Forest Science*, 50, 225, 2004.
41. Shi, H., Laurent, E. J., LeBouton, J., Racevskis, L., Hall, K. R., Donovan, M., Doepker, R. V., Walters, M. B., Lupi, F., and Liu, J., Local spatial modeling of white-tailed deer distribution, *Ecological Modeling*, 190, 171, 2006.

42. Miller, J. and Franklin, J., Explicitly incorporating spatial dependence in predictive vegetation models in the form of explanatory variables: A Mojave Desert case study, *Journal of Geographical Systems*, 8, 411, 2006.

43. Kupfer, J. A. and Farris, C. A., Incorporating spatial non-stationarity of regression coefficients into predictive vegetation models, *Landscape Ecology*, 22, 837, 2007.

44. Burrough, P. A., GIS and geostatistics: Essential partners for spatial analysis, *Environmental and Ecological Statistics*, 8, 361, 2001.

45. Syed, K. H., Goodrich, D. C., Myers, D. E., and Sorooshian, S., Spatial characteristics of thunderstorm rainfall fields and their relation to runoff, *Journal of Hydrology*, 271, 1, 2003.

46. Jerrett, M., Burnett, R. T., Kanaroglou, P., Eyles, J., Finkelstein, N., Giovis, C., and Brook, J. R., A GIS-environmental justice analysis of particulate air pollution in Hamilton, Canada, *Environment and Planning A*, 33, 955, 2001.

47. Chappell, A., The limitations of using Cs-137 for estimating soil redistribution in semi-arid environments, *Geomorphology*, 29, 135, 1999.

48. Cesaroni, D., Matarazzo, P., Allegrucci, G., and Sbordoni, V., Comparing patterns of geographic variation in cave crickets by combining geostatistic methods and Mantel tests, *Journal of Biogeography*, 24, 419, 1997.

49. Theobald, D. M. and Gross, M. D., EML: A modeling environment for exploring landscape dynamics, *Computers Environment and Urban Systems*, 18, 193, 1994.

50. Itami, R. M., Simulating spatial dynamics: Cellular-automata theory, *Landscape and Urban Planning*, 30, 27, 1994.

51. Leao, S., Bishop, I., and Evans, D., Simulating urban growth in a developing nation's region using a cellular automata-based model, *ASCE Journal of Urban Planning and Development,* 130, 145, 2004.

52. Xie, Y. C., A generalized model for cellular urban dynamics, *Geographical Analysis*, 28, 350, 1996.

53. He, C., Zhang, Q., Li, Y., Li, X., and Shi, P., Zoning grassland protection area using remote sensing and cellular automata modeling: A case study in Xilingol steppe grassland in northern China, *Journal of Arid Environments*, 63, 814, 2005.

54. Yassemi, S., Dragicevic, S., and Schmidt, M., Design and implementation of an integrated GIS-based cellular automata model to characterize forest fire behaviour, *Ecological Modeling*, 210, 71, 2008.

55. Ahmad, S. and Simonovic, S. P., Spatial system dynamics: New approach for simulation of water resources systems, *Journal of Computing in Civil Engineering*, 18, 331, 2004.

56. Wheeler, D. J., Commentary: Linking environmental models with geographic information systems for global change research, *Photogrammetric Engineering and Remote Sensing*, 59, 1497, 1993.

57. Fan, F. L., Wang, Y. P., and Wang, Z. S., Temporal and spatial change detecting (1998–2003) and predicting of land use and land cover in Core corridor of Pearl River Delta (China) by using TM and ETM+ images, *Environmental Monitoring and Assessment*, 137, 127, 2008.

58. Acevedo, M. A. and Restrepo, C., Land-cover and land-use change and its contribution to the large-scale organization of Puerto Rico's bird assemblages, *Diversity and Distributions*, 14, 114, 2008.

59. Ding, H., Wang, R. C., and Wu, J. P., Quantifying land use change in Zhejiang coastal region, China using multi-temporal landsat TM/ETM plus images, *Pedosphere*, 17, 712, 2007.

60. Alves, D. S. and Skole, D. L., Characterizing land cover dynamics using multi-temporal imagery, *International Journal of Remote Sensing*, 17, 835, 1996.

61. Barreto-Neto, A. A. and de Souza Filho, C. R., Application of fuzzy logic to the evaluation of runoff in a tropical watershed, *Environmental Modeling & Software*, 23, 244, 2008.

62. Karssenberg, D., de Jong, K., and van der Kwast, J., Modeling landscape dynamics with Python, *International Journal of Geographical Information Science*, 21, 483, 2007.
63. van der Perk, M., Burrough, P. A., and Voigt, G., GIS-based modeling to identify regions of Ukraine, Belarus and Russia affected by residues of the Chernobyl nuclear power plant accident, *Journal of Hazardous Materials*, 61, 85, 1998.
64. De Vasconcelos, M. J., Goncalves, A., Catry, F. X., Paul, J. U., and Barros, F., A working prototype of a dynamic geographical information system, *International Journal of Geographical Information Science*, 16, 69, 2002.
65. Yu, H. and Shaw, S. L., Exploring potential human activities in physical and virtual spaces: A spatio-temporal GIS approach, *International Journal of Geographical Information Science*, 22, 409, 2008.
66. Neutens, T., Van de Weghe, N., Witlox, F., and De Maeyer, P., A three-dimensional network-based space-time prism, *Journal of Geographical Systems*, 10, 89, 2008.
67. Miller, H. J., Measuring space-time accessibility benefits within transportation networks: Basic theory and computational procedures, *Geographical Analysis*, 31, 1, 1999.
68. Kwan, M. P., Gender and individual access to urban opportunities: A study using space-time measures, *Professional Geographer*, 51, 210, 1999.
69. Geertman, S., Hagoort, M., and Ottens, H., Spatial-temporal specific neighbourhood rules for cellular automata land-use modeling, *International Journal of Geographical Information Science*, 21, 547, 2007.
70. Wagner, D. F., Cellular automata and geographic information systems, *Environment and Planning B*, 24, 219, 1997.
71. Worboys, M., Event-oriented approaches to geographic phenomena, *International Journal of Geographical Information Science*, 19, 1, 2005.
72. Brown, D. G., Page, S., Riolo, R., Zellner, M., and Rand, W., Path dependence and the validation of agent-based spatial models of land use, *International Journal of Geographical Information Science*, 19, 153, 2005.

11 Keynote Paper
Spatial Scale and Neighborhood Size in Spatial Data Processing for Modeling the Natural Environment

A-Xing Zhu

CONTENTS

OVERVIEW

This chapter presents and discusses two issues related to spatial data processing for modeling the natural environment. The first issue is the mismatch between the spatial scale of processes and the neighborhood size used in spatial analysis for characterizing those processes. The second is scale incompatibility and its impact on the characterization of spatial joint distributions used in the modeling of the natural environment. It is shown in this chapter that information characterized through spatial analysis is very sensitive to the neighborhood size used. It is recommended that spatial analysis should employ a neighborhood size comparable to the spatial scale of the process. This can be achieved by altering the default neighborhood size in spatial analysis algorithms. It is also shown that scale incompatibility of geographic data can lead to significant mischaracterization of spatial joint distributions of geographic factors. It is suggested that scale transformation of spatial data is needed to mitigate the impact of scale incompatibility on the characterization of spatial joint distributions.

11.1 INTRODUCTION

The assumption that the accuracy of modeling the natural environment will increase simply as the resolution of spatial data increases may not always hold. The complication is that processes in the natural environment operate at certain spatial scales and are greatly impacted by the joint distribution of their environmental factors. The spatial scale at which a particular process operates may not be the same as the spatial resolution of geographic data. For example, the spatial scale of soil-forming processes is not necessarily the same as the spatial resolution of digital elevation data used to characterize the topography. Although the spatial resolution of geographic data may increase as our ability to capture and record these detailed variations in geographic data increases, the spatial scale of processes does not change in response to the increase in spatial resolution. In addition, many of the methods used to model the natural environment often take multiple spatial data sets as inputs in characterizing the joint variation of geographic factors over space (referred to as spatial joint distribution, such as the joint spatial variation of soil depth and slope gradient). These data sets are often produced for different purposes and at different resolutions. Mixing spatial data of different resolutions can cause incorrect characterization of the spatial joint distribution of geographic or environmental factors. This chapter illustrates these problems and describes approaches to mitigate them. The following section presents the basic concepts related to the spatial scales of geographic processes. Section 11.3 discusses the impact of neighborhood size of spatial analysis and approaches to mitigate the negative impacts. Section 11.4 illustrates the issues related to the characterization of spatial joint distributions involving spatial data of different resolutions or map scales, and provides recommendations for minimizing the chance of improperly characterizing spatial joint distributions. Section 11.5 highlights the key points made in this chapter and suggests opportunities for further research.

11.2 BASIC CONCEPTS

11.2.1 SPATIAL SCALE, GRAIN SIZE, SPATIAL RESOLUTION

Spatial scale is a fundamental issue in many geographic analyses [1–5]. Depending on the context, spatial scale can refer to the grain size or the extent [6–9]. The *grain size* refers to the spatial detail, or minimum areal unit over which a particular process should be studied. The *extent* refers to the size of the study area. Extent and grain are like a fishnet: the extent represents the size of the whole net and the grain represents the hole size in the net. Extent and grain together determine the upper and lower limits of the study [1]. Geographic analysis at large spatial scales often involves a large spatial extent and coarser grain size [1,10]. In this chapter, spatial scale refers to the grain size.

Spatial resolution refers to the spatial detail or the spatial unit over which spatial data are collected or represented. There is not necessarily any connection between the grain size and spatial resolution of the data because the resolution at which spatial data are captured or represented may not be the same as the spatial detail at which a given spatial process needs to be studied, unless these two are coordinated. For example, the spatial detail at which an ecosystem process should be studied may not be the spatial resolution of remote sensing data (such as pixel size of Landsat Thematic Mapper [TM] imagery), unless specific efforts are made to match the sensor resolution to the spatial detail needed for studying particular processes, and this is often not the case.

11.2.2 EFFECTIVE NEIGHBORHOOD OF SPATIAL PROCESSES

The effects of spatial processes on geographic patterns and vice versa manifest themselves over a certain area. The interaction of geographic factors is a process of exchanging energy and matter, and this exchange requires space (or a neighborhood) to manifest. This neighborhood is referred to as the *effective neighborhood*. For example, the growth of a tree depends on the conditions of the area immediately surrounding it, often referred to as the spatial niche. This spatial niche has to be of a certain size, otherwise the tree will not receive enough water and nutrient supplies to survive. The size of this niche is the effective neighborhood of this ecological process. For another example, topography, as an important factor controlling the redistribution of energy and matter at a local level, plays a key role in soil-forming processes. The soil characteristics at a point are not purely dependent on the topographic conditions at this particular point, but rather dependent on the topographic conditions over a certain area around this point because the redistribution of energy and matter needs an area of certain size to play out. Effective neighborhoods are different for different spatial processes due to variations in the spatial scales of those processes. For example, the effective neighborhood of topography on plant growth may be different from that of topography on soil formation.

The effective neighborhood can be treated as the grain size. In many geographical analyses, the spatial detail at which a given process should be studied should not be smaller than the size of the effective neighborhood. The spatial process will not become meaningful once the grain size is smaller than the size of the effective neighborhood. For example, characterizing the effect of topography on soil formation at the grain size of 10 cm is not necessary because soil property values are not impacted by the conditions over an area of 10 cm in size. In fact, studying spatial processes at the grain size smaller than the size of the effective neighborhood will lead to erroneous conclusions being drawn because the processes at that grain size may not have much to do with the processes impacting the phenomenon of concern.

There are two ways to consider the effective neighborhood in geographic analysis. The first is to incorporate the effective neighborhood size in models describing geographic processes; and the second, which this chapter focuses on, is to consider the effective neighborhood size when processing spatial data that are used to drive models of geographic processes.

11.2.3 SPATIAL JOINT DISTRIBUTION OF GEOGRAPHIC FACTORS

Spatial joint distribution refers to the variation in the configuration of geographic conditions over space. For example, if one traverses from a broad summit to a wide valley bottom, one would experience changes in the combination of elevation, slope gradient, and soil depth in the following way: from the combination of high elevation, gentle slope, thick soil at the summit; through the combination of high elevation, steeper slope, thin soil at the shoulder position; to the combination of lower elevation, steep slope, thin soil at the back slope; to the combination of low elevation, gentler slope, thicker soil at the foot slope; and finally to the combination of low elevation, gentle to flat slope, and thick soil at the valley bottom.

Spatial joint distributions are associated with two different contexts. The first is the interaction of geographic factors, for example, a change in slope gradient over space causes the change in soil depth: a steeper slope induces stronger soil erosion and leads to shallower soil. The second is the independent distribution of geographic factors. For example, the change of slope gradient is independent of a change in precipitation. Nevertheless, they do exhibit a spatial joint distribution.

Geographic factors are an integral part of geographic (spatial) processes, and the spatial joint distribution of geographic factors determines the outcome of spatial processes. The correct characterization of spatial joint distribution of geographic factors is therefore critical for accurate modeling of the natural environment.

11.3 NEIGHBORHOOD SIZE ISSUES IN MODELING THE NATURAL ENVIRONMENT

There is a difference between the effective neighborhood size and the neighborhood size over which geographic attributes are computed in a geographic information system (GIS). The former refers to the spatial scale at which geographic processes manifest their influence, whereas the latter merely refers to the domain that is needed for determining the value of a geographic attribute. For example, the effective neighborhood size for slope gradient to manifest its influence on soil forming is different from that needed for a gradient calculation algorithm to determine the slope gradient value at a location.

There are two interrelated issues in relation to neighborhood size in computing geographic attributes for modeling the natural environment, particularly when high resolution spatial data sets are used. The first is the issue of default neighborhood and the second is the modifiable areal unit problem (MAUP).

11.3.1 Default Neighborhood

The default neighborhood refers to the neighborhood associated with a given computation algorithm. For example, many algorithms for calculating terrain attributes (such as slope gradient, slope aspect, and surface curvature) use a 3 × 3 pixel window as the neighborhood over which the terrain attributes are computed [11,12]. This 3 × 3 pixel window is referred to as the default neighborhood for these algorithms. The default neighborhood is often tied to the resolution of the input data set: as the spatial resolution increases, the default neighborhood decreases. For example, when computing the slope gradient for a given location, a common approach is to use a 3 × 3 pixel moving window over a gridded digital elevation model (DEM), such as in ESRI ArcGIS and TAPES [13]. When the resolution of the input DEM increases (i.e., the pixel size decreases), the neighborhood over which the slope gradient is computed decreases.

Tying the default neighborhood size to spatial resolution of input data is problematic for modeling the natural environment [14,15], particularly when data of high spatial resolution are used. Sometimes the default neighborhood can be smaller than the effective neighborhood, which makes the computed value less relevant to the phenomena under study. For example, when the resolution of the DEM increases to 0.1 m, the neighborhood used to compute the slope gradient becomes an area of 0.3 m × 0.3 m. Clearly, in studying the influence of slope gradient on soil formation at the landscape level (1:24,000 map scale), the slope gradient over such a small neighborhood has little relevance. In addition, small neighborhood sizes often make the computation of geographic attributes highly susceptible to noise or local variation, which have little geographic meaning.

11.3.2 The Modifiable Areal Unit Problem (MAUP)

11.3.2.1 The Concept of MAUP

The MAUP [16–18] refers to the problem that occurs when different statistical results (characterizations) are obtained from the same set of data as a result of different grouping of area units. This can be illustrated using Simpson's paradox for the unemployment–ethnicity relationship [19]. As shown in Table 11.1, the unemployment rate is not related to ethnicity within Area A and Area B, respectively. However, when the two areas are aggregated, unemployment rates are very different between ethnic groups. This is a result of the aggregation of spatial units and represents a classic example of MAUP.

There are two types of effects associated with MAUP: the scale effect and the zoning effect (Figure 11.1) [16,18]. The scale effect refers to the problems associated with the aggregation of smaller units into larger units (difference in size). The case referred to in Table 11.1 is an example of the scale effect. The zoning effect refers to

TABLE 11.1

Illustration of Simpson's Paradox for the Unemployment/Ethnicity Relationship

Ethnicity	Unemployment	Population	Unemployment Rate (%)
Area A			
White	90	900	10
Asian	10	100	10
Area A total	100	1000	10
Area B			
White	100	500	20
Asian	100	500	20
Area B total	200	1000	20
Aggregated Area (Area A and Area B)			
White	190	1400	13.57
Asian	110	600	18.33
Total	300	2000	15

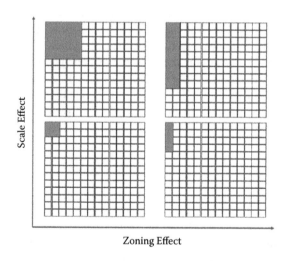

Zoning Effect

FIGURE 11.1 Two effects of the modifiable areal unit problem: scale effect and zoning effect.

a situation when the units are the same size but the different results are related to how smaller units are arranged or grouped to form the larger ones (a difference in grouping) [16]. Both effects will manifest themselves in modeling the natural environment when detailed spatial data are aggregated. In the context of neighborhood size, this chapter focuses on the scale effect.

11.3.2.2 The Scale Effect of MAUP

We will use the sensitivity of terrain attribute calculations to neighborhood size to illustrate the scale effect of MAUP on the modeling of the natural environment. To examine the effect of different neighborhood sizes (the scale effect), Zhu et al. [20] employed the variable neighborhood size concept by Wood [15] and developed a user-defined neighborhood size method for computing terrain derivatives. This method first creates a least-squares regression polynomial to produce a filtered (generalized) terrain surface over a user-defined neighborhood (see Shary et al. [21] and Schmidt et al. [22] for a discussion on polynomial methods). As is standard practice, a second-degree polynomial is used here [23]:

$$z = rx^2 + ty^2 + sxy + px + qy + u \qquad (11.1)$$

The coefficients p, r, s ... u are found by moving a window of user-specified size across the DEM and minimizing the squared difference between the polynomial and the elevation values within this window (or neighborhood area). This procedure is repeated for every elevation point, and thus z is considered a local polynomial. At every point the polynomial is differentiated analytically to obtain slope, curvature, and any other required values. This technique suppresses short-range variation at spatial scales smaller than the neighborhood size, regardless of DEM resolution. Allowing the user to specify the neighborhood size provides control over the amount of short-scale variation in the analysis. In this implementation, the neighborhood size is defined as the distance from the center of the center pixel to the window edge (it is similar to a radius rather than a diameter). Other studies have shown that this method produces more accurate terrain derivatives than other common methods (e.g., Florinsky [24]).

The above method was applied to a watershed in Dane County, Wisconsin. To examine the effect of neighborhood size, five different resolutions (10, 20, 30, 40, and 50 ft) of DEM were used. For each DEM resolution, a set of neighborhood sizes (ranging from 10 ft to 300 ft) were used to compute slope gradient, profile curvature, and contour curvature. It should be noted that the resolution of the DEM determines the smallest neighborhood size that can be applied to a specific DEM. For example, a 30 ft or coarser resolution DEM does not permit a neighborhood size of 20 ft or smaller.

The scale effect of neighborhood size on the computed terrain conditions can be examined in two ways. The first is to examine the difference between computed terrain conditions under different neighborhood sizes and those observed in the field by domain experts. In this illustration slope gradient is used as the terrain variable, and slope gradient values observed by field soil scientists are used as the observed values. A total of 81 field observations were made and the root mean squared error (RMSE) was used to measure differences. It can be observed from Figure 11.2 that the smallest RMSE, representing the best match between the computed slope gradient and the observed slope gradient, corresponds to neighborhood size somewhere between 100 ft and 110 ft, significantly different from the smallest neighborhood size available at each DEM. This means that over the study area, the spatial scale at which field soil scientists examine the impact of slope gradient on soil formation is about 100 ft and not the smallest neighborhood size available at any given DEM resolution. Table 11.2

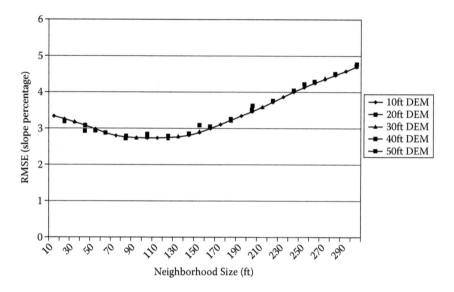

FIGURE 11.2 Root mean squared error between slope gradients calculated at different neighborhood sizes and those observed in the field.

lists the difference measured as mean error (ME) and RMSE between the gradient values computed using the default neighborhood size in ArcGIS and those observed in the field. The larger ME and RMSE further suggest that slope gradient values computed at the default neighborhood size, which is tied to DEM resolution, are not a good approximation (larger mean error and RMSE) to that used by field soil scientists.

The second method to examine the scale effect of neighborhood size on the computed terrain conditions is to analyze sensitivity of the computed terrain conditions to neighborhood size. We use two indices to compare the sensitivity to neighborhood size across different terrain variables (slope gradient, profile curvature, planform curvature): standardized magnitude and relative change [20]. *Standardized magnitude* is defined as the ratio of the individual value to the mean and measures the individual's deviation from the mean; the standardized magnitude for slope gradient at a given point for a given neighborhood size is the ratio of the gradient value at that point for the current neighborhood size over the mean of the slope gradient values at that point over all neighborhood sizes. Values of standardized magnitude far from unity mean greater deviation from the mean. Because the standardized magnitude is dimensionless, values for one terrain variable can be compared with those of other terrain variables, providing a means of assessing the relative sensitivity of different terrain derivatives to neighborhood size. It must be pointed out that standardized magnitude may not be appropriate for locations where the mean value approaches zero. *Relative change* is the difference in standardized magnitude between two consecutive neighborhood sizes at a point. It therefore provides an alternative way to characterize the sensitivity across neighborhood size by allowing us to identify neighborhood sizes to which the terrain conditions are more or most sensitive.

TABLE 11.2

Difference in Slope Gradient Values between Default Neighborhood Size in ArcGIS and User-Defined Neighborhood Size (NS)

DEM Resolution	10 ft		20 ft		30 ft	
Methods	ArcGIS	100 ft NS	ArcGIS	100 ft NS	ArcGIS	90 ft NS
ME	0.94	0.075	0.923	0.041	0.864	0.197
RMSE	3.422	2.726	3.188	2.764	3.28	2.739

Note: DEM = digital elevation model; ME = mean error; RMSE = root mean squared error.

Different terrain variables exhibit dramatic differences in sensitivity. Figure 11.3 shows the standardized magnitude for the three terrain variables (slope gradient, profile curvature, and contour [planform] curvature) over neighborhood size at two field points. This figure clearly shows that curvature measures are much more sensitive to neighborhood size than slope gradient. At some neighborhood sizes the computed curvature values are significantly higher than the overall mean, while the computed gradient values are about the same as the overall mean. Although there is some variability in sensitivity from location to location, the fact remains that curvature measures are more sensitive to neighborhood size than slope gradient.

In addition, terrain variables are more sensitive to neighborhood size at small neighborhood sizes than at large neighborhood sizes. This is seen in Figure 11.4, which shows the relative change across different neighborhood sizes for the two field points. The figure has two important features. The first is that the sensitivity is much stronger at small neighborhood sizes and generally decreases as neighborhood size increases. The second is that relative change for the two curvature variables fluctuates much more than that for slope gradient across neighborhood size. However, the general pattern of sensitivity across neighborhood size is that terrain variables are more sensitive to neighborhood size when neighborhood size is small, and less sensitive when neighborhood size is large. This makes the selection of neighborhood size much more critical for applications that require terrain information over small spatial scales. To obtain a good approximation of terrain information at a small spatial scale, one almost needs an exact match between the neighborhood size used and the desired spatial scale. At larger spatial scales this exact match may not be necessary.

11.3.3 Impact on Modeling

The combined effect of default neighborhood and MAUP on modeling the natural environment can be illustrated through a digital soil mapping example. Digital soil mapping is a predictive approach to model soil spatial variation based on the relationship between soil and its environmental conditions [25–27]. Among the many predictive digital soil-mapping approaches, SoLIM (soil land inference model) is one of the few digital soil mapping approaches that can be used in the production of soil surveys [26]. SoLIM couples GIS/remote sensing techniques with artificial

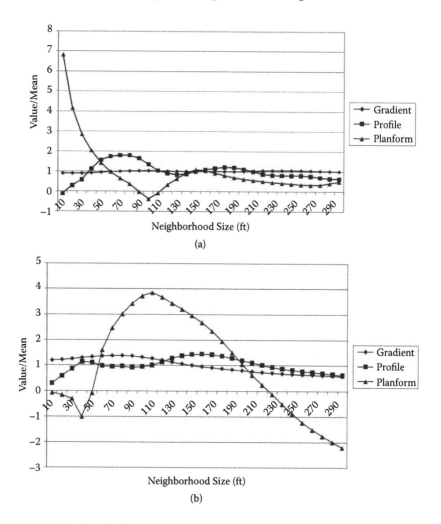

FIGURE 11.3 Standardized magnitude variation over neighborhood size (digital elevation model resolution fixed at 10 ft): (a) at Field Point 49; (b) at Field Point 105.

intelligence techniques to map spatial distributions of soil characteristics using fuzzy logic. It uses GIS/remote sensing to characterize the environmental conditions, which covary with soil conditions, and uses artificial intelligence to extract relationships between the characterized environmental conditions and soils [25,26,28,29]. A set of inference techniques developed under fuzzy logic is then used to combine the characterized environmental conditions with the extracted soil–environment relationships to predict soil spatial variation [30].

In this illustration we use the SoLIM approach as a means of examining the impact of neighborhood size on digital soil mapping. Terrain attributes (slope gradient, profile curvature, and contour curvature) were used together with other environmental variables (nonterrain data, such as geology) as inputs to SoLIM for soil mapping. To examine the impact of neighborhood size on digital soil mapping we

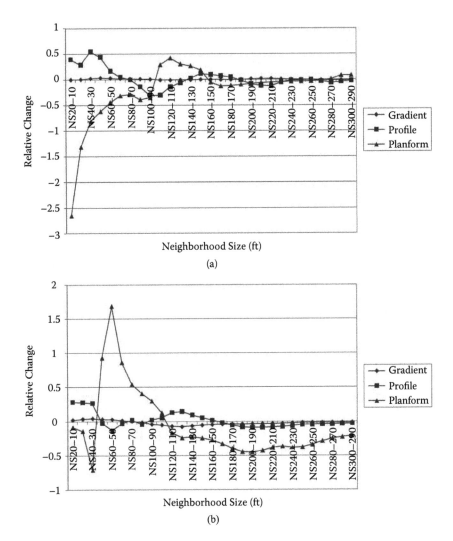

FIGURE 11.4 Relative change in standardized magnitude over neighborhood size (with digital elevation model resolution fixed at 10 ft): (a) at Field Point 49; (b) at Field Point 105.

held constant both the knowledge-base and the nonterrain data. We changed only the neighborhood size, which in turn gave varying terrain derivatives. Thus with each neighborhood size for each DEM resolution, we produced a version of the soil map using the SoLIM approach based on terrain derivatives that were generalized at that neighborhood size. For this exercise, we employed neighborhood sizes ranging from 10 ft to 180 ft for DEM resolutions of 10 ft, 15 ft, and 30 ft. Field soil samples were collected on a hill slope in Dane County, Wisconsin, to assess the accuracy of SoLIM-predicted soils under each neighborhood size. In this way, we obtained information about the impact of neighborhood size on the accuracy of digital soil mapping.

The current implementation of SoLIM requires the knowledge on soil–environment relationships to be translated manually into a digital representation [31,32]. The translated knowledge needs to be verified, which is done through subjective verification of preliminary inference results by soil scientists. If changes are suggested by soil scientists, the translated knowledge is revised based on these suggestions. The purpose of this validation is to ensure that the knowledge base used for prediction agrees with the scientist's conception of soil–environment relations. This process continues until the soil scientists are satisfied with the preliminary inference result. In this study we used results from the 10 ft DEM with 10 ft neighborhood size for verification. Although the field validation data set is not used in the verification process, it is expected that predictions using the 10 ft DEM are more accurate than those based on other DEM resolutions.

The relationship between the accuracy of digital soil predictions and neighborhood size is shown in Figure 11.5. It is clear that neighborhood size has a profound impact on the accuracy of the soil map. The difference in accuracy between different neighborhood sizes can be quite substantial, with the accuracy at one neighborhood size doubling at another. It is important to note that the most accurate soil map is not obtained at the smallest neighborhood size. The accuracy peaks at a neighborhood size of around 100 ft. Note that the somewhat high accuracy for the 10 ft DEM at the 10 ft neighborhood size is related to the verification process. Still, this accuracy is lower than that using a 100 ft neighborhood size. The finding here further suggests that the removal of certain fine-scale variations in DEMs is important for digital soil mapping because these fine scale details do not contribute to the differentiation of soil at the scale of interest for soil scientists. This supports the findings reported by Smith et al. [31].

11.3.4 CHOOSING NEIGHBORHOOD SIZE

The previous analysis and discussion have clearly illustrated that the default neighborhood associated with algorithms for spatial analysis does not necessarily capture the spatial scale of the geographic process under study. Choosing the correct neighborhood size is of paramount importance in processing geographic data for

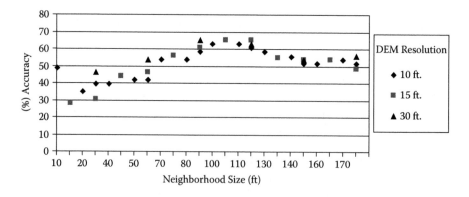

FIGURE 11.5   Accuracy of digital soil mapping and neighborhood size.

modeling the natural environment. The first step in choosing proper neighborhood size is to develop methods that permit a user-specified neighborhood, which in turn will allow the neighborhood size for spatial analysis to be separated from the spatial resolution of input data. In this regard, techniques for user-defined neighborhood size have been developed in terrain analysis [15,24,31,33,34]. The second step is to determine the spatial scale at which the spatial process of concern operates so that a neighborhood size comparable to this spatial scale can be selected for processing the geographic data. The third step is to perform sensitivity analysis of the geographic attribute to neighborhood size. This step is especially important for analyses performed at fine spatial scales, because geographic attributes are often more sensitive to neighborhood size at fine spatial scale than at coarse spatial scale, as demonstrated in the analysis above. This not only helps to select a proper neighborhood size, but also provides some insights into the potential impact of neighborhood size on the modeled results.

11.4 SCALE INCOMPATIBILITY AND SPATIAL JOINT DISTRIBUTION OF GEOGRAPHIC FACTORS

Characterizing the spatial joint distribution of multiple geographic factors is an essential part of modeling the natural environment. This characterization is commonly accomplished through the spatial analysis of geographic data of different sources and at different map scales or spatial resolutions. These data are often incompatible due to difference in sources and in scale or resolution. This incompatibility often leads to mischaracterization of spatial joint distribution of geographic factors and, as a result, leads to incorrect conclusions being drawn when modeling the natural environment.

11.4.1 Geographic Presentation, Geographic Representation, Maps, and Data Layers

Within the context of GIS there is a need to distinguish two important schemes of the manifestation of geographic information: geographic representation and geographic presentation. Although both terms imply manifestation of geographic information, geographic representation refers to a process of depicting as much information as accurately and as truthfully as possible about a given geographic feature/phenomenon for the purpose of information storage. The focus is on the detail and the richness of information about a given geographic feature or phenomenon. Geographic representation is limited only by data collection abilities and the conceptual model used to retain the collected data. Resolution is often associated with geographic representation. For example, the detail and accuracy of Band 1 of a Landsat TM scene depends on the sensitivity and the spatial resolution of the sensor taking the image. Another example is that the accuracy of a digital transportation network for a city created from global positioning system (GPS) observation depends on the accuracy of the GPS equipment and the density of readings taken in the field.

In contrast, geographic presentation refers to a process of showing selected aspects of certain geographic features or phenomena for the purpose of visualization. The focus is on the selection of information that is considered to be important or interesting to the audience. It is limited by the tools for visualization and by the sensitivity of human perception. Data for geographic presentation are often generalized and map scale is often associated with geographic presentation. For example, the accuracy and detail of a city's transportation network displayed on a computer screen are highly dependent on the size of the screen and the ability of human eyes to discern information on screen. It may be true that the actual information available on the transportation network is far more detailed than that shown on the screen, but due to the limitation of screen size, the information must be generalized before being displayed on screen in a way that people can interpret.

In summary, the level of spatial detail of geographic information is often high for geographic representation, and often low for geographic presentation. Therefore, geographic information associated with geographic representation is by nature incompatible with geographic information associated with geographic presentation. This is one of the major sources of scale incompatibility among geographic data in a GIS environment.

Maps, on one hand, are a means of geographic presentation used to communicate geographic knowledge and patterns. On the other hand, maps are also a geographic representation used to store geographic information. However, the ability of maps to store geographic information is limited by the same limitations associated with geographic presentation because the information must be shown on paper (or screen). Thus, generalization is needed to scale the content to fit on a piece of paper or screen, as well as to ensure that it is discernable by the human eye. Data layers in a GIS environment can take the form of either geographic representation, such as the Landsat TM imagery and the GPS-generated digital transportation network mentioned earlier, or geographic presentation, such as the data layer created from digitization of a paper map and other data layers that have gone through a generalization process for visualization purposes. Both forms of data (geographic representation and presentation) coexist within a GIS environment. Even with the provision of metadata, they are rarely distinguished in practice, which leads to scale incompatibility when they are used together in modeling the natural environment.

11.4.2 SCALE INCOMPATIBILITY AND CHARACTERIZATION OF SPATIAL JOINT DISTRIBUTION

Scale incompatibility can easily occur when modeling the natural environment using GIS due to the mixing of data at the geographic representation level with those at the geographic presentation level. The default neighborhood problem in characterizing spatial variation of individual geographic attributes could also add to scale incompatibility when such generated spatial data are used together. Scale incompatibility can lead to the mischaracterization of spatial joint distributions, which in turn can lead to incorrect conclusions being drawn regarding the operation of processes in the natural environment. Figure 11.6 illustrates the mischaracterization of the spatial joint distribution of slope gradient, slope aspect, and soil A-horizon depth along a transect in the Lubrecht watershed in western Montana. Figure 11.6a shows the spatial joint distribution of soil A-horizon depth from a conventional soil map (at 1:24,000) with slope gradient and slope aspect derived from a DEM at 30 m resolution. The following relationship can be discerned from Figure 11.6a: soil A-horizon depth is related to slope aspect (thick

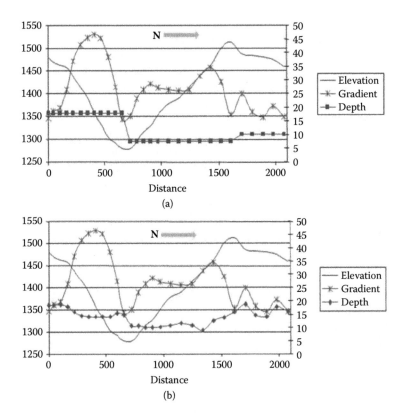

FIGURE 11.6 Impact of scale incompatibility on the characterization of spatial joint distribution: (a) spatial joint distribution between A-horizon depth and slope gradient using a conventional soil map and detailed digital terrain data; (b) spatial joint distribution between A-horizon depth and slope gradient using soil information from digital soil map and detailed digital terrain data.

on north facing slopes, thin on south facing slopes, and medium on plateau) but is not related to slope gradient. Clearly this conclusion is very different from well-known relationships between soil erosion and slope gradient. Figure 11.6b shows the spatial joint distribution along the same transect, but with soil information from a detailed soil map that is compatible with the 30 m DEM. The conclusion based on this joint distribution is that the soil A-horizon depth is related to slope aspect (as above) and slope gradient plays an important role in soil A-horizon depth, but the relationship is complex. For instance, on mountain slopes, soil A-horizon depth is related negatively to slope gradient, but on the plateau, soil A-horizon depth is positively related to slope gradient. This conclusion fits quite well with the commonly understood impact of slope gradient on soil formation; that is, for areas with a higher slope gradient, soil erosion is stronger and soil A-horizon depth is often shallower than on areas with a lower slope gradient. On the plateau, soil formation is limited by drainage conditions, and areas of higher slope gradient have better drainage conditions, which leads to deeper soil A-horizon compared to areas with lower slope gradient.

The drastic difference between the two joint spatial distributions shown in Figure 11.6 is the result of scale incompatibility between soil information derived from the conventional soil map and terrain information derived from detailed terrain data. The conventional soil map is a form of geographic presentation, and the information content is limited at the level of geographic presentation; whereas the terrain information derived from a 30 m DEM is at the level of geographic representation. There is a substantial difference in spatial scale and level of detail between the two types of data. Thus, the use of these incompatible data can lead to mischaracterization of the joint spatial distribution of geographic factors. On the other hand, the soil information used in Figure 11.6b was derived from a digital soil mapping approach and is at the level of geographic representation, thus is compatible with information derived from digital terrain analysis and remote sensing techniques [26]. The combination of these data sets can correctly characterize the spatial joint distribution of geographic factors.

11.4.3 MINIMIZING SCALE INCOMPATIBILITY

Correct characterization of spatial joint distributions requires the spatial information of the involved geographic factors to be compatible in scale. The following procedures should be followed to minimize scale incompatibility in modeling the natural environment. First, one should always ensure that all the spatial data involved are either of geographic presentation or of geographic representation schemes, and at the same time that these data should be at similar scales or spatial resolutions. If the spatial data sets involved are of different schemes (mixing of geographic representation and geographic presentation), or at very different scales or resolutions, one should perform scale transformation on the data sets so that they are of the same scheme and at a similar spatial scale. Quinn et al. [35], through their work on scale incompatibility and ecological modeling, discovered that the impact of scale incompatibility is at its minimum when the modeling scale approaches the scale of the coarsest resolution data set among those involved in the analysis. Based on their work, it is recommended here that spatial data should be transformed to the scale that is the coarsest among the spatial data involved. For example, if data are limited by geographic presentation, then cartographic generalization techniques [36] should be applied to convert data at the geographic representation level to the geographic presentation level. The second procedure recommended is that once data are of the same scheme (geographic representation or geographic presentation) and at the same scale or resolutions, one should use a neighborhood size that matches the spatial scale of the process under concern to process the spatial data and characterize the spatial joint distribution.

11.5 CONCLUSIONS

The analysis and management of the natural environment requires information on spatial variations in environmental or geographic factors. Through the discussion of effective neighborhood size and spatial joint distribution of geographic factors, this chapter explores the impacts of neighborhood size and scale incompatibility on the characterization of spatial variation of the environmental factors needed in modeling the natural environment, and highlights possible approaches to mitigate these impacts. To mitigate the negative impact of neighborhood size, this chapter suggests the development of spatial analysis techniques that allow the specification of a user-defined neighborhood so that the spatial scale at which data are processed matches the spatial scale of the process under investigation. It is further recommended that when the spatial scale of the process to be modeled is small, spatial data should be processed at multiple neighborhood sizes that are close to the spatial scale so that sensitivity of model results to neighborhood size can be examined and the use of improper neighborhood sizes can be avoided. To reduce the impacts of scale incompatibility, this chapter suggests that the spatial data used to characterize spatial joint distributions of geographic factors should be transformed into similar representation schemes and compatible scales or similar resolutions. Scale transformation can be achieved through cartographic generalization and analysis of spatial data at an appropriate neighborhood size.

Opportunities for future research activities in this area include the exploration of the exact form of the relationship between the neighborhood size of a spatial analytical technique (such as a high pass filter) and spatial scale. Although this chapter establishes that there is a connection between neighborhood size and spatial scale, it cannot be said that this represents a one-to-one relationship. Additional studies are also required to support the observation that the coarsest scale among a set of data layers should be used as the scale for characterizing spatial joint distribution of geographic factors needed for modeling the natural environment.

ACKNOWLEDGMENTS

The research reported in this chapter was supported financially by the Chinese Academy of Sciences International Partnership Project Human Activities and Ecosystem Changes (CXTD-Z2005-1), the National Basic Research Program of China (2007CB407207), and the One Hundred Talent Program of the Chinese Academy of Science.

REFERENCES

1. Wiens, J. A., Spatial scaling in ecology, *Functional Ecology*, 3(4), 385, 1989.
2. Wiens, J. A., Stenseth, N. C., Van Horne, B., and Ims, R. A., Ecological mechanisms and landscape ecology, *Oikos*, 66, 369, 1993.
3. Wiens, J. A., Predicting species occurrences: Progress, problems, and prospects, in *Predicting Species Occurrences: Issues of Accuracy and Scale*, Scott, J. M., Heglund, P. J., Morrison, M. L., Haufler, J. B., Raphael, M. G., Wall, M. A., and Samson, F. B., Eds., Island Press, Washington, 739, 2002.
4. Jenerette, G. D. and Wu, J., On the definitions of scale, *Bulletin of Ecological Society of America*, 81, 104, 2000.
5. Levin, S. A., The problems of pattern and scale in ecology, *Ecology*, 73, 1943, 1992.
6. Turner, M. G., Dale, V. H., and Gardner, R. H., Predicting across scales: Theory development and testing, *Landscape Ecology*, 3, 245, 1989.
7. MacNally, R. and Quinn, G. P., Symposium introduction: The importance of scale in ecology, *Austmtian Journal of Ecology*, 23, 1, 1998.

8. Houston, M. A., Critical issues for improving predictions, in *Predicting Species Occurrences: Issues of Accuracy and Scale*, Scott, J. M., Heglund, P. J., Morrison, M. L., Haufler, J. B., Raphael, M. G., Wall, M. A., and Samson, F. B., Eds., Island Press, Washington, 7, 2002.

9. Wu, J., Effects of changing scale on landscape pattern analysis: Scaling relations, *Landscape Ecology*, 19, 125, 2004.

10. Noon, B. R. and Dale, V. H., Broad-scale ecological science and its application, in *Applying Landscape Ecology in Biological Conservation,* Gutzwiller K. J., Ed., Springer, New York, 3, 2002.

11. Horn, B. K. P., Hill shading and the reflectance map, *Proceedings of the IEEE*, 69, 14, 1981.

12. Zevenbergen, L. W. and Thorne, C. R., Quantitative analysis of land surface topography, *Earth Surface Processing and Landforms*, 12, 47, 1987.

13. Moore, I. D., Terrain analysis program for environmental sciences (TAPES), *Agricultural Systems and Information Technologies*, 4, 37, 1992.

14. Hodgson, M. E., What cell size does the computed slope/aspect angle represent? *Photogrammetric Engineering and Remote Sensing*, 61, 513, 1995.

15. Wood, J., Scale-based characterisation of digital elevation models, in *Innovations in GIS 3*, Parker, D., Ed., Taylor & Francis, London, 163, 1996.

16. Openshaw, S. and Taylor, P., A million or so correlation coefficients: Three experiments on the modifiable areal unit problem, in *Statistical Applications in the Spatial Sciences,* Wrigley, N., Ed., Pion, London, 1979, 127.

17. Openshaw, S. and Taylor, P. J., The modifiable areal unit problem, in *Quantitative Geography: A British View,* Wrigley, N. and Bennett, R. B., Eds., Routledge and Kegan Paul, London, 1981, 60.

18. Openshaw, S., The modifiable areal unit problem, *CATMOG 38,* GeoBooks, Norwich, 1984.

19. Green, M. and Flowerdew, R., New evidence on the modifiable areal unit problem, in *Spatial Analysis: Modelling in a GIS Environment*, Longley, P. and Batty, M., Eds., GeoInformation International, Cambridge, 41, 1996.

20. Zhu, A. X., Burt, J. E., Smith, M., Wang, R., and Gao, J., The impact of neighbour-hood size on terrain derivatives and digital soil mapping, in *Advances in Digital Terrain Analysis,* Zhou, Q., Liu, X., and Lees, B., Eds., in press.

21. Shary, P. A., Sharaya, L. S., and Mitusov, A. V., Fundamental quantitative methods of land surface analysis, *Geoderma*, 107, 1, 2002.

22. Schmidt, J., Evans, I. S., and Brinkmann, J., Comparison of polynomial models for land surface curvature calculation, *International Journal of Geographical Information Science*, 17, 797, 2003.

23. Evans, I. S., An integrated system of terrain analysis and slope mapping, *Zeitschrift fur Geomorphologie, Suppl. Bd.*, 36, 274, 1980.

24. Florinsky, I. V., Accuracy of local topographic variables derived from digital elevation models, *International Journal of Geographical Information Science*, 12, 47, 1998.

25. Zhu, A. X., Band, L. E., Dutton, B., and Nimlos, T. J., Automated soil inference under fuzzy logic, *Ecological Modelling*, 90, 123, 1996.

26. Zhu, A. X., Hudson, B., Burt, J., Lubich, K., and Simonson, D., Soil mapping using GIS, expert knowledge, and fuzzy logic, *Soil Science Society of America Journal*, 65, 1463, 2001.

27. McBratney, A. B., Mendonca Santos, M. L., and Minasny, B., On digital soil mapping, *Geoderma*, 117, 3, 2003.

28. Zhu, A. X., A personal construct-based knowledge acquisition process for natural resource mapping, *International Journal of Geographical Information Science*, 13, 119, 1999.

29. Qi, F. and Zhu, A. X., Knowledge discovery from soil maps using inductive learning, *International Journal of Geographical Information Science*, 17, 771, 2003.
30. Zhu, A. X. and Band, L. E., A knowledge-based approach to data integration for soil mapping, *Canadian Journal of Remote Sensing*, 20, 408, 1994.
31. Smith, M. P., Zhu, A. X., Burt, J. E., and Styles, C., The effects of DEM resolution and neighbourhood size on digital soil survey, *Geoderma*, 137, 58, 2006.
32. Liu, J. and Zhu, A. X., Mapping with words: A new approach to automated digital soil survey, *International Journal of Intelligent Systems*, in press.
33. Schmidt, J. and Hewitt A. E., Fuzzy land element classification from DTMs based on geometry and terrain position, *Geoderma*, 121, 243, 2004.
34. Schmidt, J. and Andrew, R., Multi-scale landform characterization, *Area*, 37, 341, 2005.
35. Quinn, T., Zhu, A. X., and Burt, J. E., Effects of detail soil spatial information on watershed modeling across different model scales, *International Journal of Applied Earth Observation and Geoinformation*, 7, 324, 2005.
36. Slocum, T. A., McMaster, R. B., Kessler, F. C., and Howard, H. H., Eds., *Thematic Cartography and Geographic Visualization*, Prentice Hall, New Jersey, 518, 2005.

12 Invited Paper
Toward an Algebra for Terrain-Based Flow Analysis

David G. Tarboton and Matthew E. Baker

CONTENTS

OVERVIEW

Topography is an important land surface attribute for hydrology that, in the form of digital elevation models (DEMs), is widely used to derive information for the modeling of hydrologic processes. Much hydrologic terrain analysis is conditioned upon an information model for the topographic representation of downslope flow derived from a DEM, which enriches the information content of digital elevation data. This information model involves procedures for removing spurious sinks, deriving a structured flow field, and calculating derivative surfaces. We present a general method for recursive flow analysis that exploits this information model for calculation of a rich set of flow-based derivative surfaces beyond current weighted flow accumulation approaches commonly available in geographic information systems through the integration of multiple inputs and a broad class of algebraic rules into the calculation of flow-related quantities. This flow algebra encompasses single and multidirectional flow fields, various topographic representations, and weighted accumulation algorithms, and enables untapped potential for a host of application-specific functions. We illustrate the potential of flow algebra by presenting examples of new functions enabled by this perspective that are useful for hydrologic and environmental modeling. Future opportunities for advancing flow algebra functionality could include the development of a formulaic language that provides efficient implementation and greater access to these methods. There are also opportunities to take advantage of parallel computing for the solution of problems across very large input data sets.

12.1 INTRODUCTION

The land surface plays a crucial role in the hydrologic cycle by controlling the partitioning of precipitation into various components of runoff, infiltration, storage, and evapotranspiration. Topography is arguably the most important land surface attribute for hydrologic applications since it serves to define watersheds, the most basic hydrologic model element. Beven and Kirkby's TOPMODEL [1] has enjoyed widespread success as one of the first hydrologic models to take advantage of digital representations of topography. Terrain analyses based on digital elevation data are increasingly used in hydrology (e.g., Wilson and Gallant [2]).

Digital representation of topography is usually through one of three data structures [2]: (1) regular grids, (2) triangulated irregular networks (TINs), and (3) contours (Figure 12.1). Square grid digital elevation models (DEMs) have emerged as the most widely used data structure because of their simplicity and ease of computer implementation. Triangulated irregular networks have also found widespread use [3–5] because they can be adapted to the scale or detail of terrain information. The contour-based stream tube concept, first proposed by Onstad and Brakensiek [6], has also been used in hydrology to avoid the bias associated with grid data structures [7–12]. Despite their potential, contour-based methods have not seen widespread application, perhaps due to their complexity, with current implementations requiring careful handling of special cases [2]. The specific work reported in this chapter relies on grid digital DEMs, although many of the concepts are generic and extend to TIN or contour/flow tube elements.

Since the first grid-based DEMs appeared in the late 1980s, there has been rapid ongoing improvement of DEM data available to the hydrologic community, including the US National Elevation Dataset, which provides seamless coverage across the United States (at 10 m resolution in many locations). Worldwide, the Shuttle Radar Topography Mission (SRTM) data provides 90 m resolution coverage globally, with higher resolution data available in some places. DEM acquisition techniques based on light detection and ranging (LIDAR) are producing centimeter accuracy high-resolution DEM data sets. We stand at a threshold of improvement in surface topography

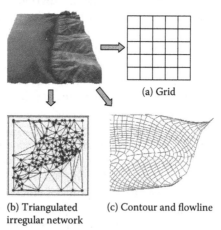

(a) Grid

(b) Triangulated
irregular network

(c) Contour and flowline

FIGURE 12.1   Models for the digital representation of terrain.

precision due to LIDAR that provides both opportunities and computing challenges. Rapid expansion of digital elevation applications is also driven by increasing power available in personal computers and the capability to rapidly download and process DEM data. This is leading to increased incorporation of terrain derivatives into analysis in many fields, including hydrology and environmental modeling. This chapter contributes to methods for development of flow-related terrain derivatives that might enhance such analyses.

Information science includes the precise representation of physical environments using data models that enhance the capability for analysis and integration of information. This chapter examines data models for the representation of flow over terrain in geographic information systems (GIS) and presents a new formalism for deriving flow-based information useful for hydrologic and environmental modeling. A basic underlying assumption is that water and its associated constituents move downhill. Terrain-based flow models enrich the information available from a DEM by deriving a structured digital representation of the flow field, which serves as the foundation for calculation of a wide range of flow-related quantities, the most basic of which is contributing area. The algorithm for calculating contributing area can be generalized to include additional information and rules, and to produce additional spatial fields of interest. Here we review methods for calculating terrain-based flow fields and existing algorithms for efficient derivation of flow-based information. Collectively, these methods form the conceptual basis for the encompassing formalism of flow algebra. Flow algebra provides a general approach for the incorporation of rules into flow-related calculations that encompass existing flow accumulation methods as special cases while allowing for the development of additional applications. Although derived with the basic downhill assumption in mind, flow algebra is not limited solely to the movement of water over terrain. The formalism applies to any noncirculating (nonlooping) flow field, and flow directions used in flow algebra can be derived from any potential surface. Flow fields derived from the gradient in any potential field (such as topographic slope in a gravitational field) are noncirculating because flow is from high to low potential. Flow algebra concepts thus have broad application for modeling the natural environment.

This chapter is organized as follows. We first describe the terrain-based flow data model. This is a review of existing work from a data modeling perspective and presents the digital representation of the terrain flow field as the foundation for recursive flow analysis, as described in Section 12.3, and flow algebra presented in Section 12.4. The section on recursive flow analysis reviews how the digital representation of the flow field supports the calculation of derivative flow related surfaces. This then leads in to the section on flow algebra as a generalization of recursive flow analysis that encompasses the use of algebraic rules in recursive calculations of flow-related derivative surfaces. We then present, in Section 12.5, examples that illustrate the capability of flow algebra and conclude in Section 2.6 with some thoughts on future directions for the development and use of flow algebra in terrain analysis for hydrologic and environmental modeling.

12.2 THE TERRAIN-BASED FLOW DATA MODEL

The terrain-based flow data model comprises a digital representation of terrain (Figure 12.1) and a representation of the flow field that connects adjacent model elements, enabling the routing of flow over a terrain surface and providing the basis for terrain-based flow calculations. This section reviews existing methods for the construction of the terrain-based flow model comprising drainage correction and calculation of the flow field representation for grid DEMs, and some of the hydrologic and environmental modeling work that has exploited this model.

In grid DEMs, sinks comprised of grid cells surrounded by higher-elevation neighbors occur due to deficiencies in DEM production processes and generalization in the representation of terrain [13,14]. Drainage correction that removes sinks is an important, but not essential, first step in the development of the terrain-based flow information model. Drainage correction is the processes of altering (correcting) the DEM to remove these sinks, and a DEM that has had all sinks removed is referred to as hydrologically correct. Care needs to be exercised not to correct nonspurious sinks or alter the DEM surface so much as to introduce further error into hydrologic analyses. The choice as to whether to remove sinks or not, therefore, needs to be based upon the physical use and interpretation of the results.

Several efficient implementations of sink filling have been developed [15,16]. Breaching or carving alterations to the DEM to allow drainage through barriers has also been suggested using either a 3-4 grid cell search [17,18], or by tracing downward from the pour point until an elevation lower than the sink is found, then carving a path from the sink to the lower elevation [19]. Soille [20] developed a logical integration of the sink filling and carving approaches that minimizes overall modification of the DEM by optimizing between raising the elevation of terrain within sinks and lowering the elevation of terrain along sink outflow paths. This approach provides a hydrologically correct DEM that is as close as possible to the original DEM data. Grimaldi et al. [21] suggested a physically based method that employs solutions from a landscape evolution model to remove sinks. A concern with this approach is that it favors the landscape-evolution model over real terrain data, in some cases altering the original DEM even more than a filling approach in order to bring the DEM surface into conformity with model solutions.

The most common procedure for routing flow over a terrain surface represented by a grid DEM is the eight-direction method (D8) first proposed by O'Callaghan and Mark [22]. In this model, the direction of steepest descent toward one of the eight (cardinal and diagonal) neighboring grid cells is used to represent the flow field [13,14,22–27]. In cases where the steepest descent cannot be determined, a broader search radius or random selection from among ties may be used. Garbrecht and Martz [18] presented a method for the routing of flow across flat surfaces, both away from higher terrain and toward lower terrain, that improved over prior methods. However, the D8 approach is limited because it can assign flow to only one of eight possible directions, each separated by 45° in a square grid [28–30].

Multiple flow direction methods [30–33] have been suggested as an attempt to solve the limitations of D8. Multiple flow direction methods proportion the outflow from each element between one or more downslope elements. They thus introduce dispersion (spreading out) into the flow with the goal to represent downslope flow in an average sense. A challenge in developing multiple flow direction approaches using grid DEMs involves balancing the introduction of dispersion against bias from routing flow along grid directions. The D-infinity (D∞) multiple flow direction model [30] represents flow direction as a vector along the direction of steepest downward slope on eight triangular facets centered at each grid cell. Flow from a grid cell is shared between the two downslope grid cells closest to the vector flow angle based on angle proportioning. Siebert and McGlynn [33] introduced an extension to D∞ called MD∞ that combines ideas from Tarboton [30] with Quinn et al. [31]. The MD∞ approach calculates slopes on triangular facets, but then proportions the flow between multiple downslope directions on triangular facets, thereby accounting for divergent situations where flow between more than two downslope grid cells is likely. MD∞ introduces more dispersion than D∞, but reduces some of the grid bias that D∞ creates in divergent situations. Figure 12.2 illustrates the representation of flow on a plane surface by single and multiple flow direction methods.

All flow field methods assign or proportion flow from each grid cell to one or more of its adjacent neighbors. In grid DEMs the basic model element is a grid cell, but the same concepts can be applied to any set of topologically connected model elements (Figure 12.3). Grid, TIN, and contour-flow-tube-element flow field assignments are all subject to the general condition that the proportions assigned to each downslope element are positive and should satisfy the conservation constraint

$$\sum_j P_{ij} = 1 \qquad (12.1)$$

where P_{ij} is the proportion of flow going from element i to a neighboring element j, and the sum is over all the neighboring elements. For the D8 grid model these proportions are either 1 (connected) or 0 (not connected). For the multiple flow direction models these proportions fall between 0 and 1 for each neighboring element. There is also a requirement that flow is noncirculating such that no portion of flow leaving one element ever returns to the same element after passing through one or more of its neighbors.

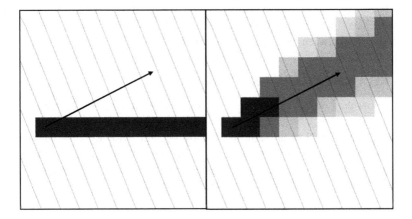

FIGURE 12.2   Flow across a plane surface represented by (a) single flow direction approach and (b) multiple flow direction approach.

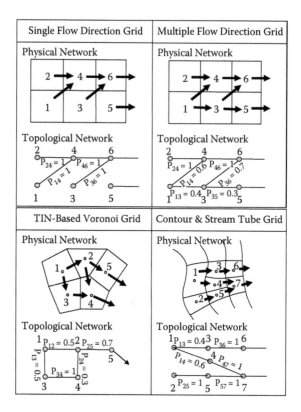

FIGURE 12.3   Downslope flow apportioning among topologically connected model elements using different flow field assignments and terrain representations.

Many measures useful in hydrologic and environmental modeling have been derived from this flow model. Without being comprehensive, these include the wetness index [1], a quasi-dynamic wetness index [34], terrain stability [35–39], erosion [40–44], contaminant transport [45,46], and riparian buffers [47–49]. Typically these measures have involved combining existing fields (e.g., slope) with outputs of an accumulation operation (e.g., specific contributing area).

12.3 RECURSIVE FLOW ANALYSIS

Once a flow data model comprising a set of flow proportions for each model element is defined, it may be used to evaluate contributing area and other accumulation derivatives across a DEM domain. In the most general sense, the flow field derived from a DEM defines the surface connectivity between any two parts of a landscape. Given a flow field, the general accumulation function is defined by an integral of a weight or loading field $r(\underline{x})$ over a contributing area, CA.

$$A(\underline{x}) = A[r(\underline{x})] = \int_{CA} r(\underline{x})d\underline{x} \qquad (12.2)$$

In this expression, $\underline{x}$ represents the location of an arbitrary point in the domain, $A(\underline{x})$ represents the result of the accumulation function evaluated at that arbitrary point, and A[.] denotes the accumulation operator, which operates on $r(\underline{x})$ to get the result $A(\underline{x})$. Figure 12.4 illustrates this concept. For a direct contributing area calculation, the weighting field, $r(\underline{x})$, is set equal to 1. In an example calculation of streamflow from excess rainfall, the weighting field would be set equal to rainfall minus infiltration.

Mark [25] presented a recursive algorithm for evaluation of accumulation in the D8 case that was extended to multiple flow direction methods by Tarboton [30]. Numerically, flow accumulation is evaluated recursively for each element as

$$A_i = A(\underline{x}_i) = r(\underline{x}_i)\Delta + \sum_{\{k:P_{ki}>0\}} P_{ki}A(\underline{x}_k) \qquad (12.3)$$

where $\underline{x}_i$ is a location in the field represented numerically by a model element such as grid cell in a DEM, and $A_i = A(\underline{x}_i)$ represents the accumulation at that element. The model element area is Δ, and the notation $\{k:P_{ki}>0\}$ denotes that summation is over the set of k values such that $P_{ki}>0$ (i.e., summing the contribution from neighboring elements k to element i). In other words, accumulated flow at any model element is the sum of flow arising from that element and flow arising from all contributing neighboring elements, each weighted according to the proportion of flow it contributes. This is a recursive definition because the accumulated flow for any model element depends upon the accumulated flow of adjacent upslope elements. The recursive definition includes a requirement that in tracing each path upstream, one must eventually arrive at a source element that has no other elements draining into it. This "termination requirement" is satisfied as long as the flow field is noncircular. Contributing area, as we have defined it in Equation 12.2 and Equation 12.3, is ill-posed for any flow field that includes looping. Appendix A presents pseudocode for the general upslope recursive algorithm for evaluating Equation 12.3 that

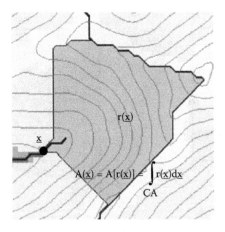

FIGURE 12.4 Physical definition of general flow accumulation function.

can be used for any flow field expressed in terms of the proportion of flow between elements, P_{ki}.

Figure 12.5 illustrates the contributing area computed using D8 flow directions. In this case P_{ki} is either 1 or 0 and is assigned to the neighboring element in the direction of steepest downward slope. The streaks aligned with grid directions illustrate the grid bias of the D8 approach. Figure 12.6 illustrates the contributing area computed using D∞ flow directions. In this case the proportions P_{ki} are shared among downslope neighbors, thus reducing grid bias and providing a contributing area result that is smoother, due to the dispersion, and appears to be a better reflection of the topography indicated by the contour lines. Tarboton [30] evaluated the differences between D8 and D∞ for theoretical surfaces where the contributing area is known, and showed that the D∞ calculations had smaller bias and mean square error.

The recursive algorithm presented earlier is an upslope recursion because it examines all the elements upslope from the element at which the quantity of interest is being evaluated. Tarboton [50] presents a number of other functions that exploit upslope recursion for the development of hydrologically useful quantities, such as downslope influence, decaying accumulation, and concentration limited accumulation. *Downslope influence*, illustrated in Figure 12.7, represents a special case of weighted flow accumulation from any target set of elements y within a given domain so that

$$I(\underline{x}|\underline{y}) = A[\underline{i}(\underline{x}|\underline{y})] \qquad (12.4)$$

where A[.] is the weighted accumulation operator presented in Equation 12.3. Isolation of the contribution from the target zone y is accomplished with the condition that $r(\underline{y}) = 1$ for x ∈ y and $r(\underline{x}) = 0$ elsewhere, denoted by a (1,0) indicator function $i(\underline{x}|\underline{y})$ on the set y. $I(\underline{x}|\underline{y})$ is the contribution (influence) from the set of elements y at each element x in the map. Downslope influence is useful in hydrology, water quality analysis,

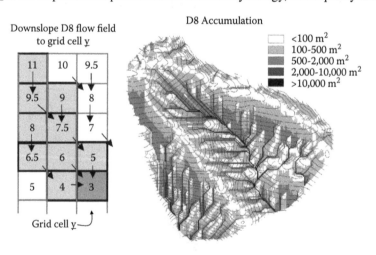

FIGURE 12.5 Flow field and contributing area from the eight-direction (D8) method. Numbers on left panel are elevations used to compute flow field.

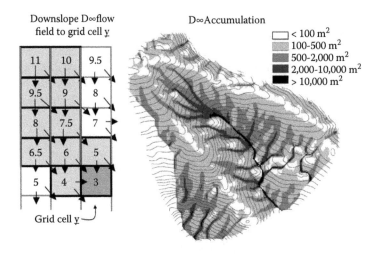

FIGURE 12.6 Flow field and contributing area from the D∞ method. Numbers on left panel are elevations used to compute flow field.

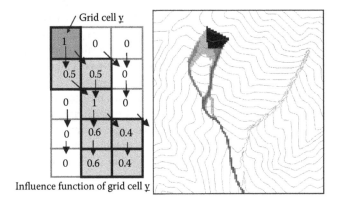

FIGURE 12.7 Downslope influence, calculated as the weighted accumulation from a target set (dark).

and land management for tracking where contaminants or sediment from a specific source are expected to move. Contributions from a set of source elements can follow several different pathways in a multidirection flow field. The level of influence along these pathways can decrease with transport distance if source contributions are spread across a greater number of receiving elements in a divergent flow field.

Recursive flow analysis can also examine elements downslope from the element at which the quantity of interest is being evaluated. A function that uses this idea [50] is the *upslope dependence* function, which is the inverse of downslope influence. Upslope dependence of a set of model elements $\underline{y}$, may be related to downslope influence by

$$D(\underline{x}\,|\,\underline{y}) = I(\underline{y}\,|\,\underline{x})$$ (12.5)

$D(\underline{x}|\underline{y})$ gives the proportion of flow from a model element $\underline{x}$ than contributes to (eventually flows through) one or more of the elements in the set $\underline{y}$. In this case, the target $\underline{y}$ is downslope rather than upslope of the elements being evaluated. Evaluation of this function requires reversal of the direction in which the flow direction field is traversed. Whereas the accumulation operator in Equation 12.3 tracks the proportion of flow from a set of elements $\underline{k}$ to a receiving element $\underline{i}$ if $P_{ki} > 0$. Here the operator moves in the opposite direction, $P_{ik} > 0$, such that

$$R_i = R(\underline{x}_i) = R[r(\underline{x}_i)] = r(\underline{x}_i)\Delta + \sum_{\{k:P_{ik}>0\}} P_{ik}R(\underline{x}_k) \qquad (12.6)$$

In this expression $R[.]$ is a general reverse weighted accumulation operator that operates on the weighting field $r(\underline{x})$ tracking the downslope amount back up the slope. The result at each model element, denoted $R_i = R(\underline{x}_i)$ is the sum of the contribution from element $\underline{x}_i$, $r(\underline{x}_i)\Delta$, and the accumulation from downslope elements, $\underline{x}_k$, according to the proportions P_{ik}. Upslope dependence of the target set $\underline{y}$ is evaluated by setting $R(\underline{x}) = 1$ for $\underline{x} \in \underline{y}$, initializing these elements as 1, setting $r(\underline{x}) = 0$ elsewhere and evaluating Equation 12.6 recursively. Pseudocode for recursive downslope, or reverse, flow accumulation that evaluates Equation 12.6 is given in Appendix B.

There is an irony in the terminology here, in that evaluation of upslope dependence requires recursion in the downslope direction. This occurs because evaluation of whether an element is upslope of a target area requires one to search downslope. Figure 12.8 illustrates how upslope dependence can be used to identify the area comprising elements that contribute some fraction of their area to the flow through a target area. Given this, the upslope dependence function can be useful for tracking the likely origins of sediment or other dissolved contaminant at a receiving location. The upslope dependence function can also be used for delineating the area draining

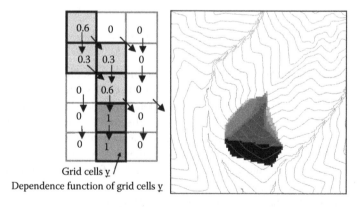

Grid cells $\underline{y}$
Dependence function of grid cells $\underline{y}$

FIGURE 12.8 Upslope dependence quantifies the proportion of flow in a domain (dark and light gray) that contribute to a target set (black). Note cells with fractional contributions along the margin.

to a watershed outlet. It should be noted that in contrast to single direction contributing areas, multiple flow direction approaches allow a single model element to contribute to both the target set (i.e., $i(\underline{x}|\underline{y}) = 1$) as well as elements outside the target set (i.e., $i(\underline{x}|\underline{y}) = 0$), or to more than one catchment outlet. Thus, to identify discrete watersheds draining to separate outlets, a rule based on the largest upslope dependence value or an upslope dependence threshold is needed.

12.4 TOWARD A FLOW ALGEBRA

Examination of the recursive flow analysis examples presented in the previous section reveals some generality and pattern to these calculations: (1) multiple direction accumulations rely on weighted flow proportioning, whereas single direction accumulations are a special case where all flow follows one pathway; (2) flow proportioning can occur to any number of neighboring model elements, so long as it conforms to the conservation constraint; (3) recursion can occur in both upslope and downslope directions; and (4) accumulations can be weighted by additional field(s) (e.g., rainfall minus infiltration). This capability, at least for upslope recursions, is available in flow accumulation functions in general purpose GIS software. However, we suggest here that recursive flow analysis need not be limited to the incorporation of additional weight fields into flow accumulation. Rather, what is needed is the ability to involve one or more additional fields in the accumulation functions that operate during the recursion according to a set of mathematical rules. We call these general rules for flow-related calculations *flow algebra*. Because these general rules encompass all existing flow-related procedures, what we present here comprises a unifying approach for understanding past, present, and future flow-related calculations. By exposing the generality of flow-field related calculations, we hope to suggest a direction for software development that will enable and stimulate generation of additional flow-derived measures useful in hydrology and environmental modeling.

Flow algebra logic exploits the recursive evaluation methodology illustrated in Equation 12.3. Recursion serves to simplify the evaluation of a flow algebra function from its global or zonal integral definition, such as in Equation 12.2, to a local evaluation where the function value at an element depends only on variables at that element and at *either* of the elements immediately upstream or downstream in the flow network, but not both at the same time. Flow algebra also generalizes the capability of zonal integral functions, enabling the evaluation of quantities that could not be defined in terms of a zonal integral because the result depends on the flow field as well as local rules or additional value fields. We distinguish within flow algebra between simple input variables and variables with recursive dependence. Simple input variables or fields, denoted $\underline{\gamma}(\underline{x})$, are fully quantified before the evaluation of a flow algebra expression. Variables that have recursive dependence on the flow field, denoted $\underline{\theta}(\underline{x})$, are quantified during the course of evaluating a flow algebra expression.

In general, a flow algebra expression may be written as

$$\underline{\theta}(\underline{x}_i) = f(\underline{\gamma}(\underline{x}_i), \underline{P}_{ki}, \underline{\theta}(\underline{x}_k), \underline{\gamma}(\underline{x}_k)) \tag{12.7}$$

for an upstream function, or

$$\underline{\theta}(\underline{x}_i) = f(\underline{\gamma}(\underline{x}_i), \underline{P}_{ik}, \underline{\theta}(\underline{x}_k), \underline{\gamma}(\underline{x}_k)) \qquad (12.8)$$

for a downstream function. The function f(.) may include any mathematical opera-tor, such as $+$, $-$, $\div$, $\times$, summation, conditional, logical, trigonometric, and math-ematical functions. In this expression, $\underline{\theta}(\underline{x}_i)$ is a list (of dimension m) of the recursive variables being evaluated at location i by the expression. $\underline{\gamma}(\underline{x}_i)$ is a list (of dimension q) of all simple input variables. $\underline{P}_{ik}$ or $\underline{P}_{ki}$ is a vector giving the proportion of flow from the first subscript element to the second subscript element, defined over all k for which $P_{(xx)}$ is nonzero. $\underline{P}_{ik}$ or $\underline{P}_{ki}$ is of dimension $\underline{n}$ where $\underline{n}$ represents the number of connected neighbor nodes. $\underline{\theta}(\underline{x}_k)$ is a list of all recursive variables evaluated at each neighbor location $\underline{k}$. It has dimension m $\times$ n. $\underline{\gamma}(\underline{x}_k)$ is a list of simple input variables at each neighbor node k. It has dimension q $\times$ $\underline{n}$.

The recursive variables, $\underline{\theta}(\underline{x})$, appear on the right-hand side of the expression because evaluation of the expression at location $\underline{x}$ depends on the values for these variables at adjacent logical network nodes, either upstream or downstream. With this structure, not only can $\underline{\gamma}(\underline{x})$ be applied as a weight, both $\overline{\gamma}(\underline{x})$ and $\underline{\theta}(\underline{x})$ fields can be applied during the calculation of any quantity with recursive dependence. A flow algebra expression is either of type upstream (e.g., contributing area, downslope influence) or downstream (e.g., upslope dependence, reverse accumulation), depend-ing on whether the functional dependence is on upstream or downstream quantities. Recursive dependence upon both upstream and downstream variability in the same expression is not allowed because such recursions would not terminate. Appendix C gives general pseudocode for the implementation of an upstream flow algebra function. The similarity of this to flow accumulation (Appendix A) is apparent. Downstream flow algebra is obtained by reversing $\underline{P}_{ki}$ to $\underline{P}_{ik}$. Upstream and down-stream flow algebra are similar in all other respects.

Flow algebra expands upon the concept of map algebra available in popular GIS by the inclusion of flow field operations. Map algebra involves point-by-point (cell-by-cell) mathematical operations between spatial fields. Flow algebra adds to this capability by incorporating operations based on the flow field and algebraic or functional descriptions of how the quantity being modeled is related to, and involved with, the flow field.

Because flow algebra encompasses multidirectional flow algorithms, it is applica-ble to any numerical representation of a flow field, including single or multiple flow direction grids, Voronoi polygons based upon a TIN discretization, or flow net model elements based upon contour and flow line discretization. Each flow field representa-tion has an underlying logical network structure defining the connectivity between elements (Figure 12.3). This may be implicit (as in the case of grids) or explicit (for Voronoi polygons and flow net model elements). Flow algebra elements could also be topographically delineated catchments. For example, the Arc Hydro data model [51] provides connectivity between stream reaches and stream reach catchments (the area draining directly to a stream reach) within a stream network, and implements accumulation functions using reach catchments as model elements.

12.5 EXAMPLES OF FUNCTIONS CONSTRUCTED USING FLOW ALGEBRA

This section gives examples to illustrate how flow algebra may be used to extend the functional capability of recursive flow analysis through the incorporation of rules into the recursive evaluation methodology. The examples have an increasing level of complexity so as to develop basic concepts using simple functions and then, by gradually adding modifications, illustrate potential for more specific applications.

A natural measurement derived from any flow field is that of *distance along a flow pathway*. Specifically, we consider here the distance in a downslope direction from each model element to a target set, such as a stream or catchment outlet, though upslope distances may also be defined using an upslope recursion. In hydrologic analyses, flow lengths have been used to characterize geomorphologic instantaneous unit hydrographs [52], estimate water residence times [53], contrast geomorphologic versus hydrodynamic attenuation/dispersion [54], and characterize water quality [55,56]. A variety of ecological analyses have used flow path distances to understand the influence of the spatial arrangement of watershed attributes on water quality and biotic responses [57–60].

In the D8 model, flow can only proceed to a single downslope element. D8 flow length calculations are consequently relatively straightforward and comprise accumulation of cardinal (Δx, Δy) or diagonal $\left(\sqrt{\Delta x^2 + \Delta y^2}\right)$ cell traverses, where Δx and Δy are element dimensions. In a multiple direction flow model, the distance from any model element $\underline{x}_i$ to another element $\underline{x}_j$ is not uniquely defined. Flow that originates at element $\underline{x}_i$ may arrive at $\underline{x}_j$ by a number of distinct pathways, and flow length is thus defined by a distribution rather than a single number. Bogaart and Troch [61] proposed calculating the average of this length distribution by weighting by the fraction of flow directed along a particular flow pathway. In the following, we present a general implementation using flow algebra. Practically speaking, the full length distribution cannot be accumulated easily over large domains due to excessive computational demands; however, distance functions that retain the longest and shortest paths may also be defined.

For the evaluation of average distance using flow algebra, the vector of simple inputs, $\underline{y}(\underline{x})$, is comprised of the coordinates of the center of each element and a target set indicator $\underline{y}$ (e.g., $\underline{y}_i = 1$ on the stream and 0 off the stream). The vector of recursive variables, $\theta(\underline{x})$, comprises the average distance to the target set from element $\underline{x}_i$, denoted ad($\underline{x}_i$). Average distance is calculated using a downslope recursion with flow algebra expression f(.), Equation 12.8, defined as:

if $\underline{y}_i = 1$ (if on the indicator set):

$$ad(\underline{x}_i) = 0$$

else

$$ad(\underline{x}_i) = \sum_{\{k:P_{ik}>0\,\&\,ad(\underline{x}_k)\geq0\}} P_{ik}(dist(\underline{x}_i,\underline{x}_k)+ad(\underline{x}_k))\Big/ \sum_{\{k:P_{ik}>0\,\&\,ad(\underline{x}_k)\geq0\}} P_{ik} \qquad (12.9)$$

The extra condition $ad(\underline{x}_k) \geq 0$ is placed in the summation to accumulate only those elements for which the average distance is defined, because distance is not defined for those elements with no downslope elements in the target set. Division by the sum of proportions is to account for partial contribution of a model element to downslope elements for which distance is defined. In most cases, the denominator will be equal to 1 except, for example, when a downslope element flows into a neighboring catchment and out of the domain, in which case $ad(\underline{x}_k)$ will be undefined. The function $dist(\underline{x}_i, \underline{x}_k)$ evaluates the geometric distance between the center of elements $\underline{i}$ and $\underline{k}$.

Similarly, the longest distance to the target set from each element $\underline{x}_i$, denoted $ld(\underline{x}_i)$, is calculated using a downslope recursion:

if $\underline{y}_i = 1$

$$ld(\underline{x}_i) = 0$$

else

$$ld(\underline{x}_i) = \underset{\{k:P_{ik}>0\,\&\,ld(\underline{x}_k)\geq 0\}}{Max}(dist(\underline{x}_i, \underline{x}_k) + ld(\underline{x}_k)) \tag{12.10}$$

where for each downslope neighbor ($P_{ik} > 0$), the function selects the maximum of the longest distance from that neighbor plus the distance to that neighbor. The shortest distance, $sd(\underline{x}_i)$, is calculated as:

if $\underline{y}_i = 1$

$$sd(\underline{x}_i) = 0$$

else

$$sd(\underline{x}_i) = \underset{\{k:P_{ik}>0\,\&\,sd(\underline{x}_k)\geq 0\}}{Min}(dist(\underline{x}_i, \underline{x}_k) + sd(\underline{x}_k)) \tag{12.11}$$

It may be of practical interest to weight the flow distance to calculate distance differently across a set of element values. A *weighted flow distance* may be calculated by adding a weight field, $w(\underline{x}_i)$, to the input vector $\underline{y}(\underline{x})$. A flow algebra expressions for weighted flow distance, similar to Equation 12.9 is

$$ad(\underline{x}_i) = \cfrac{\displaystyle\sum_{k:P_{ik}>0} P_{ik}\left(\cfrac{w(\underline{x}_i) + w(\underline{x}_k)}{2} dist(\underline{x}_i, \underline{x}_k) + ad(\underline{x}_k)\right)}{\displaystyle\sum_{\{k:P_{ik}>0\,\&\,ad(\underline{x}_k)\geq 0\}} P_{ik}} \tag{12.12}$$

The weights associated with the originating and receiving elements are averaged and multiplied by the distance between elements in this calculation.

Weighted distances have recently been applied to the problem of calulating filtering effects of streamside forests and wetlands, which have been observed to reduce concentrations of dissolved nutrients along field-to-stream transects. Baker et al. [49] used distances measured from row crop agriculture to streams weighted by the presence of forest or wetlands along each flow pathway to characterize the extent of riparian filtering across catchments. In this calculation, croplands are identified from a land cover raster as potential nutrient sources, whereas potential sinks (buffers) include forest and wetlands along flowpaths between each crop element and the stream (Figure 12.9a and Figure 12.9b). Forests and wetlands occuring adjacent to the stream but not on a flow path between a nutrient source and a stream *are not considered* in the analysis because they are assumed not to be involved in nutrient transport or filtering. A similar approach was recently used to understand how stream map resolution or seasonal expansion and contraction of stream networks might influence estimation of source–sink connectivity and relative nutrient uptake in streamside forests versus headwater streams [62]. Figure 12.9 also illustrates how flow length and connectivity estimates may be altered through the use of single (Figure 12.9c and d) versus multidirectional (Figure 12.9c and d) flow fields. In some cases, alternate pathways

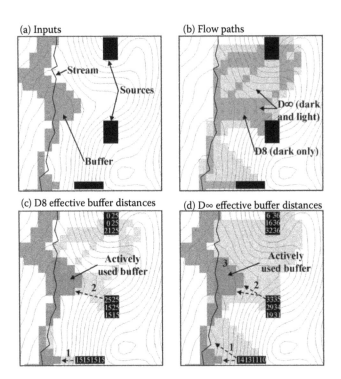

FIGURE 12.9   Weighted flow length-to-stream measures used in buffer analyses for water quality modeling in tributaries of Chesapeake Bay, Maryland.

identifed by a multidirectional flow field may be less (e.g., label 1 in Figure 12.9c and d) or more (e.g., label 2 in Figure 12.9c and d) buffered compared to single-direction paths. In every case where multidirectional flow dispersion occurs (e.g., label 3 in Figure 12.9d), estimates of the area of the potential buffer used in buffering will be necessarily greater than when using single directional estimates.

A simple extension of the above recursion, the *drop function*, is defined for any model element as the elevation difference from a location $\underline{x}_i$ on the land surface to a target region downslope, usually the stream or catchment outlet. In this case, a DEM serves as an additional input field, $\chi(\underline{x})$, providing the value z. McGuire et al. [53] used MD∞ to accumulate flow in their study of water residence time, but were limited to using D8 for flow distance, flow gradient (drop/distance), and gradient-to-distance ratios. Next we present a flow algebra solution to this problem. Given a multiple flow direction field with flow out of each element being proportioned between downslope model elements, there is no single pathway by which flow from any $\underline{x}_i$ reaches a set of downslope elements $\underline{y}$. The drop function may therefore be defined in terms of the maximum drop:

$$\underset{\{j:Q_{ij}>0\}}{Max}\left(z(\underline{x}_i)-(z(\underline{y}_j))\right) \tag{12.13}$$

the minimum drop:

$$\underset{\{j:Q_{ij}>0\}}{Min}\left(z(\underline{x}_i)-(z(\underline{y}_j))\right) \tag{12.14}$$

or the average drop:

$$z(\underline{x}_i)-\sum_{\{k:Q_{ij}>0\}}Q_{ij}\,z(\underline{y}_j) \tag{12.15}$$

As in flow distance calculations, the target region to which drop is being measured is indicated by the set of elements $\underline{y}$. These may be quite a long way from the element $\underline{x}_i$. A subset of these receive flow from the element $\underline{x}_i$. Q_{ij} denotes the proportion of flow from element $\underline{x}_i$ that eventually gets to $\underline{y}_j$ in the set $\underline{y}$. The maximum drop formula evaluates the elevation drop to the lowest point where flow from element $\underline{x}_i$ enters $\underline{y}$. The minimum drop formula evaluates the elevation drop to the highest location where any flow from element $\underline{x}_i$ enters $\underline{y}$. The average drop formula weights the drop based on the proportion of flow entering element $\underline{y}_j$ at each location. Numerically, these equations are evaluated using a downslope recursion based on the multiple flow proportions P_{ik} giving flow from grid cell $\underline{i}$ to grid cell $\underline{k}$. The maximum drop is calculated as

$$mxdrp(\underline{x}_i)=\underset{\{k:P_{ik}>0\}}{Max}\left(z_i-z_k+mxdrp(\underline{x}_k)\right) \tag{12.16}$$

This adds the drop from i to neighbor k to the longest drop from neighboring element k. The maximum is over all the neighbors that receive a positive proportion of the flow, $P_{ik} > 0$. The minimum drop is similarly calculated as

$$mndrp(\underline{x}_i) = \underset{\{k:P_{ik}>0\}}{Min} (z_i - z_k + mndrp(\underline{x}_k)) \qquad (12.17)$$

and the average drop as

$$avdrp(\underline{x}_i) = \sum_{\{k:p_{ik}>0\}} P_{ik}(z_i - z_k + avdrp(\underline{x}_k)) / \sum_{\{k:P_{ik}>0\ \&\ avdrp(\underline{x}_k)\geq 0\}} P_{ik} \qquad (12.18)$$

These recursive definitions have the escape condition that $mxdrp(\underline{x}_k)$, $mndrp(\underline{x}_k)$, and $avdrp(\underline{x}_k)$ are 0 for model elements $\underline{x}_k$ that belong to the set of target elements $\underline{y}$.

Similarly, minimum, maximum, and average *rise to ridge functions* (rtr) from any element $\underline{x}_i$ may be defined, essentially just by switching i and k in Equation 12.16 to Equation 12.18 to switch from downslope to upslope recursion, and renaming the functions

$$minrtr(\underline{x}_i) = \underset{\{k:P_{ki}\geq T\}}{Min} (z_k - z_i + minrtr(\underline{x}_k)) \ \ if \ \sum P_{ki} > 0, \ 0 \ Otherwise \quad (12.19)$$

$$maxrtr(\underline{x}_i) = \underset{\{k:P_{ki}\geq T\}}{Max} (z_k - z_i + maxrtr(\underline{x}_k)) \ \ if \ \sum P_{ki} > 0, \ 0 \ Otherwise \quad (12.20)$$

$$artr(\underline{x}_i) = \frac{\sum_{\{k:P_{ki}\geq T\}} P_{ki}(z_k - z_i + artr(\underline{x}_k))}{\sum_{\{k:P_{ki}\geq T\}} P_{ki}} \ \ if \ \sum P_{ki} > 0, \ 0 \ Otherwise \quad (12.21)$$

In the rise to ridge functions, the escape condition for the recursion is $\Sigma P_{ki} > 0$ that defines ridge elements as elements that do not have any upslope elements. In Equations 12.19 to 12.21 we also introduced the option for a user input threshold, T, to control upslope paths from neighbors k that enter element $\underline{x}_i$ that are considered to be upslope.

Transport limited accumulation is a flow algebra function that introduces further rules into flow-related calculations. This function is designed to calculate the transport of sediment that may be limited by *both* the sediment supply and the capacity of the flow field to transport sediment. This is an example of an algorithm not currently available to general GIS users without the functionality of flow algebra. We have framed the calculation in a general way with supply and transport capacity fields as inputs (components of $\underline{y}(\underline{x})$), so as to apply to any transport process where there is both distributed supply of a substance and a limited capacity for transport of that substance. This function accumulates substance flux subject to the rule that transport out of any model element is the minimum between supply and transport capacity. The total supply is calculated as the sum of transport in to the element from upslope elements plus the supply contribution from the element. This is again a recursive definition, since it depends upon the transport flux from upslope elements. Specifically,

$$T(\underline{x}_i) = Min(C(\underline{x}_i), \sum_{\{k:P_{ki}>0\}} P_{ki}T(\underline{x}_k) + S(\underline{x}_i)) \tag{12.22}$$

where $C(\underline{x}_i)$ is the transport capacity associated with model element $\underline{x}_i$, $S(\underline{x}_i)$ the supply (e.g., erosion potential) at model element $\underline{x}_i$, and $T(\underline{x}_i)$ gives the resulting transport limited accumulation flux. If $C(\underline{x}_i)$ exceeds transport to the element plus local supply, then the flux is supply limited and the second term in the Min is chosen. If the available substance from the sum of influx plus local supply exceeds $C(\underline{x}_i)$, then the flux is transport limited and the outflux is the transport capacity, $C(\underline{x}_i)$. Both transport capacity and local supply fields ($C(\underline{x}_i)$ and $S(\underline{x}_i)$) are inputs and thus components of $\gamma(\underline{x})$; whereas the resultant transport limited accumulation flux is the result of recursion on the flow field and thus an element of the vector $\underline{\theta}(\underline{x})$. Another part of $\underline{\theta}(\underline{x})$ and a by-product of this calculation is the deposition $D(\underline{x}_i)$ at any point, calculated as total supply minus actual transport,

$$D(\underline{x}_i) = \sum_{\{k:P_{ki}>0\}} P_{ki}T(\underline{x}_k) + S(\underline{x}_i) - T(\underline{x}_i) \tag{12.23}$$

$D(\underline{x}_i)$ is 0 at supply limited elements, whereas at transport limited elements it quantifies the excess of total supply over transport capacity. Comparison of $D(\underline{x}_i)$ to $S(x_i)$ is required to distinguish deposition of substance from local supply, versus substance that is transported into an element from another upslope element. This model for accumulation of a substance subject to supply and transport capacity limits is consistent with sediment transport and erosion theory involving the separate processes of detachment and transport [63,64]. Figure 12.10 illustrates transport limited accumulation. The supply field may be based on erodibility from soil surveys, whereas transport capacity in this example is based on slope–area relationships [35,65]. Reductions in sediment delivery ratios as drainage area increases are naturally modeled by this function due to the trapping of sediment at locations where transport capacity is limited.

Calculation of an *avalanche runout zone* provides another, more comprehensive opportunity to illustrate the generality and potential of flow algebra for calculations involving multiple terrain and flow fields. In this application, avalanche source zones, identified manually using expert knowledge and visual interpretation of maps, are used as input (although there is clearly potential for modeling avalanche source zones based upon topographic attributes as has been done for landslides [36,37,66]). The rule for identifying runout zones is that all locations downslope from a source zone are potentially affected up until the energy from the avalanche is depleted. This depletion point is estimated when the slope between the source and the affected area is less than a threshold angle (alpha). The alpha angle is calculated using the distance from the highest point in the source zone to points within the potential runout zone (Figure 12.11). Distance may be measured either along a straight line or along a flow path. This alpha-angle model is a simple model for avalanche or debris flow runout that is used in practice to evaluate potential hazards (e.g., Schaerer [67], McClung and Schaerer [68], Iverson [69], and Toyos et al. [70]). Because evaluation of the runout zone requires looking upslope, flow algebra with upslope recursion is used.

For the avalanche application using a multidirectional flow field, it may be desirable to exclude model elements from the runout zone that receive only a small fraction of flow from the avalanche source. We therefore specify a threshold, T, supplied by the user, that must be exceeded before an element is counted as contributing to a downslope neighbor for the purposes of defining the avalanche runout zone and calculating alpha angle (e.g., $P_{ki} > T$ where $T = 0.2$). T may be input as 0 if all fractional contributions to a downslope element, no matter how small, are to be counted. The

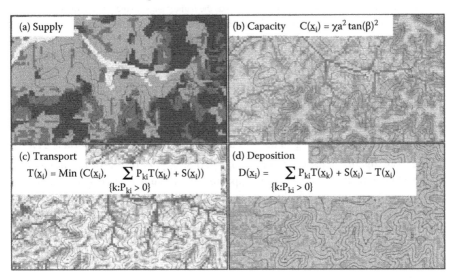

(a) Supply

(b) Capacity $\quad C(\underline{x}_i) = \chi a^2 \tan(\beta)^2$

(c) Transport
$$T(\underline{x}_i) = \text{Min}\ (C(\underline{x}_i),\ \sum_{\{k:P_{ki} > 0\}} P_{ki}T(\underline{x}_k) + S(\underline{x}_i))$$

(d) Deposition
$$D(\underline{x}_i) = \sum_{\{k:P_{ki} > 0\}} P_{ki}T(\underline{x}_k) + S(\underline{x}_i) - T(\underline{x}_i)$$

FIGURE 12.10 Transport limited accumulation is a function of distributed supply and transport capacity.

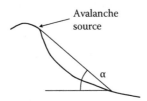

Avalanche source

α

FIGURE 12.11 Alpha (α) angle from point in avalanche runout to avalanche source.

TABLE 12.1

Variables in Avalanche Runout Flow Algebra Function

Symbol	Description
	Simple Input Variables: $\gamma(\underline{x})$
T	Flow proportion threshold
α	Alpha angle
as	Avalanche source set
x_i, y_i	Coordinates of the center of each element
z_i	Elevation of the center of each element
	Recursive Variables: $\underline{\theta}(\underline{x})$
rz	A runout zone indicator with value 0 to indicate that this grid cell is not in the runout zone and value > 0 to indicate that this grid cell is in the runout zone. Since there may be information in the angle to the associated source site, this variable will be assigned the angle to the source site, denoted as β here (in degrees).
xm, ym	X and Y locations of the source site that has the highest angle to the point in question
zm	Elevation of the source site that has the highest angle to the point in question
dm	Flow distance from the source site that has the highest angle to the point in question. This is included to allow evaluation of source angles using either straight-line or flow-path distances.

avalanche source zone is input as an indicator set $\underline{as}$ ($\underline{as}_i = 1$ in avalanche source zone, and 0 otherwise). The simple and recursive variables involved in avalanche runout calculation cast in terms of the general flow algebra construct are listed in Table 12.1.

The flow algebra expression $f(\gamma(\underline{x}_i), \underline{P}_{ki}, \underline{\theta}(\underline{x}_k), \gamma(\underline{x}_k))$ for $\underline{\theta}(\underline{x}_i)$ at element $\underline{x}_i$ is evaluated by the pseudocode in Appendix D. The suite of inputs and calculated fields in this function exceeds the capacity of currently available accumulation operators, but is relatively straightforward within the flow algebra construct. Figure 12.12 illustrates the avalanche runout from three potential source zones computed using $\alpha = 22°$ for a snow avalanche prone area in Logan Canyon, Utah.

12.6 FUTURE DIRECTIONS

The example flow algebra functions presented have been programmed for use with grid DEM data using the D∞ multiple flow direction model and included as part of the Terrain Analysis Using Digital Elevation Models (TauDEM) software distributed by the first author [71]. Code that implements the recursion is in a C++ library that has been wrapped with a Visual Basic graphical user interface callable from the ESRI ArcGIS software package as an ArcMap toolbar or geoprocessing toolbox, as well as from the open source Mapwindow™ GIS [72]. Source code and compiled executables for a personal computer are distributed using an open source license. However, such implementations, though based on flow algebra concepts, do not provide the full capability we envision. The recursive algorithms, though compact in terms of coding and efficient in terms of model element evaluations (each element is

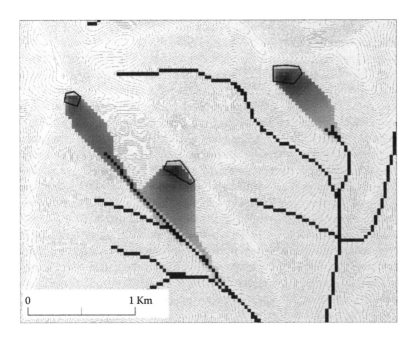

FIGURE 12.12 Avalanche runout zones for Wood Camp Hollow in Logan Canyon, Utah, computed using $\alpha = 22°$. Contour interval is 10 m. The intensity of color is scaled by the angle to source, β, subject to the constraint $\beta > \alpha$.

visited only once), can be inefficient in terms of memory requirements (because at each recursion step the function state is saved on a stack), and are not implemented to take advantage of parallel processing. Broad-scale application of these methods to large data sets will require work to address these limitations. Another step in the implementation process will involve the development of text parsing software for translating user inputs into process-specific recursive accumulations. This software would provide an interface that enables users to design their own combinations of $\gamma(\underline{x})$ and $\underline{\theta}(\underline{x})$, specifying their own algebraic and logical rules for custom flow algebra functions.

Despite rapid advances in computer technology, there remains a considerable gap among digital representations of terrain, flow fields, and real-world observations. As a result, geographic and hydrologic models lag behind current hydrologic theory in their representation of physical processes. Computational modeling frameworks are required that enable the implementation and rapid evaluation of new theories and field-based concepts. Flow algebra provides a formalism for thinking about and modeling spatial processes that are related to, or occur embedded within, a flow field. We hope that flow algebra serves to fill some of this gap through the terrain-based flow analyses it enables.

This chapter has framed an existing information model for the analysis of flow over terrain in GIS. This model establishes a flow field through (1) drainage correction involving the removal of sinks, followed by (2) definition of the flow field through a general multidirectional proportioning of flow from each element among downslope neighbors. The flow field is required to be noncirculating and, as such, is suitable for representation of flow derived from the gradient of any potential field. Flow proportions arising from any model element should sum to one to ensure conservation. Once this flow field is defined, a broad class of upstream and downstream recursive functions may be constructed using the formalism of flow algebra. We have presented some examples for exploiting this capability, including new techniques for addressing the measurement of flow distances, elevation drops, sediment transport, and avalanche runouts. The new techniques have already been utilized in several distinct applications, and they serve to illustrate a small portion of the untapped potential of the recursive flow algebra approach. Although the examples we present have been developed using grid data structures, the logic of flow algebra is applicable for any set of logically connected elements defining flow in a noncirculating flow field. Many advances in hydrologic modeling have not made their way to GIS applications for the simple reason that they did not work well within a grid data structure, or suffered from limitations due to single flow direction approaches. Advances have also been hampered by the difficulty associated with implementation of rules and logic within flow field-related calculations. It is our hope that flow algebra will provide a more inclusive modeling framework for moving across data structures in hydrologic modeling of the natural environment.

ACKNOWLEDGMENTS

Funding and support for this project was provided by CICEET, the Cooperative Institute for Coastal and Estuarine Environmental Technology. A partnership of the National Oceanic and Atmospheric Administration (NA06NOS4190167) and the University of New Hampshire, CICEET develops tools for clean water and healthy coasts nationwide. The authors also acknowledge partial research support through Utah State University's new faculty seed grant program.

REFERENCES

1. Beven, K. J. and Kirkby, M. J., A physically based variable contributing area model of basin hydrology, *Hydrological Sciences Bulletin*, 24, 43, 1979.
2. Wilson, J. P. and Gallant, J. C., *Terrain Analysis: Principles and Applications*, John Wiley & Sons, New York, 2000.
3. Jones, N. L., Wright, S. G., and Maidment, D. R., Watershed delineation with triangle-based terrain models, *Journal of Hydraulic Engineering*, 116, 1232, 1990.
4. Nelson, E. J., Jones, N. L., and Berrett, R. J., Adaptive tessellation method for creating TINs from GIS data, *ASCE Journal of Hydrologic Engineering*, 4, 2, 1999.
5. Tucker, G. E., Lancaster, S. T., Gasparini, N. M., Bras, R. L., and Rybarczyk, S. M., An object oriented framework for distributed hydrologic and geomorphic modeling using triangulated irregular networks, *Computers and Geosciences*, 27, 959, 2001.
6. Onstad, C. A. and Brakensiek, D. L., Watershed simulation by the stream path analogy, *Water Resources Research*, 4, 965, 1968.
7. O'Loughlin, E. M., Saturation regions in catchment and their relations to soil and topographic properties, *Journal of Hydrology*, 53, 229, 1981.
8. O'Loughlin, E. M., Prediction of surface saturation zones in natural catchments by topographic analysis, *Water Resources Research*, 22, 794, 1986.

9. Moore, I., O'Loughlin, E. M., and Burch, G. J., A contour based topographic model for hydrological and ecological applications, *Earth Surface Processes and Landforms*, 13, 305, 1988.

10. Moore, I. D. and Grayson, R. B., Terrain-based catchment partitioning and runoff prediction using vector elevation data, *Water Resources Research*, 27, 1177, 1991.

11. Grayson, R. B., Moore I. D., and McMahon, T. A., Physically based hydrologic modeling 1: A terrain-based model for investigative purposes, *Water Resources Research*, 28, 2639, 1992.

12. Dawes, W. R. and Short, D., The significance of topology for modeling the surface hydrology of fluvial landscapes, *Water Resources Research*, 30, 1045, 1994.

13. Jenson, S. K., Applications of hydrologic information automatically extracted from digital elevation models, *Hydrological Processes*, 5, 31, 1991.

14. Jenson, S. K. and Domingue, J. O., Extracting topographic structure from digital elevation data for geographic information system analysis, *Photogrammetric Engineering and Remote Sensing*, 54, 1593, 1988.

15. Planchon, O. and Darboux, F., A fast, simple and versatile algorithm to fill the depressions of digital elevation models, *Catena*, 46, 159, 2001.

16. Arge, L., Chase, J., Halpin, P., Toma, L., Vitter, J., Urban, D., and Wickremesinghe, R., Efficient flow computation on massive grid terrain datasets, *Geoinformatica*, 7, 283, 2003.

17. Garbrecht, J. and Martz, L., TOPAZ, An automated digital landscape analysis tool for topographic evaluation, drainage identification, watershed segmentation, and subcatchment parameterization, NAWQL, 95-1, National Agricultural Water Quality Laboratory, USDA, ARS, Durant, OK, 1995.

18. Garbrecht, J. and Martz, L. W., The assignment of drainage direction over flat surfaces in raster digital elevation models, *Journal of Hydrology*, 193, 204, 1997.

19. Soille, P., Vogt, J., and Colombo, R., Carving and adaptive drainage enforcement of grid digital elevation models, *Water Resources Research*, 39, 1366, 2003.

20. Soille, P., Optimal removal of spurious pits in grid digital elevation models, *Water Resources Research*, 40, W12509, 2004.

21. Grimaldi, S., Nardi, F., Benedetto, F., Istanbulluoglu, E., and Bras, R. L., A physically-based method for removing pits in digital elevation models, *Advances in Water Resources*, 30, 2151, 2007.

22. O'Callaghan, J. F. and Mark, D. M., The extraction of drainage networks from digital elevation data, *Computer Vision, Graphics and Image Processing*, 28, 328, 1984.

23. Marks, D., Dozier, J., and Frew, J., Automated basin delineation from digital elevation data, *Geo-Processing*, 2, 299, 1984.

24. Band, L. E., Topographic partition of watersheds with digital elevation models, *Water Resources Research*, 22, 15, 1986.

25. Mark, D. M., Network models in geomorphology, in *Modelling in Geomorphological Systems*, Anderson, M. G., Ed., John Wiley, 73, 1988.

26. Morris, D. G. and Heerdegen, R. G., Automatically drained catchment boundaries and channel networks and their hydrological applications, *Geomorphology*, 1, 131, 1988.

27. Martz, L. W. and Garbrecht, J., Numerical definition of drainage network and subcatchment areas from digital elevation models, *Computers and Geosciences*, 18, 747, 1992.

28. Fairfield, J. and Leymarie, P., Drainage networks from grid digital elevation models, *Water Resources Research*, 27, 709, 1991.

29. Costa-Cabral, M. and Burges, S. J., Digital elevation model networks (DEMON): A model of flow over hillslopes for computation of contributing and dispersal areas, *Water Resources Research*, 30, 1681, 1994.

30. Tarboton, D. G., A new method for the determination of flow directions and contributing areas in grid Digital Elevation Models, *Water Resources Research*, 33, 309, 1997.

31. Quinn, P., Beven, K., Chevallier, P., and Planchon, O., The prediction of hillslope flow paths for distributed hydrological modeling using digital terrain models, *Hydrological Processes*, 5, 59, 1991.
32. Freeman, T. G., Calculating catchment area with divergent flow based on a regular grid, *Computers and Geosciences*, 17, 413, 1991.
33. Seibert, J. and McGlynn, B. L., A new triangular multiple flow-direction algorithm for computing upslope areas from gridded digital elevation models, *Water Resources Research*, 43, W04501, 2007.
34. Barling, R. D., Moore, I. D., and Grayson, R. B., A quasi-dynamic wetness index for characterizing the spatial distribution of zones of surface saturation and soil water content, *Water Resources Research*, 30, 1029, 1994.
35. Montgomery, D. R. and Dietrich, W. E., A physically based model for the topographic control on shallow landsliding, *Water Resources Research*, 30, 1153, 1994.
36. Pack, R. T., Tarboton, D. G., and Goodwin, C. N., The SINMAP approach to terrain stability mapping, 8th Congress of the International Association of Engineering Geology, Vancouver, British Columbia, Canada, September 1998.
37. Pack, R. T., Tarboton, D. G., and Goodwin, C. N., Terrain stability mapping with SINMAP, technical description and users guide for version 1.00, Report Number 4114-0, Terratech Consulting Ltd., Salmon Arm, British Columbia, Canada, 1998, http://www.engineering.usu.edu/dtarb/, accessed May 2008.
38. Pack, R. T., Tarboton, D. G., and Goodwin, C. N., Assessing terrain stability in a GIS using SINMAP, 15th annual GIS conference, GIS 2001, Vancouver, British Columbia, February 2001.
39. Borga, M., Fontana, G. D., and Cazorzi, F., Analysis of topographic control on shallow landsliding using a quasi-dynamic wetness index, *Journal of Hydrology*, 268, 56, 2002.
40. Roering, J. J., Kirchner J. W., and Dietrich, W. E., Evidence for nonlinear, diffusive sediment transport on hillslopes and implications for landscape morphology, *Water Resources Research*, 35, 853, 1999.
41. Jones, R., Algorithms for using a DEM for mapping catchment areas of stream sediment samples, *Computers and Geosciences*, 28, 1051, 2002.
42. Istanbulluoglu, E., Tarboton, D. G., Pack, R. T., and Luce, C., A probabilistic approach for channel initiation, *Water Resources Research*, 38, 1325, 2002.
43. Istanbulluoglu, E., Tarboton, D. G., Pack, R. T., and Luce, C., A sediment transport model for incising gullies on steep topography, *Water Resources Research*, 39, 1103, 2003.
44. Cochrane, T. A. and Flanagan, D. C., Representative hillslope methods for applying the WEPP model with DEMS and GIS, *Transactions of the ASAE*, 46, 1041, 2003.
45. Ning, S. K., Jeng, K. Y., and Chang, N. B., Evaluation of non-point sources pollution impacts by integrated 3S information technologies and GWLF modelling, *Water Science And Technology*, 46, 217, 2002.
46. Endreny, T. A. and Wood, E. F., Watershed weighting of export coefficients to map critical phosphorous loading areas, *Journal of the American Water Resources Association*, 39, 165, 2003.
47. Tomer, M. D., James, D. E., and Isenhart, T. M., Optimizing the placement of riparian practices in a watershed using terrain analysis, *Journal of Soil and Water Conservation*, 58, 198, 2003.
48. McGlynn, B. L. and Seibert, J., Distributed assessment of contributing area and riparian buffering along stream networks, *Water Resources Research*, 39, 1082, 2003.
49. Baker, M. E., Weller, D. E., and Jordan, T. E., Improved methods for quantifying potential nutrient interception by riparian buffers, *Landscape Ecology*, 21, 1327, 2006.

50. Tarboton, D. G., Terrain analysis using digital elevation models in hydrology, 23rd ESRI International Users Conference, San Diego, California, July 2003.

51. Maidment, D. R., Ed., *Arc Hydro GIS for Water Resources*, ESRI Press, Redlands, California, 2002.

52. Rodriguez-Iturbe, I. and Valdes, J. B., The geomorphologic structure of hydrologic response, *Water Resources Research*, 15, 1409, 1979.

53. McGuire, K. J., McDonnel, J. J., Weiler, M., Kendall, C., McGlynn, B. L., Welker, J. M., and Seibert, J., The role of topography on catchment-scale water residence time, *Water Resources Research*, 41, W05002, 2005.

54. White, A. B., Kumar, P., Saco, P. M., Rhoads, B. L., and Yen, B. C., Hydrodynamic and geomorphologic dispersion: Scale effects in the Illinois River basin, *Journal of Hydrology*, 288, 237, 2004.

55. Alexander, R. B., Smith, R. A., and Schwarz, G. E., Effect of stream channel size on the delivery of nitrogen to the Gulf of Mexico, *Nature*, 403, 758, 2000.

56. Soranno, P. A., Hubler, S. L., and Carpenter, S. R., Phosphorous loads to surface waters: A simple model to account for spatial pattern of land use, *Ecological Applications*, 6, 865, 1996.

57. King, R. S., Baker, M E., Whigham, D. F., Weller, D. E., Jordan, T. E., Kazyak, P. F., and Hurd, M. K., Spatial considerations for linking watershed land cover to ecological indicators in streams, *Ecological Applications*, 51, 137, 2005.

58. Frimpong, E. A., Sutton, T. M., Lim, K. J., Hrodey, P. J., Engel, B. A., Simon, T. P., Lee, J. G., and Le Master, D. C., Determination of optimal riparian forest buffer dimensions of stream biota—landscape association models using multimetric and multivariate responses, *Canadian Journal of Fisheries and Aquatic Sciences*, 62, 1, 2005.

59. King, R. S., Beaman, J. R., Whigham, D. F., Hines, A. H., Baker, M. E., and Weller, D. E., Watershed land use is strongly linked to PCBs in white perch in Chesapeake Bay subestuaries, *Environmental Science and Technology*, 38, 6546, 2004.

60. Van Sickle, J. and Johnson, C. B., Parametric distance weighting of landscape influence on streams, *Landscape Ecology*, 23, 427, 2008.

61. Bogaart, P. W. and Troch, P. A. Curvature distribution within hillslopes and catchments and its effect on the hydrological response, *Hydrology and Earth System Sciences*, 10, 925, 2006.

62. Baker, M. E., Weller D. E., and Jordan, T. E., Effects of stream map resolution on measures of riparian buffer distribution and nutrient retention potential, *Landscape Ecology*, 22, 973, 2007.

63. Hairsine, P. B. and Rose, C. W., Modeling water erosion due to overland flow using physical principles 1. Sheet flow, *Water Resources Research*, 28, 237, 1992.

64. Hairsine, P. B. and Rose, C. W., Modeling water erosion due to overland flow using physical principles 2. Rill Flow, *Water Resources Research*, 28, 245, 1992.

65. Dietrich, W. E., Wilson, C. J., Montgomery, D. R., McKean, J., and Bauer, R., Erosion thresholds and land surface morphology, *Geology*, 20, 675, 1992.

66. Tarolli, P. and Tarboton, D. G., A new method for determination of most likely landslide initiation points and the evaluation of digital terrain model scale in terrain stability mapping, *Hydrology and Earth System Sciences*, 10, 663, 2006.

67. Schaerer, P. A., Avalanches, in *Handbook of Snow*, Gray, D. M. and Male, D. H., Eds., Pergamon Press, Willowdale, Canada, 1981.

68. McClung, D. and Schaerer, P., *The Avalanche Handbook*, 2nd ed., Mountaineers Books, 1993.

69. Iverson, R. M., The physics of debris flows, *Reviews of Geophysics*, 35, 245, 1997.

70. Toyos, G., Oramas Dorta, D., Oppenheimer, C., Pareschi, M. T., Sulpizio, R., and Zanchetta, G., GIS-assisted modelling for debris flow hazard assessment based on the events of May 1998 in the area of Sarno, Southern Italy: Part I. Maximum run-out, *Earth Surface Processes and Landforms*, 32, 1491, 2007.
71. TauDEM: Terrain Analysis Using Digital Elevation Models, http://www.engineering.usu.edu/dtarb/taudem, accessed May 2008.
72. MapWindow™ GIS, http://www.mapwindow.org, accessed May 2008.

APPENDIX A: PSEUDOCODE FOR RECURSIVE UPSLOPE FLOW ACCUMULATION

Global variables A_i, $r(\underline{x}_i)$, P_{ij}, Δ
Function FlowAccumulation(x_i)
  if A_i is known
  then
      no action
  else
      for each neighbor location $\underline{x}_k$ indexed by k
          if($P_{ki} > 0$)then
              call FlowAccumulation($\underline{x}_k$)
              //This is the recursive call to calculate area for the neighbor
      Next k
      // At this point all the neighboring A_k inputs are available

$$A_i = r(\underline{x}_i)\Delta + \sum_{\{k:P_{ki}>0\}} P_{ki}A_k$$

return

APPENDIX B: PSEUDOCODE FOR RECURSIVE DOWNSLOPE OR REVERSE FLOW ACCUMULATION

Global variables R_i, $r(\underline{x}_i)$, P_{ij}, Δ
Function ReverseAccumulation($\underline{x}_i$)
  if R_i is known
  then
      no action
  else
      for each neighbor location $\underline{x}_k$ indexed by k
          if($P_{ik} > 0$)then
              call ReverseAccumulation($\underline{x}_k$)
              //This is the recursive call to the downslope neighbor

Next k
// At this point all the neighboring R_k inputs are available

$$R_i = r(\underline{x}_i)\Delta + \sum_{\{k:P_{ik}>0\}} P_{ik} R_k$$

return

APPENDIX C: GENERAL PSEUDOCODE FOR UPSTREAM FLOW ALGEBRA EVALUATION

Global variables γ, $\underline{\theta}$, P_{ij}
Function FlowAlgebraUpstream($\underline{x}_i$)
   if $\underline{\theta}(\underline{x}_i)$ is known
   then
       no action
   else
       for each neighbor location $\underline{x}_k$ indexed by k
           if($P_{ki}>0$)then
               call FlowAlgebraUpstream($\underline{x}_k$)
               //This is the recursive call to traverse to an upslope neighbor
       Next k
       // At this point all the necessary inputs are available
       Evaluate Algebraic expression $\underline{\theta}(\underline{x}_i) = f(\gamma(\underline{x}_i), \underline{P}_{ki}, \underline{\theta}(\underline{x}_k), \gamma(\underline{x}_k))$
return

APPENDIX D: GENERAL PSEUDOCODE FOR AVALANCHE RUNOUT ZONE EVALUATION

Global variables γ, $\underline{\theta}$, $\underline{P}_{ij}$
Function AvalancheRunout(x_i)
   if $\underline{as}_i > 0$ (if in source zone)
       $rz_i = \alpha$
       $xm = x_i$
       $ym = y_i$
       $zm = z_i$
       $dm = 0$
   else
       initialize $rz_i = $ nodata
       For each k with $P_{ki} > T$
           if $rz_k >= \alpha$ (neighbor k is in the runout zone)
           if path distance

$d = dm_k + \text{dist}(x_i, y_i, x_k, y_k)$ (This is the total distance along flow paths through a neighbor k to the element i)

else

$d = \text{dist}(x_i, y_i, xm_k, ym_k)$ (This is the horizontal distance from element i to the element with maximum angle on the upslope flow path ending at neighbor k)

$zd = zm_k - z_i$ (This is the elevation difference from the source on a path coming through neighbor k to cell i)

$\beta = \text{atan}(zd/d)*180/\pi$ (This is the angle in degrees from a source on a path coming through neighbor k to cell i)

if $\beta >= \alpha$ and $\beta > rz_i$ (The set of assignments below assign the vector $\underline{\theta}(\underline{x}_i)$ using the flow path from a neighbor k for which the angle to the source on that flow path, β, is a maximum)

$rz_i = \beta$

$xm = xm_k$

$ym = ym_k$

$zm = zm_k$

$dm = d$

Next k

Return

13 Spatial Terrain Modeling
A Hierarchical Approach Toward 3-D Geospatial Data Set Merging

Sagi Dalyot and Yerach Doytsher

CONTENTS

OVERVIEW

Correct, accurate, and updated representation and modeling of our natural environment is recognized as an invaluable resource. One of the more widespread and applicable data type representations is achieved by using digital terrain model (DTM) data sets. A variety of applications, such as visualization or terrain analysis, in the fields of mapping and geoinformation (or geophysics) can benefit from using up-to-date DTM data sets. A key issue still to be addressed when working with DTM data set representation is data merging: integrating data from different sets. Various factors cause global systematic errors as well as local random ones, which reflect on geometric and radiometric data representation. In this chapter a new approach for merging DTM data sets is presented, which analyzes local inconsistencies inherent in geospatial data sets prior to actual data integration. The concept of implementing a hierarchical approach is introduced, in which global geometric discrepancies are monitored, and then used locally for accurate spatial modeling. This approach

195

leads to a more qualitative and reliable representation of the natural environment, thus offering control over the various levels of errors. As part of the proposed merging solution, the extraction of a new DTM look-alike database is introduced, which stores data that represents local discrepancies in the form of affine transformation parameters of the whole integrated area. The hierarchical approach produces a singular, unified, and spatially continuous surface representation of the terrain relief, achieving a more accurate modeling result of the terrain than any of the original data sets.

13.1 INTRODUCTION

As a result of developments in data acquisition and data processing in recent years, as well as in computer technologies, the capabilities of analysis and computation procedures have improved tremendously. Consequently, new data collection technologies yield frequent updating of outdated geospatial data sets. Digital terrain relief data is among the most important sources of information for the representation, characterization, and modeling of the earth, and its natural and nonnatural (man-made) environmental processes. DTM data sets are today one of the main resources for a wide range of applications concerned with terrain relief research, such as for the spatial sciences community, and the geographic information systems [GIS] in particular. Moreover, these data sets represent a key tool for a variety of analysis and research purposes, such as visualization, management, and spatial analysis.

Considered as a continuous and usually smoothed surface representation, a DTM has a grid structure that stores in each of its nodes position and height, commonly referred to as height rasters. It can therefore provide several terrain attributes, including gradient, curvature, slope, and aspect. New data collection technologies and techniques for producing DTMs such as laser scanning and radar interferometry result in a frequent need for updating of outdated geospatial data sets. The updated data sets will usually display more accurate and up-to-date representation of the terrain relief, and frequently in a much denser data structure. The general assumption is, therefore, that by updating inferior DTMs with these updated data sets, there is high potential for achieving enhanced and more accurate topographic representation of the terrain. An update process will usually merge (integrate) data to achieve enhanced and improved product quality and reliability. This in turn enables the support of better geospatial operations as well as decision-making processes.

When the task of merging or change detection is at hand, one might need to compare and analyze the data derived from different DTMs. The fact that these data sets present geospatial data from different sources, which consists of various geometries, scales, resolutions, types, accuracies, and epochs (i.e., dates acquired), might lead to observing global as well as local discrepancies inherent in the different DTMs. These discrepancies may occur due to natural causes such as underground activities, landslides, and earthquakes, or human activities that took place during the data acquisition epochs, as well as inherent errors occurring during the observations or production (object modeling) stages [1]. These various nonuniform factors present global systematic errors as well as local random ones, which reflect on different scales of geometric differences. Therefore, prior to integrating or merging different DTM geospatial data sets, a thorough investigation of different algorithms designed to deal with these various factors is mandatory. Ignoring the topographic discrepancies and integrating the data by one of the common mechanisms, such as replacing the less accurate data with the more accurate data (i.e., cut and paste; height averaging; or even height smoothing and mosaicking), will probably result in erroneous representation and spatial gaps in the terrain relief representation. These mechanisms address only the height representation issue of the terrain, and not its characteristics—topology and morphological structures [2]. The result of implementing these types of mechanisms will be expressed in visible terrain relief representation discontinuities; the pattern of topographic entities will be represented truncated and broken close to the borders of the mutual coverage area. The characteristics of the integrated terrain are not preserved, and may even result in a final product that is inferior to any of the original data sets. A correct preliminary positioning of the updated patch within the outdated one is therefore essential to achieve correct modeling and updating processes of the geospatial data sets, which will prevent discontinuities in the terrain representation. Figure 13.1 shows an example of two DTMs representing the same coverage area, where both were produced on different epochs and with different production techniques. It is clear that there are substantial irregularities in the topographic representation between the different data sets. These irregularities must be solved via morphologic and accuracy adjustments. Otherwise, when merged, the merged data set will not preserve both topographies, and the seam line on the mutual coverage area's border will become a distinctive discontinuity line in the merged DTM that can be visually characterized as a topographic wall. Moreover, a generic solution is needed when two DTMs that represent similar accuracies are to be merged; that is, replacing or averaging is not an option. An appropriate monitoring algorithm for resolving topographic differences must be implemented prior to the integration process.

The reference point in recent work, which addresses the problem of integration of multiple DTM data sets with various accuracies, such as Podobnikar [3] or Frederiksen et al. [4], is that all the data sets utilized for the task were already mutually georeferenced and that there are no morphological incongruities between them.

(a)   (b)

FIGURE 13.1 Shaded relief representation of two data sets representing the same coverage area; substantial planimetric and altimetric topographic discrepancies are visible.

This reference point is the result of considerable *a priori* work, which is not always possible, as in the case of real-time and near real-time applications. Heipke [5], Koch and Heipke [6], and Walter and Fritsch [7], among others, have addressed the issue of integrating DTM data sets with other types of data structures, such as 2-D (two-dimensional) and 2.5-D vector data. However, this work was carried out mainly for the purpose of GIS data integration, and analysis was principally focused on semantic visualization. Furthermore, although vector data represent entities, such as networks or discrete data structures (points, polylines, and polygons), the hypothesis of DTM geospatial data sets is that they represent continuous reality, i.e., terrain. Katzil and Doytsher [8] have addressed the issue of the seam line between two adjacent DTMs and its close surrounding by using a new conflation algorithm based on homologous (three-dimensional) 3-D polylines to achieve a continuous strip of DTM data sets. It is worth noting that this chapter has addressed the issue of the seam line (similar to the height smoothing mechanism), and not the issue of a complete integration and updating between two (or more) data sets, which involves the extraction of the inner inherent topological characteristics and their relations.

It is evident that different DTMs have different sensitivity to computed terrain attributes and representation, which is derived directly from the fact that the data sets present heterogeneous geospatial data [9]. Furthermore, DTMs only partly describe the reality of a continuous terrain relief, mainly because of its discrete representation. So based on the discrete DTM's nature of data structure for terrain representation, integration of two or more sources can significantly improve the quality of the merged DTM, and thus represent more adequately the natural environment. However, an appropriate and thorough analysis of the different DTMs' inherent discrepancies must be carried out prior to integration to achieve a reliable and accurate solution. This solution must work on the spatial domain, and hence preserve the morphologic and topographic connections and representations that exist in each data set.

13.2 PREVIOUS RESEARCH

A merging procedure where no implicit data correlation between different data sets is known can rely on geometrical- and topological-based techniques. This requires utilizing complete and accurate sets of data relations that exist within the entire mutual coverage area. The use of these data relations will enable precise modeling, which is essential for an accurate merging of the mutual data. A merging process can be carried out according to three main stages:

1. Preintegration, that is, georegistration, where a common schema of both geospatial data sets is chosen while relying on sets of selective unique homologous features (objects). This is a crucial stage since no implicit information between the geographical data is available—knowledge that is crucial for qualitative and precise modeling.
2. Matching, which is based on geometric schema specification analyses (known also as conflation) and is essential for achieving a precise reciprocal modeling framework between the two data sets.

3. Merging, which uses the matching modeling relations and the data that exists in both data sets for data fusion, hence achieving an enhanced and accurate terrain representation.

Stage 1 implies that no prior knowledge on the reference frames exists. Hence, the preintegration process is designated to acquire this reference frame by registering homologous unique features identified in both data sets. This is a mandatory stage since, otherwise, the subsequent processes, namely, matching and modeling, will result in an inaccurate outcome. The extraction of the approximated correspondence, that is, initial georeferencing between the mutual coverage data of the data sets, is carried out. A rough translation (an expression of this reference frame) can be extracted by constraining a spatial-shift transformation based on selected unique homologous features that represent the same real-world object and can be identified in both data sets. Brown [10] specifies different types of registrations that deal with a variety of available cases, which mostly occur due to differences in the acquisition processes, resulting in image misalignment, perspective distortions, and scale and scene changes. Though Brown mostly discusses the registration between images, some registrations can easily be translated into affine transformation in 3-D grid-points space.

Rusinkiewicz and Levoy [11] showed that obtaining prior knowledge regarding the geometric spatial relations, that is, georegistration, between the data sets is crucial to ensure a successful and nonbiased matching process. Zhengyou [12] also discussed this issue, and stated that because the matching algorithm converges to a local minimum, the "motion" extraction (as he characterized it) between the two merged data sets—especially in the case of a large motion—must use an initial registration method before the matching procedure can be implemented. Schenk and Csatho [13] discuss the importance of building a reciprocal reference working frame—knowledge that must be given to the integration stage. As a result, synergy between the different data sets will give a more accurate and complete representation of the natural environment. The nonrigid affine registration can be achieved by various schemas, such as the clustering approach [14] or invariant property [15]. Here, we suggest the working scheme of the forward Hausdorff distance mechanism [16], mainly because of its algorithmic simplicity and quickness, its compatibility to discrete features such as points, and the fact that no prior knowledge of the reference coordinates systems is needed for its implementation.

Spatial geometric data set matching can be achieved by one of several processes, including conflation and mosaicking. These matching algorithms mainly depend on the geometric types of the objects to be matched, their topological relations, the volumes of the data sets, and the semantic attributes [17]. A robust and qualitative matching process suitable for the data characteristics at hand is the iterative closest point (ICP) algorithm [18]. The ICP algorithm is designated for 3-D point cloud matching by nearest neighbor criteria when using an iterative least squares matching (LSM) process [19]. This procedure utilizes *a priori* georegistration values, needed to exclude local minima solutions that might occur otherwise.

Several researchers have addressed the problem of ensuring the continuity of surface merging and modeling when a matching procedure is obtained. Feldmar and

Ayache [20] presented a framework for nonrigid surface registration which ensured a semantic and geometric representation of free-form surfaces. The authors have based their research on the assumption that the matched surfaces represent the same object (mostly medical related surfaces). However, in the case of DTM surface representation, distortions, deformations, and major displacements are quite common, so the term "same" becomes ambiguous here. Furthermore, though local affine transformation was implemented, it was based on the same "best" global transformation extracted for the whole data set. Walter and Fritsch [7] also discussed the importance of integration methods, and described a relational matching approach for integrating spatial data from different sources. Though the same methodology is used in the work presented here, the first registration stage in Walter and Fritsch's research was actually carried out manually before the integration stage, and no implementation of postlocal adjustments on that value during the subsequent integration stages was done.

A data sets merging procedure that relies on geometric matching parameters and ensures a spatial continuity of surface modeling can be based on a "reverse engineering" procedure. This process involves a simultaneous quantification of the DTMs', geometric shape for the reconstruction of the merged DTM 3-D model. Still, because of the discrete nature of the data and the fact that DTMs can represent different resolutions, a smooth interpolation must be used. Doytsher and Hall [21] have described a bidirectional third-degree parabolic interpolation algorithm of a DTM, ensuring robust, smooth, and no-gaps topography modeling and representation. This calculation enables a smooth transition between two DTMs, though a weighting average interpolation on the transformation values has to be taken into account.

13.3 PROPOSED HIERARCHICAL APPROACH

The hierarchical approach presented here suggests the implementation of two working levels of topographic frames: (1) global preintegration, and (2) local matching and merging. The motivation is to be able to monitor the nonunified global zonal discrepancies of both DTMs, and hence extract the accurate corresponding georegistration values per zonal area. This is followed by precise localized ICP matching and merging processes, enabling the calculation of a spatially continuous merged surface representation and model of the terrain relief. Figure 13.2 presents a block diagram of the suggested procedure. Various mathematical processes, as well as geometric concepts designed for correct and accurate DTM discrepancy monitoring, were implemented in this research, mainly to address the diverse issues discussed earlier. This hierarchical concept is in contrast to working with the entire data as a "global bundle," which clearly might lead to overlooking localized topographic trends. These two working levels are essential to ensure qualitative initial georegistration parameter extraction, enabling an accurate, localized, constrained ICP matching process, which leads to precise modeling and integration of the data sets.

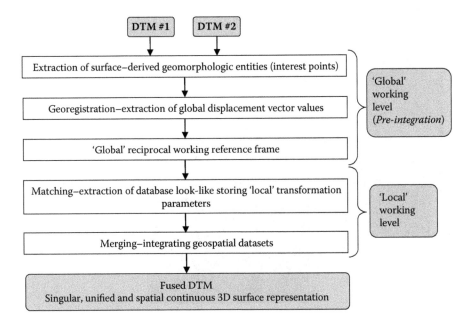

FIGURE 13.2 Block diagram describing the workflow and its relevant stages.

13.3.1 GLOBAL WORKING LEVEL (PREINTEGRATION)

13.3.1.1 First Order Division

As the size of each data set designated for merging is unknown, a first order division on each data set is mandatory. The division is carried out ensuring full topographic overlap between patches in the different data sets. Each medium-size patch (*msp*) is cut with a preliminary known size, as depicted in Figure 13.3. Tests and analyses on the quality of the solution in respect to different patch sizes have shown that a preliminary size of approximately 100 km^2 per patch proved statistically efficient and accurate. This value is mainly derived from the number of interest points that was sufficient for a qualitative georegistration process; this size can be altered if needed. The division is required for extracting global-discrepancy values that exist between the two data sets. The stages of this section are carried out on the zonal patches data: extraction of unique local geomorphologic points and calculation of initial georegistration values that correspond to each overlapping *msp*.

13.3.1.2 Interest Points Extraction

It is evident that geometric, topologic, and topographic conditions define unique surface-derived geomorphologic points, such as mountains or hill peaks. Relying on a designated registration process performed on these extracted unique points, which represent the same reality in each DTM, will satisfy the need for calculating local discrepancies, that is, georegistration parameters between overlapping *msps*.

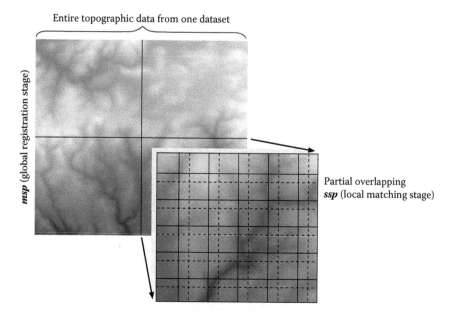

FIGURE 13.3 Two working topographic zoning levels: global registration (medium-size patches); local matching (small-size patches).

It is worth noting that the mathematical approach shown here uses topographic maxima only and not minima. This is due to the fact that local minima points are very rare topographically, which necessitates building an additional set of geometric and statistical rules to those already implemented. The small number of existing topographic minima proved that relying only on maxima points was still statistically sufficient for the proposed georegistration approach.

A successful extraction of surface-derived geomorphologic points requires the examination of the topological conditions that exist in the neighborhood of each DTM grid point. A new computational approach was devised to correctly define geomorphologic interest points. It is based on statistical tests and topologic definitions and constraints according to a set of geometric rules. This computational approach is subdivided into five steps:

1. Extracting four perpendicular second-degree polynomials. Each polynomial is derived from the height (Z) and L, which denotes the local axis coordinates in each of the four principal directions (north, east, south, and west). It was found that choosing six consecutive discrete points in each direction for the calculation of the polynomials, as depicted in Figure 13.4, gave a satisfying and precise definition of the generalized topography description of the surroundings of each grid point. A common DTM resolution is usually a few dozens of meters, which corresponds to the hypothesis presented here concerning the number of points that will enable the correct extraction of unique topographic entities. It is worth emphasizing that step 4, as will be discussed, ascertains the exact position of a unique entity even in high

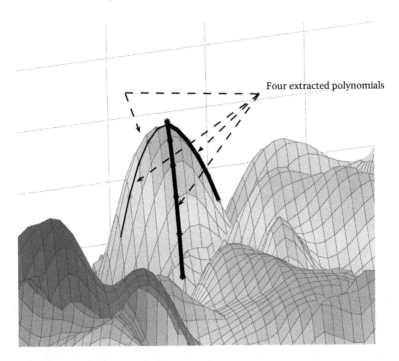

Four extracted polynomials

FIGURE 13.4 Four perpendicular second-degree polynomials define the grid-point topographic neighborhood.

resolution DTMs. Hence, the value of L(i, j) is between zero (for the examined grid point j = 1), and five times the resolution in X and Y, respectively (for the farthest point j = 6), as depicted in Equation 13.1. A least squares adjustment process on each polynomial ensures the extraction of the polynomial's three coefficients: a, b, and c. These twelve coefficients—three for each polynomial—quantitatively define the topographic neighborhood of each examined grid point, as depicted in Figure 13.5.

$$Z_{(i,j)} = a_i + b_i \cdot L_{(i,j)} + c_i \cdot L^2_{(i,j)} \tag{13.1}$$

where i denotes the polynomial index ($i \in [1–4]$); j denotes the point index ($j \in [1–6]$); a_i, b_i, and c_i are the polynomial i coefficients; and $Z_{(i,j)}$ and $L_{(i,j)}$ denote the local axis coordinates.

2. Calculating the integral (area) of the four polynomials in Z direction relative to the height of the farthest point, as depicted in Figure 13.5.

3. Statistical tests on the extracted geometric values will ensure a preliminary qualitative consideration of the examined grid point as one of interest. Two of the polynomial coefficients—b and c—are tested according to statistical thresholds, as well as the polynomial integral value. These tests inspect the polynomials' topological behavior, define their type (ascending or descending), and scales the height magnitude of the examined grid point in respect to its surroundings.

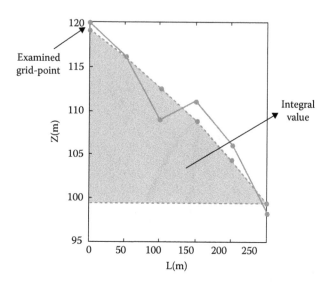

FIGURE 13.5 Extraction of second-degree polynomial (step 1), and calculation of integral (step 2): solid line represents existing topography; dashed line represents extracted polynomial.

4. The three previous steps result in a preliminary evaluation of interest points. Local clustering on these points is carried out, aimed at finding an interest area, which will enable the identification of a grid point that represents the maximum of this interest area. A predefined number of points and search distance criteria, derived from surface characteristics, are stated to qualitatively define an interest area. This process is finalized with selection of the cluster's highest grid point, as depicted in Figure 13.6

5. To ascertain that in each cluster the highest topographic location is chosen, a local bidirectional interpolation within the cluster is carried out, as depicted in Figure 13.6. This process ensures the precise calculation of the highest topographic location, thus achieving planimetric subresolution accuracy. This calculation is done by extracting local polynomials that intersect at the location of the highest grid point found in step 4 in X and Y directions: Z_x and Z_y, respectively. First derivative geometric constraint enables the calculation of the shift values with respect to the highest grid point in the cluster: S_x and S_y, shown in Equation 13.2. This shift vector will point to the precise topographic location of the required unique surface-derived geomorphologic point position.

$$S_x = -\frac{a_1}{2 \cdot a_2}; S_y = -\frac{a_4}{2 \cdot a_5};$$

$$X_{interest\_point} = X_{highest} + S_x; \quad Y_{interest\_point} = Y_{highest} + S_y$$

(13.2)

where a_1 and a_2 denote polynomial coefficients of Z_x; a_4 and a_5 denote polynomial coefficients of Z_y; and S_x and S_y denote the shift values in direction X and Y, respectively.

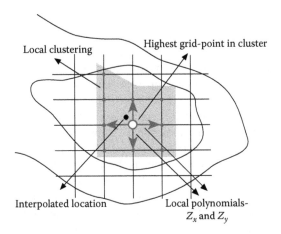

FIGURE 13.6 Precise location of an interest point: clustering process (step 4), and bidirectional local interpolation (step 5).

13.3.1.3 Calculation of Initial Shift Vectors

The extraction of interest points in each overlapping *msp* enables the calculation of a correct zonal topographic spatial displacement. The registration process is performed on interest points present in both data sets. The process proposed identifies and couples-up homologous features existing in both data sets in order to extract a rough reference frame. It relies on the forward Hausdorff distance algorithm, which does not require any constraints or prior knowledge of the points' dispersal or the topologic relations between the sets. Given two sets of points: $A = \{a_1, \ldots, a_m\}$ and $B = \{b_1, \ldots, b_n\}$, the forward Hausdorff distance (h) measures the degree of mismatch between the two sets, as defined in Equation 13.3. This equation identifies point $a \in A$ that is farthest from any point in B, and then measures the distance from a to its nearest neighbor b in B. This distance gives an initial estimation of the global geo-registration value, which is statistically evaluated from the correspondence it obtains between all the points in sets A and B.

$$h(A, B) = \max_{a \in A} \min_{b \in B} \|a - b\| \qquad (13.3)$$

The minimum number of required paired-up points is derived from the registration model, in this case an affine transformation model. Here we refer only to the 2-D translation vector (ignoring the height dimension) while requiring a minimum number of pairs that will obtain a good standard deviation values evaluation of the registration. Hence, the output is a vector set of initial registration values for the global zonal area, that is, overlapping *msp: dx^0, dy^0*, used as *a priori* data for the matching process.

13.3.2 Local Working Level

13.3.2.1 Constrained ICP Matching

After establishing the global reciprocal working reference frame, a second order division on the data is carried out. In contrast to the implementation of one global matching procedure on the entire mutual data (which assumes no local trends exist—a hypothesis that is in contrast to the fundamental concept stated here), we suggest dividing the entire mutual coverage area covered by each *msp* into homologous separate mutual smaller frames. By implementing a separate and independent ICP matching process on each of these smaller frames, several segregated localized monitoring procedures are achieved. This enables a more accurate characterization of local phenomena in comparison with an ambiguous global one. Every *msp* is subdivided with a preliminary known size into partial-overlapped small-size patches (*ssp*), as depicted in Figure 13.3. It was found that to ensure that the merged DTM is unified and continuous throughout the area, and with respect to the data sets' resolution, an *ssp* size of 1 km² per patch is efficient, resulting in qualitative merged terrain representation. This size is resolution-dependent and can be altered if needed. Based on the assumption that inherent local discrepancies exist between the DTMs, it is clear that small zonal patches could be fitted more accurately than large patches. Hence, the matching process will introduce approximately the same topographic matching values in neighboring small patches. This is in contrast to large variations and truncated values that are the consequence of large patches, which introduce large deviations and disorder in their correspondence. Statistically, smaller patches show homogeneous and unified topographic characteristics that allow better evaluation of the correct registration results. This ensures that an ICP process implemented on *ssp*s will produce more accurate matching results in reference to the existing relations between the 3-D grid points. In this research a constrained ICP process was implemented, which was derived from the constraints that the data characteristics and problem imposed.

In general, a matching process is aimed at finding the best geometric correspondence between two data sets, described here by two 3-D point clouds denoted by $f(x,y,z)$ and $g(x,y,z)$. The magnitude of the correspondence of the two data sets is derived from an error vector denoted by $e(x,y,z)$. This error vector (e) describes the relations of the two data sets, which can be denoted by $\{f(x,y,z) - g(x,y,z)\}$. Vector e includes local random errors as well as global systematic ones (which have been modeled in the preintegration stage). Thus, it is obvious that the matching of small patches will monitor more effectively and accurately these various types of errors. The error vector extraction is achieved by minimizing the target function, defined here by a transformation model, extracting the best possible correspondence between data sets f and g.

A constrained ICP process is implemented on each overlapping *ssp*, which suggests a nearest neighbor search criteria process according to the three constraints outlined in Equation 13.4 (first, second, and third row, respectively):

1. The coordinates of the paired-up nearest neighbor i in data set g ($X^g i$, $Y^g i$, $Z^g i$), which correspond to point i in data set f, fit geometrically a local cell-plane in data set g (cell-plane model is defined by a bilinear interpolation).
2. The line equation, derived from the coordinates of point i transformed from data set f to data set g with the best known transformation parameters (denoted by x'_f, y'_f, and z'_f), and the paired-up nearest neighbor i in data set

g (Xgi, Ygi, Zgi), is perpendicular to the local cell-plane in data set *g* in the X direction (achieved by first order derivative).

3. This uses the same constraint outlined in step 2, only here the line equation is perpendicular to the local cell-plane in data set *g* in the Y direction (achieved by first order derivative).

$$Z_i^g = \frac{h_1}{D} \cdot X_i^g + \frac{h_3}{D} \cdot Y_i^g + \frac{h_4}{D^2} \cdot X_i^g \cdot Y_i^g$$

$$Z_i^g = -\frac{h_4 \cdot y_f'}{D^2} \cdot X_i^g + \frac{h_3}{D} \cdot Y_i^g + \frac{h_4}{D^2} \cdot X_i^g \cdot Y_i^g + \left(z_f' - \frac{h_3 \cdot y_f'}{D} \right) \qquad (13.4)$$

$$Z_i^g = \frac{h_1}{D} \cdot X_i^g - \frac{h_4 \cdot x_f'}{D^2} \cdot Y_i^g + \frac{h_4}{D^2} \cdot X_i^g \cdot Y_i^g + \left(z_f' - \frac{h_1 \cdot x_f'}{D} \right)$$

where *i* denotes the grid-node's index; h_1 to h_4 are calculated from the height of local grid's cell corners in data set *g*: Z_1 to Z_4 ($h_1 = Z_1 - Z_0$, $h_2 = Z_2 - Z_0$, $h_3 = Z_3 - Z_0$, $h_4 = h_2 - h_1 - h_3$); *D* denotes the grid's spacing; *g* and *f* denote the data sets; (X_i^g, Y_i^g, Z_i^g) denote the paired-up nearest neighbor; and (x_f', y_f', z_f') denote the transformed point *i* from data set *f*.

Because both DTMs represent reality in true scale, it can be assumed that the two DTMs represent the terrain relief with approximately the same scale factor (S). Hence, the transformation model implemented here was modeled according to six parameters: three translation parameters (*dx, dy,* and *dz*) and three rotation angles (φ, κ, and ω) shown in Equation 13.5. A scale factor (S) can easily be added to the model if required. Because linearization is needed to solve this transformation model, initial transformation values per *ssp* are required. The initial shift vector used for each *ssp* ICP matching is the one that corresponds to its higher-level *msp* (i.e., dx_0, dy_0, and $dz_0 = 0$), which was extracted in the preintegration stage. The assumption outlined in the beginning of this paragraph, and the fact that the DTMs are close to being orthogonal projections of the terrain, coerces the diagonal values of the rotation matrix (R) to be close to 1. Thus, the initial rotation angles values—φ_0, κ_0, and ω_0—were evaluated initially as zero degrees.

$$\begin{bmatrix} X_g - X_g^M \\ Y_g - Y_g^M \\ Z_g - Z_g^M \end{bmatrix} = R(\varphi,\kappa,\omega) \bullet \begin{bmatrix} X_f - X_f^M \\ Y_f - Y_f^M \\ Z_f - Z_f^M \end{bmatrix} + \begin{bmatrix} dx \\ dy \\ dz \end{bmatrix} \qquad (13.5)$$

where *g* and *f* denote the data sets; (*X, Y,* and *Z*) denote data set coordinates; *R* denotes the rotation matrix; *dx, dy,* and *dz* denote three translation parameters; φ, κ, and ω denote three rotation angles; and *M* denotes the center coordinates of each congruent *ssp*.

For each point in data set f, a nearest point from data set g is paired up as long as the criteria outlined in Equation 13.4 are fulfilled. Consequently, with all the point pairs extracted, a local six-parameter matching is achieved for each mutual ssp. The ICP process on each ssp is carried out iteratively and independently until predefined statistical criteria are achieved, which are based on the difference of the six parameters' values in each consecutive iteration, the number of iterations, the number of point pairs, and height difference criteria. The process yields a better localized spatial matching calculation, thus ensuring topographic continuity of the entire area, as well as excluding a local minima solution for the ICP process, and minimizing the computation time. In addition, a new concept can be pointed out: a novel DTM look-alike database extraction, which is the direct product of the matching stage. Figure 13.7 illustrates this database, which stores in its nodes the six-parameter geo-registration values corresponding to the center of each congruent ssp. This database can contribute to the effectiveness of the merging process carried out in the next stage, as well as to DTM seaming procedures. It also enables full monitoring capabilities, which support a better statistical analysis and investigation of the spatial relations existing between the DTMs.

13.3.2.2 Merging

Since the local accurate topographic relations between all the local $ssps$ are stored in the novel georegistration database, which include the local random errors as well as the global systematic ones, this correct modeling enables the implementation of a merging process. Merging is achieved via a reverse engineering mechanism, which uses the known spatial correspondence between the two DTMs that is stored in the georegistration database grid-nodes—while relying on the DTMs', data—thus ensuring spatial continuity of surface modeling. A merged DTM is calculated with

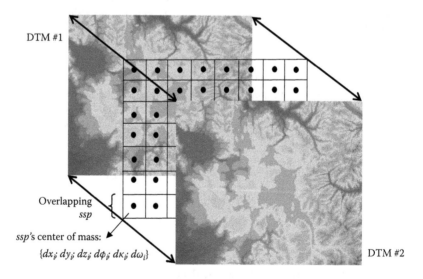

FIGURE 13.7   Digital terrain model look-alike database storing the corresponding six-parameter georegistration, that is, transformation, values for overlapping mutual ssp zones.

respect to the two original DTMs. It can be described spatially, as if the merged DTM is situated in the space between the two original DTMs. It should be emphasized that the georegistration database and the original DTMs have different resolution. Usually the original data set stores grid points in a resolution of a few meters up to several dozens of meters. The georegistration database resolution, however, is derived from the local patch size that was chosen in the matching process, thus storing the parameters in a resolution of a few hundred meters. To achieve the required DTM's resolution in the merging stage, and hence ensure continuity of the calculated values, an interpolation algorithm on the discrete transformation values stored in the georegistration database is essential. This process is divided into two main stages:

1. The localized corresponding six-parameter georegistration values of each grid point in the merged DTM are needed for the "reverse" transformation. These registration values will be used respectively for the two-way transformation (merged DTM toward each of the data sets). This calculation is done by utilizing a bidirectional third-degree parabolic interpolation on the three georegistration translation values stored in the database grid nodes, as outlined in Equation 13.6. The three rotation values are interpolated using quaternions (as will be explained in stage 2). These two interpolation mechanisms ensure smooth and robust calculation of transformation values within the corners of the *ssp*s, and hence in the entire area:

$$F_1(t) = -0.5 \cdot t + 1.0 \cdot t^2 - 0.5 \cdot t^3$$

$$F_2(t) = +1.0 - 2.5 \cdot t^2 + 1.5 \cdot t^3$$

$$F_3(t) = +0.5 \cdot t + 2.0 \cdot t^2 - 1.5 \cdot t^3 \qquad (13.6)$$

$$F_4(t) = -0.5 \cdot t^2 + 0.5 \cdot t^3$$

$$Z_p = \sum_{i=1}^{4} \sum_{j=1}^{4} F_j(x) \cdot F_i(y) \cdot H(i,j)$$

where $F_1(t)$ to $F_4(t)$ denote the third-degree parabolic equations; Z_p denotes the height interpolation; t denotes the normalized coordinates $\{0 \leq t \leq 1\}$; x and y denote the inner cell normalized coordinates; $H(i,j)$ denotes the elevations of the corner points (value of six transformation parameters from the georegistration database); and i and j denote the index of 4×4 neighboring corner points.

2. Once the corresponding georegistration parameters are known, the height of the merged DTM grid point can be calculated. This is achieved via a reverse engineering procedure that calculates a weighted average of the two corresponding heights in each of the original DTMs by a two-way transformation. Nevertheless, the georegistration parameters calculated in stage 1 represent the relations of one geospatial data set to the other (source and

target). Because a transformation from the merged DTM to each of the data sets is needed, a weighted average on these georegistration parameters is essential. It is clear that a weighted average on the translation values is straightforward due to their linear nature. A problem arises when trying to do the same on the rotation values: interpolating orientation parameters represented by Euler angles will fail due to their nonlinear character (rotation on objects involves multiplication, and rotation matrices do not commute in multiplication). In order to solve this problem and satisfy a rigid and continuous calculation, two operations are carried out:

a.   Transformation of the rotation angles from 3-D space into 4-D unit hypersphere in quaternion space, as suggested by Shoemake [22]. By doing so, the dependence among the three axes that exists with Euler angles representation is solved.

b.   Execution of a spherical linear interpolation (slerp) on the quaternion values. The idea behind slerp, as explained in Watt and Watt [23], is to avoid the problem of a motion acceleration in the middle of two key orientations, which occurs when a linear interpolation is implemented. This happens because of cutting across the hypersphere and not along its surface. The implementation of slerp, shown in Equation 13.7, ensures just that: a steady rotation that guarantees that the movement along a geodesic arc passes through the two key orientations. Consequently, the interpolated orientation is calculated using both key orientations and the relative accuracy of the DTMs (denoted by t), which derives the interpolated orientation magnitude in space.

$$\theta = \cos^{-1}\left(q_i \cdot q_n\right)$$

$$slerp\left(q_i, q_n, t\right) = \frac{\sin\left(\left(1-t\right)\cdot\theta\right)}{\sin\left(\theta\right)}\cdot q_i + \frac{\sin\left(t\cdot\theta\right)}{\sin\left(\theta\right)}\cdot q_n \qquad (13.7)$$

where q_i and q_n denote the two key orientations in quaternion 4-D space, and t denotes the normalized relative accuracy of the DTMs $\{0 \le t \le 1\}$, such that in $t = 0$ the accuracy of DTM$_i$ is ∞, and the accuracy of DTM$_n$ is 0.

By implementing the bidirectional third-degree parabolic interpolation, transforming into 4-D space and executing the slerp concept, and by using the local relations extracted earlier, the merged DTM produced satisfies the preliminary requirements that were the aim of this research: obtaining a singular, unified, and spatially continuous surface representation of the terrain relief. In addition, the merged DTM introduces more accurate modeling results of the terrain than any of the original data sets, while preserving all the topologic relations and morphologic entities represented in each of them separately. In the case where both DTMs present similar accuracies, the merged DTM will show an averaging topographic representation of both DTMs with respect to the accurate relations of both. However, it is worth emphasizing that the process and algorithms implemented are fully capable of calculating the in-between correct topography as derived from the relative accuracy of both DTMs.

13.4 EXPERIMENTAL RESULTS

In order to evaluate the proposed solution and its relevant algorithms and mathematical concepts, the suggested approach was tested on various DTMs. These DTMs were produced from several data sources, presenting a variety of resolutions, datum, and accuracies. Tests were also conducted on synthetic DTMs, which were specially produced to check specific statistical aspects of the solution.

In the global working level, the interest points' extraction mechanism proved geomorphologically to be accurate and efficient. The mechanism was examined on various DTMs representing different levels of detailing, and proved to be robust and stable. The automatic process was capable of accurately defining local surface-derived extremes in the topographic relief represented by each DTM, as can be seen in Figure 13.8. Furthermore, the accuracy of the interest points' topographic positioning was of subcell resolution, that is, higher than the resolution of the DTM. This fact contributed to the calculation accuracy of the initial georegistration vectors process between the overlapped *msp*s, which has great affect on the statistical quality of the

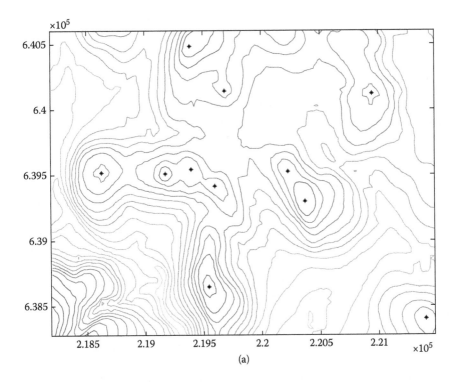

(a)

FIGURE 13.8 Contour representation showing identification of local geomorphologic surface-derived points (denoted by asterisks): (a) a 50 m resolution digital terrain model, (b) a 25 m resolution digital terrain model.

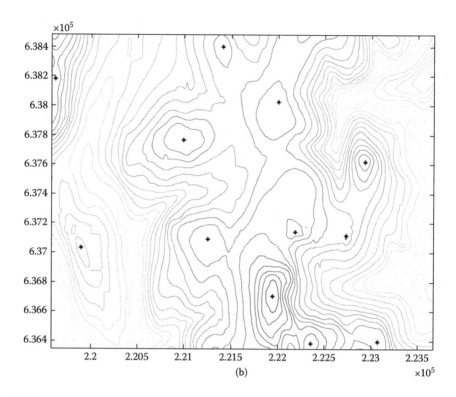

FIGURE 13.8   (Continued)

matching stage (as will be shown later). Moreover, it was concluded that the level of detailing of the DTM, which is mainly dependent on the resolution of the data set, has an effect on the number of extracted interest points: a more detailed DTM results in more interest points extracted, and thus a more reliable topographic positioning is calculated.

The forward Hausdorff distance registration algorithm proved to be robust, precise, and fast. Even when topographic discrepancies of several hundreds of meters were evident between the geospatial data sets, the algorithm was able to extract an adequate initial georegistration value. Figure 13.9 depicts a synthetic georeferencing test on a 25 m resolution patch (marked with inner frame), which covers approximately 25 km². This area was cropped and then georegistered back to a wider DTM data set, which covers approximately 100 km². Different reference systems were assigned to both DTMs, while adding spatial discrepancies and noise, resulting in an arbitrary movement of the grid points of several hundred meters. A total of 170 interest points were extracted automatically within the wide DTM coverage area (denoted by circles), while 32 were extracted in the patch (denoted by asterisks). The two sets of extracted interest points are depicted superimposed in the figure after the registration process. While several thousand pairing possibilities exist, the registration process was able to pair 23 points (marked with black circles and asterisks). The differences between the calculated registration values and the values used in the

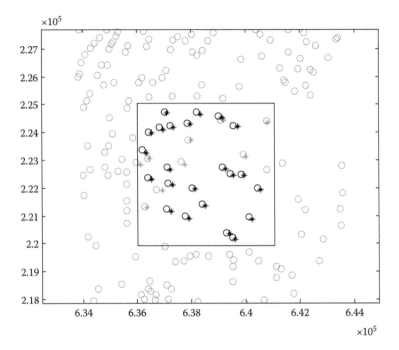

FIGURE 13.9 Superimposition of two data sets after registration (values in meters).

synthetic data test (which, as mentioned above, were several hundred meters) were less than ±5 m. The noise that was added in this test was in the range of several dozens of meters, while the resulting standard deviation was less than 10 m, hence proving that this algorithm is reliable and accurate.

Table 13.1 shows statistical values of two ICP matching processes. The aim was to ascertain the importance a correct initial georegistration shift value has on the quality of a matching process. In this case, two real DTMs (shown earlier in Figure 13.1) produced by different production techniques at different epochs were used. The second column shows the statistical results received when an ICP process utilized the extracted initial shift vector (a), whereas the third column shows the statistical results received when no prior knowledge was used for the ICP process (b).

The main conclusions arising from analysis of the numerical values in the table are:

1. The mean number of iterations needed for all the 132 processes to converge was much smaller in (a) than in (b): 3.8 compared to 15.3, and hence the computation time was shorter for (a).
2. In (a), the transformation parameters extracted—dx, dy, and dz—for the 132 *ssps* were consistent in value, were close to the initial georegistration value, and had small standard deviation (STD) values: ±0.36 m, ±0.64 m, and ±0.16 m, respectively. In contrast, the values of the transformation parameters extracted in (b) were scattered, were significantly different from the

TABLE 13.1

Statistics of the Iterative Closest Point Process Executed on 132 Small-Size Patches (Approximately 100 km²)

	Utilizing	
	(a) Initial Shift Vector Extracted ($dx_0 = 129.8$ m, $dy_0 = -50.1$ m, $dz_0 = 0$ m)	(b) No Prior Knowledge ($dx_0 = 0$ m, $dy_0 = 0$ m, $dz_0 = 0$ m)
Calculated for All 132 *ssps*		
Number of iterations$_{mean}$	3.8 (−)	15.3
dx_{mean}	124.62 m	16.16 m
dy_{mean}	−50.07 m	−9.03 m
dz_{mean}	30.02 m	28.16 m
dx_{STD}	±0.36 m	±7.59 m
dy_{STD}	±0.64 m	±4.25 m
dz_{STD}	±0.16 m	±1.60 m
z_s_{mean}	±0.23 m	±12.16 m

initial value used, and had high STD values: ±7.59 m, ±4.25 m, and ±1.60 m, respectively. This indicates that a number of matching solutions in (b) were a result of local minima convergence. Taken together, these results emphasize the importance of the registration stage on the entire procedure.

3. When the quality of the transformation parameters extracted from both ICP processes are compared, it is clear that the statistical quality of those in (a) is superior. This evaluation was done using a statistical test value denoted z_s. This number evaluates the quality of the transformation parameters by comparing the height differences between: (1) the source DTM transformed via the extracted transformation parameters, and (2) the target DTM. Evidently, this value is much smaller and closer to zero in (a): ±0.23 m, than (b): ±12.16 m. This indicates that the transformation parameters extracted in (a) are markedly more accurate.

The quality of a merged DTM can be examined and evaluated by inspecting the preservation of morphologic entities within the terrain relief represented by the original data sets. Additionally, it can be evaluated computationally by comparing the discrepancies between the original DTMs and those calculated by the merging process used. Figure 13.10 presents this phenomenon by comparing the proposed merging concept and the common height averaging mechanism. This analysis was carried out on the synthetic data, in which a patch from the 25 m resolution DTM was spatially shifted, and then merged with the original data set. It is clear that the proposed concept preserves the morphology of the topography, as opposed to the height averaging mechanism. Frame A shows one example, in which the averaging of steep terrain is smeared and smoothed, whereas the proposed mechanism preserves the topography. Frame B shows another example, in which the common mechanism creates a planar topography from merging a hill and a crevice, whereas

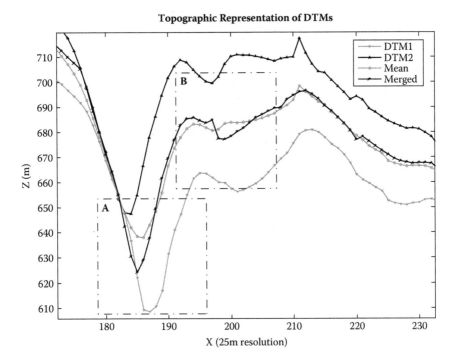

FIGURE 13.10 Side view of two digital terrain model merging procedures: (A) height averaging (mean), and (B) the proposed approach (merged).

the proposed concept preserves both morphologies. This is due to the fact that the proposed concept is implemented on the spatial domain, that is, applying the known spatial correlations and morphological adjustments, whereas the height averaging mechanism is implemented only on the vertical domain with no horizontal adjustments at all.

Figure 13.11 presents a synthetic test, in which an area of close to 25 km^2 was cropped and then merged to an original data set representing approximately 100 km^2. The DTM patch was originally copied from the original data set and was spatially shifted: a few hundred meters in X and Y (with added noise), and a few dozens meters accompanied with vertical noise in Z. Both data sets are in a 25 m resolution, which translated to approximately 40,000 and 160,000 grid nodes in the small and original data sets, respectively. As can be seen from the shaded relief representation in Figure 13.11, the updated data set is unified, spatially continuous, and free of gaps throughout the area. A careful examination of the morphologic structures shows a correct terrain relief representation with no discontinuities, including on the border of the mutual area.

Figure 13.12 presents an area of close to 40 km^2 presented by two real DTMs (used in Table 13.1)—(a) and (b) as well as the merged DTM (c) that was calculated by the approach presented here. The merged DTM presented in this 3-D representation figure is unified, spatially continuous, and has no gaps throughout its area.

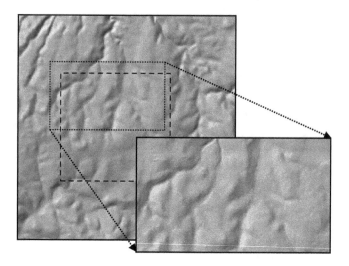

FIGURE 13.11   Complete updating procedure showing continuous terrain relief and morphological representation: dashed rectangle describes mutual coverage area; dotted rectangle describes zoomed area (on right).

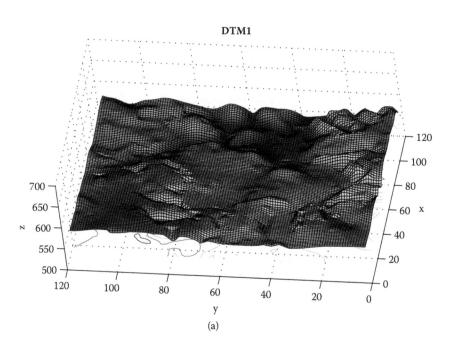

(a)

FIGURE 13.12   Two real digital terrain models (a and b), and the corresponding merged digital terrain model (c).

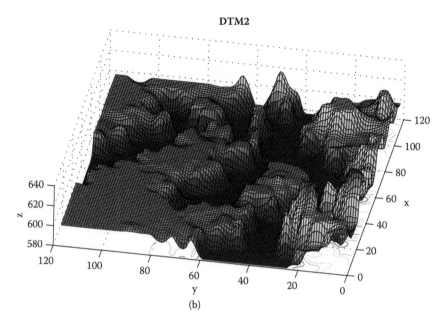

(b)

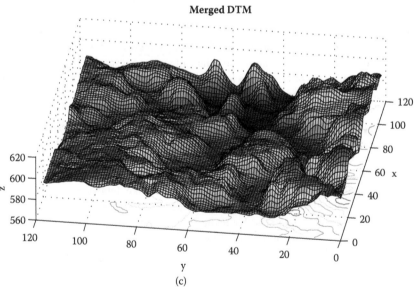

(c)

FIGURE 13.12 (Continued)

A careful examination proved that the merged data set described the surface correctly, [as it preserved] and presents the morphologic structures, such as hilltops and ravines, which exist in the DTMs used for its calculation.

In addition to the visual examination, statistical evaluations of the proposed solution were carried out on synthetic data. In one test, a spatial transformation using a sinusoidal wave height transformation with added planar shifts was carried out on a real DTM. The sinusoidal wave height was superimposed on the entire DTM area, so each zonal mutual area (*ssp*) presented a different height change from its neighboring *ssp*s (which varied from 0 to 50 m vertically). Moreover, even within the *ssp* itself, the height change was not a fixed value. These two DTMs were then merged for statistical evaluation, which was done by analyzing and comparing height gaps by: (1) implementing the proposed mechanism, and (2) imitating the height averaging mechanism. Table 13.2 shows this comparison: calculating the standard deviation of the heights difference values for all grid-point positions (X, Y) per *ssp*.

It is clear that the proposed mechanism (left column) shows relatively smaller height differences. This can be explained by the fact that the proposed algorithm takes into consideration the local spatial topographic relations that exist between the DTMs per *ssp*, which is ignored otherwise. It should be emphasized that in an additional synthetic test, in which fixed vertical shifts were used instead of the sinusoidal version (along with planar shifts), the resulting STD values of the height differences for the entire area were very close to zero.

13.5 DISCUSSION

When discussing the problem of merging geospatial DTM data sets, the characteristics of different strategies and algorithms should be considered. Generally, if one data set is considerably more accurate and detailed than the other, then in most cases the more accurate one should be chosen as the correct terrain representation. However, the more common situation when merging geospatial DTM data sets is that the two data sets have similar levels of detail and accuracy, while containing local and/or global discrepancies. In this situation, the merging procedure of the two data sets must preserve both presented morphologies, thus achieving a more accurate and reliable representation of the terrain than either of the two data sets separately.

In this chapter, a new hierarchical approach and algorithms for merging DTM data sets were introduced. Implementing this approach and algorithms ensure the preservation of local geometric features and their topological relations, while preventing distortion. Furthermore, the new DTM look-alike database, which stores the local topographic relations between the data sets, presents a conceptual approach for merging geospatial data sets. By using this database, the entirety of the spatial relations are known—in contrast to averaging or replacing only the height values. In addition, because the topographic relations of the overlapping zone are now known, a

TABLE 13.2

Standard Deviation of Vertical Heights Differences Calculated for 132 Small-Sized Patches

	Procedure	
	Proposed Mechanism	**Height Averaging Mechanism**
Height gaps *STD* value	±(0.2–0.8) m	±(3.5–5.6) m

correct smooth and continuous seaming of the two DTMs at the borders is achieved. Moreover, the implementation of separate levels of working data, as proposed here, enables local discrepancies between the different DTM geospatial data sets to be monitored. This is in contrast to using global transformations that can lead to ignoring or "smearing" of local existing geomorphologic features.

Nevertheless, under extreme geometric conditions such as large discrepancies, no correspondence, or in the case of very smooth surfaces, the attempt to extract the georegistration values might lead to incorrect results. This will probably lead to a biased ICP matching process that will divert to local minima instead of an implicit one. However, these cases are rare, as shown in the various tests reported here. Therefore, in most situations the suggested approach will result in a satisfactory solution of the merged DTM, presenting a unified, gapless, and spatial continuous representation and visualization of the terrain relief. Moreover, the proposed updating and integrating concept and algorithms can be adjusted with relatively slight modifications to data structures other than grid DTMs which are common nowadays, such as LiDAR data, which presents an irregular data structure.

REFERENCES

1. Hutchinson, M. F. and Gallant, J. C., Digital elevation models and representation of terrain shape, in *Terrain Analysis: Principles and Applications*, Wilson, J. P. and Gallant, J. C., Eds., John Wiley & Sons, New York, 2000, chap. 2.
2. Laurini, R., Spatial multi-database topological continuity and indexing: A step towards seamless GIS data interoperability, *International Journal of Geographical Information Science*, 12, 373, 1998.
3. Podobnikar, T., Production of integrated digital terrain model from multiple data sets of different quality, *International Journal of Geographical Information Science*, 19, 69, 2005.
4. Frederiksen, P., Grum, J., and Joergensen, L. T., Strategies for updating a national 3-D topographic database and related geoinformation, in *Proceedings of ISPRS XXth Congress*, Inter-Commission WG II/IV, Istanbul, Turkey, 2004.
5. Heipke, C., Some requirements for geographic information systems: A photogrammetric point of view, *Photogrammetric Engineering and Remote Sensing*, 70, 185, 2004.
6. Koch, A. and Heipke, C., Semantically correct 2.5D GIS data: The integration of a DTM and topographic vector data, in *Developments in Spatial Data Handling*, Fisher, P., Ed., Springer, Berlin, 509, 2004.
7. Walter, V. and Fritsch, D., Matching spatial data sets: A statistical approach, *International Journal of Geographical Information Science*, 13, 445, 1999.
8. Katzil, Y. and Doytsher, Y., Spatial rubber sheeting of DTMs, *Proceedings of the 6th Geomatic Week Conference*, Barcelona, Spain, 2005.
9. Wilson, J. P., Repetto, P. L., and Snyder, R. D., Effect of data source, grid resolution, and flow-routing method on computed topographic attributes, in *Terrain Analysis: Principles and Applications*, Wilson, J. P. and Gallant, J. C., Eds., John Wiley & Sons, New York, 2000, chap. 5.
10. Brown, L. G., A survey of image registration techniques, *ACM Computing Surveys (CSUR)*, 24(4), 325, 1992.
11. Rusinkiewicz, S. and Levoy, M., Efficient variants of the ICP algorithm, *Proceedings of the Third International Conference on 3D Digital Imaging and Modeling*, IEEE Computer Society Press, 145, 2001.

12. Zhengyou, Z., Iterative point matching for registration of free-form curves and surfaces, *International Journal of Computer Vision*, 13, 119, 1993.

13. Schenk, T. and Csatho, B., Fusion of LiDAR data and aerial imagery for a more complete description, *International Archives of Photogrammetry and Remote Sensing*, 34, 310, 2002.

14. Stockman, G., Kopstein S., and Benett S., Matching images to models for registration and object detection via clustering, *Pattern Analysis and Machine Intelligence*, 4, 229, 1982.

15. Lamdan Y., Schwartz J. T., and Wolfson H. J., Object recognition by affine invariant matching, *Proceedings Computer Vision and Pattern Recognition*, 335, 1988.

16. Huttenlocher, D. P., Klanderman, G. A., and Rucklidge, W. J., Comparing images using the Hausdorff distance, *IEEE Transactions on Pattern Intelligence and Machine Intelligence*, 15, 850, 1993.

17. Mount, D. M., Netanyahu, N. S., and Le Moigne, J., Efficient algorithms for robust feature matching, *Pattern Recognition,* 32, 17, 1998.

18. Besl, P. J. and McKay, N. D., A method for registration of 3-D shapes, *IEEE Transactions on Pattern Analysis and Machine Intelligence*, 14, 239, 1992.

19. Gruen, A. and Akca, D., Least squares 3D surface matching, *Proceedings of the ASPRS Annual Conference*, Baltimore, Maryland, 2005.

20. Feldmar, J. and Ayache, N., Rigid, affine and locally affine registration of free-form surfaces, *International Journal of Computer Vision*, 13, 99, 1994.

21. Doythser, Y. and Hall, J. K., Interpolation of DTM using bi-directional third-degree parabolic equations, with FORTRAN subroutines, *Computers and Geosciences*, 23, 1013, 1997.

22. Shoemake, K., Animating rotation with quaternion curves, *Computer Graphics*, 19, 245, 1985.

23. Watt, A. and Watt, M., Overview and low-level motion specification, in *Advanced Animation and Rendering Techniques Theory and Practice*, Watt, A. and Watt, M., Eds., ACM Press, New York, 1992, chap. 15.

14 Regions and Patterns of Forest Change in Brazil

A Geographically Weighted Regression

Alejandro de las Heras and Iain R. Lake

CONTENTS

OVERVIEW

Despite the acknowledged global importance of Brazilian forests, the evidence on the drivers of tree-cover change is mostly based on local analyses. The emphasis on Amazonia also overshadows tree-cover research in the rest of the biomes. This contrasts with widespread human encroachment. This chapter examines the totality of Brazil's tree-cover change from 1992 to 2001, from the viewpoint of biophysical and socioeconomic impacts. But a large-area study brings forth the issues of spatial variability of the impacts, and local or neighborhood effects. These are dealt with explicitly in our models. The additional problem of sampling, often used to allay the issue of spatial autocorrelation, is examined. The results highlight the strong and significant effect of initial forest cover in predicting subsequent tree loss. In the opposite direction, local effects seem to indicate that forest pixels in forest-only tracts are less likely to suffer tree loss. The historical southeast–northwest direction of deforestation pulses in Brazil is confirmed, and on average, higher distances to cities tend to protect tree cover. Another strong driver of tree-cover loss seems to be the density of cattle. Considering Amazonia, the Arc of Deforestation (AoD), and Southern Brazil, the significant variables are the same, pointing to similar processes. However, the strength of these variables differs regionally. Temperature and latitude also differ in sign in Amazonia and Southern Brazil. Considering the empirical spatial impacts identified by geographically weighted regression (GWR), steep slopes seem to protect tree cover in northern Amazonia and the Atlantic forest. In Eastern Brazil, higher distances to cities seemed to indicate locations that are detrimental to tree cover. Close spatial matches with the shape of the AoD are given by the density of unpaved roads, whose important role is thus confirmed, and by human population density. The best match with the AoD is given by the impact of cattle density. The foregoing results support the use of GWR in identifying and measuring empirical patterns of impact, along with spatial autoregressive regressions that measure the strength of local interactions.

14.1 INTRODUCTION

Brazil has a majority share of the world's tropical forest. Absolute forest loss there is rapid and is accelerating [1,2]. The Amazonian forest is a global switch region where modest changes in vegetation and climate can affect the earth system significantly [3]. The Cerrado is also a world biodiversity hot spot. However, Brazil-wide measurements of tree-cover change are very recent, and generalizations have to rely on local deforestation models. Recently available global data, although coarser in resolution, allow large areas and connections between heterogeneous regions to be studied [4]. In this chapter, the aim is to bring together the importance of large areas and local variation in the study of tree-cover change and its drivers. We also try to depart from *a priori* defined regions and to identify the empirical patterns of impact created by the drivers of tree-cover change.

Recent evidence on the drivers of tree-cover change is first examined, as are recent multivariate models. The relevant tree-cover data and predictors are then examined; the study regions are defined based on vegetation considerations. Multiple linear regressions, spatial autoregressive regressions, and especially geographically weighted regressions (GWR) are considered, along with sampling, spatial autocorrelation, and spatial heterogeneity issues. The results identify the average impact of the drivers in Brazil, their regional impacts controlling for neighborhood effects, and contrast *a priori* regions with empirical spatial impacts found with the GWR. The discussion deals with GWR and a framework based on living populations, attractiveness, and accessibility that was helpful in interpreting the results.

14.1.1 SOCIOECONOMIC AND BIOPHYSICAL DRIVERS

Spatial heterogeneity in tree-cover change has a natural component: Amazonian primary productivity is maximum in seasonally dry, deep-rooted, evergreen, eastern forests [22]. Similarly, change in solar radiation, temperature, and precipitation, as well as terrain, soils, and nutrient deposition explain the location of tree-cover change in recent decades [16–21]. On the human side, forest conversion to agriculture is inversely related to precipitation, and directly to drought and fire [23–26].

This clear-cut evidence contrasts, on the human side, with the changing spatial impacts of deforestation drivers. Deforestation on a large scale started in the São Paulo Atlantic Forest and tree-dominated Cerrado by 1920, expanded hundreds of kilometers to the northwest by the mid-1940s [5,6] and started in Amazonia by 1970 with in-migrant settlers [7]. At the turn of the 21st century, however, Amazonia showed a migratory picture of near-equilibrium [12]. Road impacts used to vary spatially depending on distance to markets, presence of state-promoted settlements, climate, or road type [8–11]. But by the 1990s, deforestation was at a maximum around unpaved roads and in absence of state funding [8]. The most recent deforestation wave, apparently led by forest conversion into pastures [13], is also different from one area to another. This spatial heterogeneity might explain the very large uncertainty of general conclusions: cattle is said to explain between 70% and "little deforestation" [14,15].

To aid the analysis of tree-cover change models, the significant biophysical and socioeconomic drivers of forest-cover change are discussed in the forthcoming sections. But the interpretation can also be aided by grouping the drivers as part of three spatial processes. First, an attraction process, whereby living populations (trees, humans, and cattle) try to establish themselves depending on biophysical conditions that favor natural vegetation (forest and pasture) and attract humans who grow crops or pastures [82,83]. A trial-and-error search for adequate land use cues on climate, terrain, or soils, and leads to nonrandom land conversion by humans [27,28]. Attractiveness is a relatively new concept applied in land use change; it expresses the compounded value attached to a tract of land [81]. Second, a process of progressive access to forestland limits the search for attractive land; access is dependent on terrain, road, and river networks, as well as protected areas. Accessibility of forest or agricultural markets is a key variable in deforestation studies. Third, competition for similar land tracts among living populations is due to the fact that they are attracted by similar conditions (soil, nutrients, water, and solar radiation). In turn, economic uses of land, such as agriculture, cattle ranching, and human settlements, compete for land.

14.1.2 RECENT MULTIVARIATE MODELS

Multiple linear regressions using sampling fractions of up to 20% to cope with spatial autocorrelation (SA) have yielded important insights [10,29–31]. Refined sampling has excluded points in the autocorrelation zone (the area where stronger spatial autocorrelation appears) or stratified the study into more homogeneous areas. A handful of studies have used neighborhood covariates [32], moving windows or kernels [33,34] or spatially autoregressive models [5]. Large-scale models are still the exception [10,36] and none has used a solution to SA other than sampling.

14.2 DATA

14.2.1 TREE-COVER CHANGE

Tree cover (TC) is the percentage of a grid cell that is covered by trees. Our independent variable is the percentage point tree-cover change (TCC), obtained by subtracting 1992–3 AVHRR-based TC [37] from the 2001 MODIS TC [38]. Changes in cell land-cover over time, measured by image difference, or postclassification comparison, have repeatedly been found to be suitable [77–80]. These data sets are the only [available pair of observations] comparable through time across Brazil (Figure 14.1 and Table 14.1). They are composites of images taken at different seasons thus avoiding seasonality. AVHRR and MODIS were used in each epoch because of the difficulties in obtaining images with little cloud cover over such large areas. The same problems applied to the use of high-resolution images, plus the increased burden of processing higher resolution data.

Tree-cover change, unlike dichotomic deforestation, measures hot spots of forest-loss and tree-cover gain. For example, a tree-cover loss from 80% tree cover to 15%, that is, a 65% point loss, is overlooked by the concept of deforestation, which only accounts for change beneath the 10% tree-cover threshold. Nonclassified areas in 1992–3 are displayed in white; they seem correlated with steep slopes, bare soils, and absence of vegetation in 2001, but it was decided that imputations of nonclassified areas would have introduced errors. We visually verified, using several resolutions, that forest-loss attributable to narrow, isolated roads, is visible in our data sets in both 1992–3 and 2001.

Regarding classification accuracy, the question can be divided into three problems: (1) positional and (2) classification inaccuracies, and (3) the measurement of errors themselves. We relied on visual inspection of positional and classification adequacy (e.g., deforestation of road corridors, comparison of vector and raster rivers, comparison of settlements data sets) in several characteristic places such as the Amazon River, the Arc of Deforestation, and the municipio of Alta Floresta on which additional work has been done by us. There, both location and tree cover proved satisfactory. Further measures of positional error, such as the root mean square error, are difficult due to the very large expanse needed for a good estimate, as the error is certainly not the same throughout Brazil. It was decided not to impute the dependent variable where missing, as this required unavailable ancillary data or the use of the independent variables to model the dependent one, thus artificially augmenting their correlation and the fit of the model.

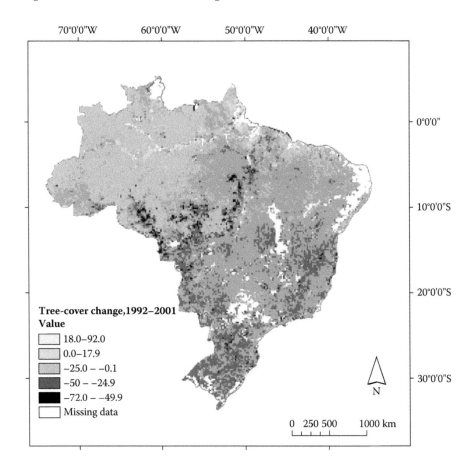

FIGURE 14.1   Map showing tree-cover change (%) in Brazil between 1992 and 2001.

14.2.2 Predictors

Most of our data sets are global, so similar studies can be replicated elsewhere. Geoprocessing generated 27 explanatory indicators. Some socioeconomic predictors are areal interpolations (cattle density, gross domestic product [GDP], and population density). We areally interpolated municipal in-migrants to avoid comparisons of municipalities with very different sizes, and to allay cross-level bias, that is, assigning the value of a unit to its subunits [43]. Some of our dependent variables have been assessed in the literature [44–46]. Consistency was visually verified for tree cover and slope; roads, rivers, and settlements; and several data sets of population and urban extents.

Considerable literature points to distance to human features as critical to deforestation. Unlike projections that faithfully represent the area, Lat Long preserves the distance between points, and so was considered suitable. The data were referenced to the WGS 1984 datum. Different resolutions in data were converted to common 20 km by 20 km and 60 km by 60 km cell resolution to avoid introducing more

TABLE 14.1

Data Sources

Variables, Units, Years	Original Resolution (at Equator)	Source
Dependent Variable		
Tree-cover change between 1992–3 and 2001 (% of each cell)	1 km (1992–3), 0.5 km (2001)	DeFries [37] and Hansen et al. [38]
Independent Variables, Biophysical		
Elevation (m)	1 km	USGS [47]
Slope (°)		
Soil total carbon (g/kg)	1:5000000	Batjes et al. [48]
Soil total nitrogen (g/kg)		
Soil cation exchange capacity (cmol₋/kg)		
Total precipitation [(mm/day)*10], 1981–1990	55 km	New et al. [49,50]
Mean temperature (°C*10), 1981–1990		
Cloud cover (%), 1981–1990		
Solar radiation (W/m²), 1981–1990		
Wet day frequency (days*10), 1981–1990		
Tree cover (% of each cell), 1992–3	1 km	DeFries [37]
Tree-cover change between 1992–3 and 2001, (% of an 8-cell neighborhood around each cell)	20 km	DeFries [37] and Hansen et al. [38]
Latitude, longitude (center of each cell)	20 km	DeFries [37]
Independent Variables, Administrative		
Natural protected areas (years-protection) during 1992–2001	1:1000000	WDPA Consortium [51]
Indigenous areas (years-protection) during 1992–2001	1:1000000	WDPA Consortium [51]
Independent Variables, Socioeconomic		
Population density (hab/km²), 1990	0.5 km	CIESIN CIAT [52]
Population density change 1990–2000 (hab/km²)		
Distance to nearest high population density point 1990 [decimal degree (dd)]		
In-migrants (persons/cell), 1990–2000	20 km	IBGE [12] and UMd [53]
Mean cattle density (10*heads/ha)	5.5 km	FAO [54]
Gross domestic product (purchasing power parity)	0.5 km	Sutton and Costanza [55]
Independent Variables, Accessibility		
Distance to nearest road (dd), 1999	1:1000000	IBGE [56]
Road density (dd), 1999		
Road dispersal in subcell quadrants [0–4 scale], 1999		

precision than actually exists in our coarsest resolution layers (soil conditions). This is in line with 50 km and 8 km resolution models [10,22]. Only the 60 km or 0.5 decimal degrees resolution is presented for the GWR, a resolution often used in earth observation data sets, therefore of interest to a wider audience.

14.2.3 STUDY AREA AND REGIONS

A regional approach to control for spatial heterogeneity of tree-cover change processes is hinted at by stratified approaches, or split approaches. Not until 2005 did collaborative efforts lead to a definition of Amazonia [57]. We define Brazilian Amazonia based on tree-cover data, and for a stable reference, the Amazon basin and adjacent catchments [58]. The 50% tree-cover threshold filtered out most of the Cerrado. The Cerrado, a savannah-forest continuum, is 70% Cerrado sensu stricto with a tree cover of 10% to 60%, at times with closed canopy [59]. Amazonia thus includes savannah islands and extends as forests into the Cerrado, Pantanal, and Caatinga biomes [60–62].

In this chapter, we also try to locate the Arc of Deforestation (AoD) accurately; first documented in 1990 [63] but with still uncertain location and shape. Here, the AoD is defined by all hot spots of forest loss (tree-cover change ≥50% in 1992–2001) in Brazil. These all appear to be within Amazonia (Figure 14.2). The AoD joins all hot spots, with a few narrow strings of lower deforestation tending to be absorbed by the AoD. With a 50% tree-cover threshold, the outline of Amazonian forest in Brazil had a good agreement with other data sources [39,40]. The AoD spatially coincided with the forest-Cerrado ecotone [41,42]. A measure of patchiness, the difference between a cell and its neighbors, showed that in the AoD tree cover was patchier than

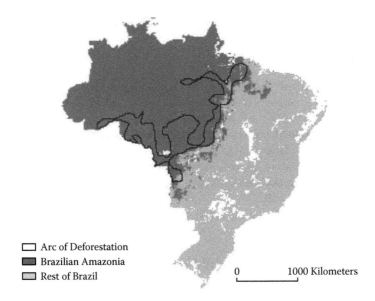

□ Arc of Deforestation
▇ Brazilian Amazonia
▢ Rest of Brazil

0   1000 Kilometers

FIGURE 14.2 Map identifying regions within the Brazil study area, highlighting the Arc of Deforestation.

other parts of Amazonia, suggesting a mixture of land uses and secondary forest of uneven ages. The southern border of the AoD was the limit of Amazonian forest in 1992. Nine years later, the forest was confined to the Amazon basin.

14.3 METHOD

Drivers of forest change are interlinked, so multiple linear regressions were necessary to measure net effects (i.e., control for the presence and level of other predictors). We simultaneously dealt with homogenizing processes among nearby areas and heterogenizing processes as distance augments.

14.3.1 DEALING WITH SPATIAL AUTOCORRELATION (SA) AND SPATIAL VARIABILITY

Spatial autocorrelation (SA) expresses the similarity of units, due to interaction and diffusion, and is usually stronger as distance decreases. SA occurs naturally but contradicts the assumption of independent cases in regressions and thus provokes biased estimates [43,64,65]. To deal with SA, virtually all current regression models, for example cokriging, spatial autoregressive models (SAR), and GWR, include neighborhood covariates or kernels. All these models are applications of the widely known multiple linear regression (MLR) [66,67]. Still, sampling is commonly used to augment distance between observations, and reduce SA. To identify the differences between these approaches, sampling, SAR, and GWR were compared.

Local variability, intrinsic to the phenomena under study, cause variations in the sign and strength of a regression coefficient. Global models (e.g., MLR or SAR) estimate an average net effect over the study area, which may be unobservable anywhere. This can be improved by regionally split SAR. Alternatively, GWR fit local estimates for each predictor and thus uncover their spatial impact patterns.

14.3.2 WHY GWR

Some regression methods that deal with SA have disadvantages. Cokriging assumes stationarity (i.e., a predictor's effect is constant across space), and it can in practice be replaced by MLR [67]. SAR models do not map varying regression estimates. Other approaches (spatial expansion method, spatially adaptive filtering, multilevel modeling, random coefficient models, kriging) are reviewed by Fotheringham et al. [66]. Three special issues of journals highlight weighting, local estimates, scale, and residuals among recent developments in spatial modeling [68–70], which arguably are already echoed in GWR.

In GWR, every point i is estimated on its own value and that of neighbors in proportion to their proximity [71]:

$$y_i = \beta_0(i) + \beta_1(i)\, x_{1i} + \beta_2(i)\, x_{2i} + \dots + \beta_n(i)\, x_{ni} + \varepsilon_i \qquad (14.1)$$

With n points to estimate, β has n sets of local parameters and is a function of X, Y, and an n by n spatial weighting matrix of distances. GWR is a continuous surface model whose results are maps showing curves of equal influence of explanatory variables.

14.3.3 CALIBRATING GWR

The first step in the GWR process is to fit a global (MLR) model, and then calibrate the local estimates (Figure 14.3). Calibrating refers to optimizing the bias-variance trade-off of the model by controlling the neighborhood size around each cell. Bias is defined as the difference between the estimate at any given cell and the estimate based on its neighbors. As neighboring observations are brought in to calculate each local estimate, bias is introduced but the estimate variance decreases [66]. A small bias is preferable to an average centered on the true value but with a variance such that the estimate might be quite far from the true value. In other words, it is more probable to be near the real value with a slightly biased GWR estimate than with a largely variant estimate. In addition, too small a neighborhood yields unpredictable results and failed estimation can occur [72]. This optimization is automated in the GWR software (Figure 14.3). However, in the presence of local collinearity in at least one location, the local estimates become unstable and overly large in absolute values. The solution is either to skip a variable or to modify the neighborhood size [73]. The options selected for neighborhood-size control are as per Figure 14.3.

Figure 14.3 shows the rationale of GWR calibration:

- The automated options optimize the bias-variance trade-off (Figure 14.3a) by minimizing the cross validation score (CV) or the corrected Akaike information criterion (AICc). CV is the sum of squared predicted errors obtained for each combination of variance and bias [66]; thus the CV should be minimized. The CV is a leave-one-out CV, omitting the cell around which the neighborhood is centered, so as to avoid an estimate based only on the center cell. This is imperative as otherwise the model "wraps itself" around each cell with $R^2 = 1$ but with null explanation as the local estimate equals the local intercept [71]. This leave-one-out CV is similar to the 8-neighbors covariate used in SAR. The AICc (AIC corrected for a sample) performs a

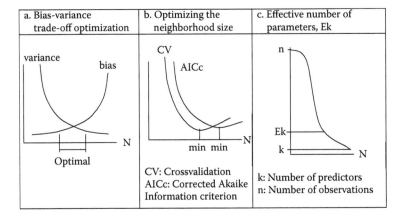

FIGURE 14.3 Calibration of the neighborhood size, N, in geographically weighted regression (GWR) [71,73].

similar task but also measures the penalty for having a model more complex than MLR. CV and AICc results were compared.

- In the presence of local multicollinearity, it is best to graphically identify the minima for CV or AICc (Figure 14.3b). In our case, local estimates with automated optimization were very high in absolute value, so manual control of neighborhood size and graphical identification of optimal neighborhood size was used.
- Effective number of parameters, Ek. This is the number of predictors k in MLR, to which the number of neighbors is added in GWR. This is an interesting statistic to look at as $k < Ek < n$, where n is the number of observations (Figure 14.3c). As the neighborhood size N diminishes, Ek tends to k and it becomes an MLR. As N augments, Ek tends to n, and the model becomes unreliable as each local estimate is calculated on a large sample with large variance.

14.3.3.1 Measuring GWR Improvement over MLR

Four indicators help ascertain GWR improvement over MLR: the explained variance, an AICc rule, an Ek rule, and an ANOVA F test. The AICc is used to penalize the increased complexity of GWR compared to MLR. An AICc difference of at least 3 corroborates a noticeable difference between models, and the lowest AICc shows which model has the least distance to the true distribution [66]. An Ek substantially lower than the number of observations dispels overfitting concerns. An ANOVA tests whether GWR improves on MLR [72].

14.3.3.2 Testing Spatial Nonstationarity

The next step in GWR is to test for variables that do not vary spatially. This avoids mapping of spatially stationary variables. Some available options are computer-intensive with long runtimes (Monte Carlo tests), some only have rough rules of thumb (SE of local estimate/SE of global estimate), or are local and do not provide an idea for the variable as a whole (local t-values). This leaves the interquartile range as an easy, albeit approximate, test whereby (2*SE (global estimate)) < (interquartile range of local estimate) indicates a spatially nonstationary relationship between the predictor and the dependent variable [66]. Under nonstationarity, the range of observed local estimates is larger than the expected range when all estimates are equal (as in the global estimate). The use of 68% of the expected distribution versus 50% of the observed estimates gives a safety margin to infer possible spatial variation of a predictor. This is illustrated in Figure 14.4.

14.4 RESULTS

The results dealt with the consequences of sampling, the identification of forest-cover change drivers, and the measurement of their relative impact. In interpreting the outcome of the models, the use of the terms "protective and risk effects" was justified, based on the epidemiological usage that the presence of a factor may worsen (risk factor) or ameliorate (protective factor) a condition. This is preferable to

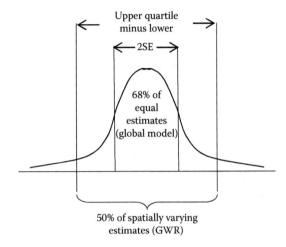

FIGURE 14.4 Interquartile range approximate test of spatial nonstationarity: interquartile range versus range of two standard errors.

"positive and negative effects," as a positive effect on deforestation may mean either an improvement in forest cover or an increase in deforestation.

14.4.1 AVERAGE EFFECTS IN BRAZIL

Low sampling fractions reduce SA [74], but small samples (low sampling fractions combined with coarse resolution) were more likely to produce inconsistencies in a set of 16 models. This primarily affected the sign of variables, which were unevenly distributed over space (paved roads, GDP). Moreover, the bias in MLR predicted by the literature was confirmed by the difference between each MLR and SAR estimate (Figure 14.5). Samples (thin lines) behave like 100% models, and differ from SAR models. Hence, it seems sampling did not relieve bias, but the SA covariate did. Whereas sampling eliminated SA by artificially creating distance between observation points, SAR views SA as a real phenomenon.

Therefore, the SAR model better represents the average effects over Brazil than the MLR. The strongest effects are tree cover in 1992 (risk effect) and neighborhood effects (protective effect). Tree-cover change is not independent from previously existing vegetation, which is targeted to be depleted. Noticeably, latitude (positive) and longitude (negative) are consistent with the historical southeast–northwest trend of Brazil's colonization and urbanization (distance to cities is a protective effect). [Solar radiation enhances plant growth, attracting humans for slash-and-burn, such that the effect of radiation is negative; it denotes areas with seasonal drought and less cloud cover.] A feedback is likely whereby deforestation reduces solar radiation absorption. A strong and significant effect of ln (natural logarithm) (cattle density) was found in our linear model; this suggested that our original variable (when ln is removed) had an exponential effect on tree-cover change. As to the pattern of slope, it corresponds to the south–north Brazilian orography profile: deteriorated Atlantic

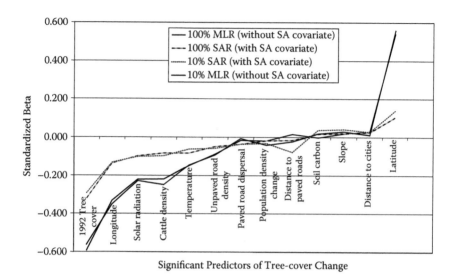

FIGURE 14.5   Multiple linear regression (MLR) and spatial autoregressive models (SAR): significant predictors of tree cover and their relative effects.

forests occupy areas with steep slopes, savannah and woodland are in the highlands, and flatter lowlands coincide with Amazonia.

14.4.2 REGIONALLY SPLIT SAR

Significant variables are the same in Amazonia, the AoD, and Southern Brazil, suggesting similar processes. However, these regional impacts varied in strength, as shown by comparing them relative to those measured in Amazonia. For instance, a value lower than one indicates a lesser impact than in Amazonia (Figure 14.6). Also, Southern Brazil seems the most sensitive region. Impacts have similar strength in Amazonia and the AoD. The largest differences between Amazonian forest and Southern Brazil are temperature and latitude; they also differ in sign. More impacts have the same sign, with slope standing out in Southern Brazil. Opposite signs show that the farther north in Amazonia, [the better for tree-cover change,] whereas in Southern Brazil, the farther north [the worse for tree cover.] Distance to cities is a protective factor in Amazonia, but a detrimental one in Southern Brazil. Distance to paved roads is a risk factor in Amazonia but has a weak effect in Southern Brazil. Finally, higher temperature is a risk factor in Amazonia but a protective one in Southern Brazil. This is because higher temperatures in Southern Brazil occur in less attractive dry or semiarid places.

Regionally split models imply problematic assumptions: first, homogeneity of processes within regions; second, interaction and diffusion causing SA occur within, not among regions; and third, interactions and diffusion exist only within 100 km from each point (within eight nearest neighbors). In the next section, GWR removes these assumptions by accounting for heterogeneity without preconceived regions and optimizing neighborhood size.

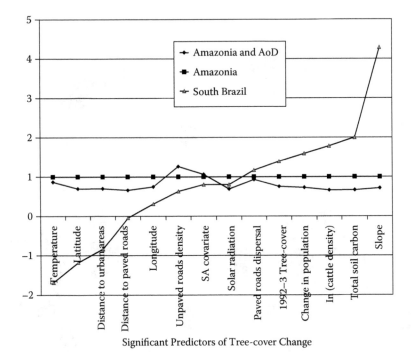

FIGURE 14.6 Regional differences in tree-cover change impacts (ratios of SAR standard-ized betas, relative to Amazonia).

14.4.3 GWR Spatial Patterns

A GWR without latitude, longitude, and the SA covariate, deemed redundant, was optimal using 358 nearest neighbors. Goodness-of-fit improved from 0.604 (MLR) to 0.818 (GWR). GWR is closer to the true distribution (AICc = 18594 vs. 20675 for MLR), and significantly improves on MLR (F0.95,180,2723 = 20.11, p < 0.001). The effective number of parameters, Ek = 192, well below the number of cells (n = 2913), dispels overfitting. An interquartile test for nonstationarity [66] showed that all significant regional SAR predictors varied spatially. The main GWR outputs are 11 maps of net local effects of each predictor (Figures 14.7, 14.8, and 14.9). The interpretation is in terms of attraction and access processes, and outcome for living populations (trees, humans, and livestock).

14.4.3.1 Attractiveness

Biophysical variables indicate agricultural suitability, that is, attractiveness. They overlap the AoD to different degrees (Figure 14.7). Slope effect is most acute near the Amazon, Tocantins, and adjacent river systems' estuaries. The steepest slopes (near the Guyana shield in northern Amazonia, and Atlantic forest in southeast Brazil) seem to protect tree cover. Temperature's impact seems weak in general, except in the southwestern AoD, quite opposite to the northeastern AoD. Soil carbon has a still smaller effect and is very localized in the central part of the AoD, northern

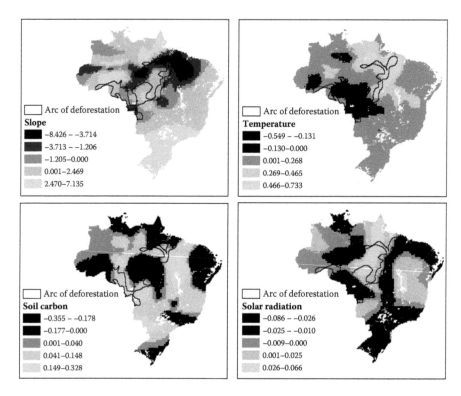

FIGURE 14.7   Geographically weighted regression (GWR) beta estimates for attractiveness/biophysical predictors of tree-cover change.

Amazonia, and the Nordeste shrubland. In the latter, soil may be very erodible or plant growth quite slow. Finally, the small range of solar radiation impact [helps comprehend] a surprising pattern for a variable expected to change gradually over space. Gradients are much sharper than in previous variables: extreme values even come in contact with one another. Solar radiation is higher wherever seasonal droughts occur and could be affected by deforestation causing higher albedo.

14.4.3.2 Accessibility

A negative effect is given by distance to urban extents (Figure 14.8). This is stronger in eastern Brazil, historically the area of early colonial settlements and now a metropolis, as well as in settlements in the southern portion of the AoD. A stronger effect is distance to paved roads with a pattern in eastern Brazil similar to distance to cities, albeit more localized. In Amazonia, the effect is almost neutral except perhaps around Santarem, the capital of Para. Paved road dispersal and density of unpaved roads both present an unexpected pattern: one of the remotest areas in westernmost Amazonia seems particularly sensitive to extant roads, despite still largely untouched forest tracts. This sensitivity could mean that if paved roads become any more disperse, and unpaved roads any more dense, they may have severe impacts there. The local importance of unpaved road density is well known, as agricultural landholdings develop along fishbone patterns;

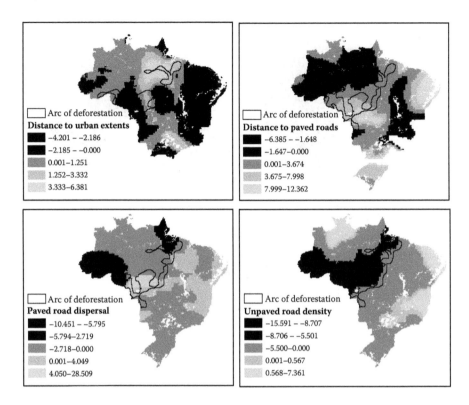

FIGURE 14.8 Geographically weighted regression (GWR) beta estimates for accessibility predictors of tree-cover change.

regionally, the density of unpaved roads has the second closest match with the shape of the AoD, with only cattle density showing a higher predictive value.

14.4.3.3 Living Populations

The best match with the AoD is given by cattle density, which has an exponential effect on tree-cover loss. Initial tree cover is a close match for the southern and eastern limits of the AoD, with two marked zones of higher impact: southern urban markets, [and in Western Amazonia a very sensitive zone,] regarded among the best preserved in 1992 (Figure 14.9). This zone coincides with [that shown by roads,] and this sensitivity could prefigure future rather than actual impacts. Possible causes for an artifact of GWR smoothing or interpolation are discussed in Section 14.5. Quite straightforward though is the impact of cattle density, which predicts accurately the shape of the AoD. Despite the absence of data on other land uses, this shows the exponential impact of cattle. Finally, change in human population density shows agreement with the shape of the AoD, but is offset to the north. Roads, initial tree cover, and increased human density are also offset in this fashion. It is possible that tree-cover change is time lagged compared to these variables showing humans settling north of the AoD, and that tree cover will suffer as a consequence in the coming years.

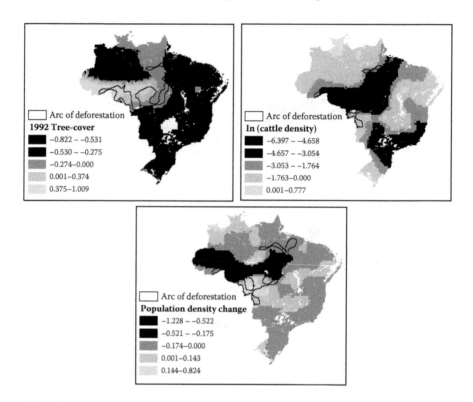

FIGURE 14.9   Geographically weighted regression (GWR) beta estimates for living populations as predictors of tree-cover change.

14.5 DISCUSSION

14.5.1 SAMPLING ISSUES

A basic principle of sampling is to avoid bias, defined as a systematic error by selecting one outcome more often that others, based on the characteristics of the observed objects or the observer's choice [84]. As a result of bias, the difference of a sample estimate to the population value becomes larger, that is, accuracy decreases [85]. We found that sampling, while reducing spatial autocorrelation, does not suppress bias. In summary, of 16 models that explored sensitivity to SA and to resolution, the 60 km resolution was not the best solution when combined with sampling. Sampling is not a reliable solution to SA. The combined effect of low sampling fractions and very coarse resolution produces small samples, hence large variation in coefficients or sign. There is also higher heterogeneity within each aggregated grid cell. However, a 20 km resolution rapidly becomes a heavy computational load.

But sampling omits context as well. This makes it unlikely to identify regional patterns such as that of enhanced tree-cover along Amazonian rivers as identified in our data set. Simple random sampling seems very likely to omit isolated features, such as roads, or a hot spot of deforestation in a patchy sample from a broad region [86]. Analyzing tree-cover change in Brazil as a whole avoided bias in the selection of the study site. However, this did not account for spatial heterogeneity, and SAR and GWR had to be carried out.

Methodologically, further work has to be carried out to fully assess the ineffectiveness, identified here, of sampling as the usual device to deal with spatial autocorrelation. Some relatively modest effort is needed to distinguish the effect on current knowledge of case studies' site-selection bias. This could easily be done by mapping known land-use or land-cover change studies onto the GWR maps obtained here. But it can already be surmised that generalizations based on too few case studies, clustered in space, are only valid for those areas. Much caution is therefore needed when generalizations are attempted on case studies or meta-analyses, or reviews are carried out without previous knowledge of overly clustered case studies and in the presence of spatial heterogeneity. Possible uses of GWR maps in this setting could take place before site selection or during analyses to measure the departure of a local site from its context.

14.5.2 GWR GRANULARITY AND SMOOTHING

Despite the many advantages of GWR in this application, we mention several possible drawbacks of GWR as used here. Granularity, defined as how fine the spatial variation of a phenomenon actually is, is assumed by GWR to be constant for all the predictors; that is, the size of the relevant neighborhood is assumed to be the same for all the drivers. Had we not included the rest of Amazonia as a context, it is possible that the use of a fixed-sized neighborhood would have been ill-adapted to model a stripe-shaped area like the AoD.

Other potential issues are possible artifacts of GWR smoothing or interpolation due to the use of large neighborhoods. Almost all the GWR maps displayed transition areas between the largest negative and positive effects, some of which could be attributable to a smoothing effect. However, this did not happen for solar radiation, despite the fact that this was expected to be a gradually varying variable.

GWR maps can also be seen as maps of susceptibility to change, and their coefficients give a clear measure of impact per unit of change in the explanatory variable. But explaining the intriguing patterns seems to require a good deal of local and regional knowledge. In particular, initial tree cover was shown to be the most important predictor in our model, and it presented an unexpected [dark shaded patch] in northwestern Amazonia, an area hitherto known as relatively untouched. Satellite data of late 2006 and early 2007 show that these are the areas that show the most marked primary productivity decrease, compared with the 1998–2004 average [87]. This area has also known the steepest decline in rainfall in recent years in Brazil [88]. The area may also be very sensitive to deforestation, as its vegetation often occurs on sandy soils [89]. It seems, therefore, that some GWR patterns, to be validated, require considerable information. But these maps also convey essential hypotheses about tree-cover change at the regional scale.

14.6 CONCLUSIONS

We carried out the first Brazil-wide models of tree-cover change. Successive methods helped lift assumptions and reduce estimation bias. There was a clear consistency between the models. In particular, the bias entailed by spatial autocorrelation in multiple linear regression did not prevent the adequate rank order of effects from being identified from this type of regression.

GWR was a technique new to the deforestation field. GWR helped cope with spatial interaction (i.e., spatial autocorrelation) and account for spatial structure (i.e., spatial variation). GWR was also an empirical alternative to deal with high variability in forest-deterioration situations that caused heteroskedasticity in MLR and SAR models. GWR did not partition the space into submodels, a solution that has conceptual and practical drawbacks, because it considers space as disjoint and rapidly encounters problems of having to fit many models with too few cases, especially if sampling is used to control for SA. On the contrary, GWR views space as continuous and does not imply *a priori* segmentation; neither does GWR assume homogeneous effects of each predictor within a region.

GWR addressed Tobler's law, which states that all things are similar, but things that are close to one another are more similar [90]. Similarity is the product of spatial interactions, found to take place in an area 358 times larger than an individual observation. Although this is a very large area, it minimizes the predicted errors and optimizes the trade-off of bias versus variance. It is also a nonarbitrary measurement of interaction in the tree-cover change layer. In addition, the spatial patterns of impacts on forests unveiled here seem one possible response to Simpson's paradox, whereby local results are reversed when pooled together. At this point we cannot conclude that a spatially varying model is always better than a global one, since this question involves a hierarchy of levels of analysis and the bottom-up versus top-down direction of effects, which might be better identified by complementary approaches.

Here GWR was used as a pattern-recognition tool. Further work will be carried out in substantial hypothesis testing as well. For instance, regarding the future of Brazilian forests, GWR can help answer a typical question linked to protected areas: Is compliance/enforcement of protection status more important than protected-area remoteness and agricultural unsuitability?

REFERENCES

1. FAO, Global Forest Resources Assessment 2000 Main report, Forestry Paper No. 140, 2001.
2. FAO, Global Forest Resources Assessment 2005, Progress towards sustainable forest management, Rome, 2005.
3. Schellnhuber, H. J., Coping with Earth system complexity and irregularity, in *Challenges of a Changing Earth: Proceedings of the Global Change Open Science Conference, Amsterdam, The Netherlands, 10–13 July 2001*, Steffen, W., Jager, J., Carson, D., and Bradshaw, C., Eds., Springer, Berlin, 2002.
4. Easterling, W. E. and Kok, K., Emergent properties of scale in global environmental modeling—Are there any?, *Integrated Assessment*, 3(2–3), 233, 2002.
5. Brannstrom, C., Coffee labor regimes and deforestation on a Brazilian frontier, 1915–1965, *Econ. Geogr.*, 76, 326, 2000.
6. Dean, W., *With Broadax and Firebrand: Destruction of Brazilian Atlantic*, University of California Press, Berkeley, 1995.
7. Fearnside, P. M., Land-tenure issues as factors in environmental destruction in Brazilian Amazonia: The case of Southern Pará, *World Dev.*, 29, 1361, 2001.

8. Câmara, G., Aguiar, A. P. D., Escada, M. I., Amaral, S., Carneiro, T., Monteiro, A. M. V., Araújo, R., et al., Amazonian deforestation models, *Science*, 307, 1043c, 2005.

9. Laurance, W. F., Fearnside, P. M., Albernaz, A. K. M., Vasconcelos, H. L., and Ferreira, L. V., Response to "Amazonian deforestation models", *Science*, 307, 1044, 2005.

10. Laurance, W. F., Albernaz, A. K. M., Schroth, G., Fearnside, P. M., Bergen, S., Venticinquez, E. M., and Da Costa, C., Predictors of deforestation in the Brazilian Amazon, *J. Biogeo.*, 29, 737, 2002.

11. Nelson, G., De Pinto, A., Harris, V., and Stone, S., Land use and road improvements: A spatial perspective, *Int. Reg. Sci. Rev.*, 27, 297, 2004.

12. IBGE, Saldo e principais fluxos migratórios 2000, in *Atlas do censo demográfico 2000*, Instituto Brasileiro de Geografia e Estatística, 2003, available at: http://www.ibge.gov.br/ home/estatistica/populacao/censo2000/atlas/pag058.pdf, accessed October 2005.

13. Cardille, J. A. and Foley, J. A., Agricultural land-use change in Brazilian Amazônia between 1980 and 1995: Evidence from integrated satellite and census data, *Rem. Sens. Env.*, 87, 551, 2003.

14. Cerri, C. E. P., Paustian, K., Bernoux, M., Victoria, R. L., Melillo, J. M., and Cerri, C. C., Modeling changes in soil organic matter in Amazon forest to pasture conversion with the Century model, *Global Change Biol.*, 10, 815, 2004.

15. Walker, R. and Qi, J., Forest cover for South America, Michigan State University's LBA Team FTP, 2002, available at: ftp://www.marajo.geo.msu.edu/lba/, accessed October 2005.

16. Baker, T. R., Phillips, O. L., Malhi, Y., Almeida, S., Arroyo, L., Di Fiore, A., Erwin, T., et al., Increasing biomass in Amazonian forest plots, *Phil. Trans. R. Soc. Lond. B, Biol. Sci.*, 359, 353, 2004.

17. House, J., Prentice, I. C., Ramamkutty, N., Houghton, R. A., and Heimann, M., Reconciling apparent inconsistencies in estimates of terrestrial CO_2 sources and sinks, *Tellus B*, 55, 345, 2003.

18. Malhi, Y., Baker, T. R., Phillips, O. L., Almeida, S., Alvarez, E., Arroyo, L., Chave, J., et al., The above-ground coarse wood productivity of 104 Neotropical forest plots, *Gloal. Change Biol.*, 10, 563, 2004.

19. Nemani, R. R., Keeling, C. D., Hashimoto, H., Jolly, W. M., Piper, S. C., Tucker, C. J., Myneni, R. B., et al., Climate-driven increases in global terrestrial net primary production from 1982 to 1999, *Science*, 300, 1560, 2003.

20. Swap, R., Garstang, M., Greco, S., Talbert, R., and Kallberg, P., Saharan dust in the Amazon Basin, *Tellus B*, 44, 133, 1992.

21. Renck, A. and Lehmann, J., Rapid water flow and transport of inorganic and organic nitrogen in a highly aggregated tropical soil, *Soil Science*, 169(5), 330, 2004.

22. Potter, C. S. et al., Regional application of an ecosystem production model for studies of biogeochemistry in Brazilian Amazonia, *Global Change Biol.*, 4, 315, 1998.

23. Chomitz, K. M. and Thomas, T. S., Determinants of land use in Amazonia: A fine-scale spatial analysis, *Am. J. Agric. Econ.*, 85, 1016, 2003.

24. Laurance, W. F., Forest-climate interactions in fragmented tropical landscapes, *Phil. Trans. R. Soc. Lond. B, Biol. Sci.*, 359, 345, 2004.

25. Nepstad, D., Lefebvre, P., Da Silva, U. L., Tomasella, J., Schlesinger, P., Soloranzo, D., Mountinho, P., et al., Amazon drought and its implications for forest flammability and tree growth: A basin-wide analysis, *Global Change Biol.*, 10, 704, 2004.

26. Steininger, M. K., Tucker, C. J., Townshend, J. R., Killeen, T. R., Desch, A., Tropical deforestation in the Bolivian Amazon, *Env. Conserv.*, 28, 2001.

27. Carpenter, F. L., Mayorga, S. P., Quintero, E. G., and Schroeder, M., Land-use and erosion of a Costa Rican Ultisol affect soil chemistry, mycorrhizal fungi and early regeneration, *Forest Ecol. Manag.*, 144, 1, 2001.

28. Lomolino, M.V. and Perault, D.R., Geographic gradients of deforestation and mammalian communities in a fragmented, temperaterain forest landscape, *Global Ecol. Biogeogr.*, 13, 55, 2004.

29. Mertens, B., Poccard-Chapuis, R., Piketty, M.-G., Lacques, A.-E.,Venturieri, A., Crossing spatial analyses and livestock economics to understand deforestation processes in the Brazilian Amazon: The case of São Félix do Xingu in south Pará, *Agric. Econ.*, 27, 269, 2002.

30. Stolle, F., Chomitz, K. M., Lambin, E. F., and Tomich, T. P., Land use and vegetation fires in Jambi Province, Sumatra, Indonesia, *Forest Ecol. Manag.*, 179, 277, 2003.

31. Verburg, P. H., Overmars, K. P., and Witte, N., Accessibility and land-use patterns at the forest fringe in the northeastern part of the Philippines, *Geo. J.*, 170, 238, 2004.

32. Geoghegan, J., Villar, S. C., Klepeis, P., Mendoza, P. M., Ogneva-Himmelberger, Y., Chowdhury, R. R., Turner, B. L., et al., Modeling tropical deforestation in the southern Yucatan peninsular region: Comparing survey and satellite data, *Agric. Ecosys. Env.*, 85, 25, 2001.

33. Munroe, D. K., Southworth, J., and Tucker, C. M., Modeling spatially and temporally complex land-cover change: The case of western Honduras, *Profess. Geo.*, 56, 544, 2004.

34. Schneider, L. C. and Pontius, R. G. J., Modeling land-use change in the Ipswich watershed, Massachusetts, USA, *Agric. Ecosys. Env.*, 85, 83, 2001.

35. Bucini, G. and Lambin, E. F., Fire impacts on vegetation in Central Africa: A remote-sensing-based statistical analysis, *Appl. Geo.*, 22, 27, 2002.

36. Grainger, A., Francisco, H. A., and Tiraswat, P., The impact of changes in agricultural technology on long-term trends in deforestation, *Land Use Pol.*, 20, 209, 2003.

37. DeFries, R. S., Hansen, M. C., Townshend, J. R., Janetos, A. C., and Loveland, T. R., A new global 1-km data set of percentage tree cover derived from remote sensing, *Global Change Bio.*, 6, 247, 2000.

38. Hansen, M., DeFries, R. S., Townshend, J. R., Carroll, M., Dimiceli, C., Sohlberg, R. A., Global percent tree cover at a spatial resolution of 500 meters: First results of the MODIS Vegetation Continuous Field algorithm, *Earth Interact.*, 7, 2, 2003.

39. Eva, H. D., A vegetation map of South America, Office for Official Publications of the European Communities, Luxembourg, 2002.

40. Walker, R., Moran, E., and Anselin, L., Deforestation and cattle ranching in the Brazilian Amazon: External capital and household processes, *World Dev.*, 28, 683, 2000.

41. WWF, Biomas Brasileiros, 2006, available at: http://www.wwf.org.br/natureza_brasileira/biomas/index.cfm, accessed May 2006.

42. INPE, CPTEC-ProVeg, Mapa da Vegetação da Amazônia Legal, 2006, available at: http://www.cptec.inpe.br/proveg/ areaest.shtml, accessed May 2006.

43. Anselin, L., Under the hood: Issues in the specification and interpretation of spatial regression models, *Agric. Econ.*, 27, 247, 2002.

44. White, M., Shaw, J., and Ramsey, R., Accuracy assessment of the vegetation continuous field tree cover product using 3954 ground plots in the south-western USA, *Int. J. Rem. Sens.*, 26, 2699, 2005.

45. Doll, C. N., Muller, J.-P., and Morley, J.G., Mapping regional economic activity from night-time light, *Ecol. Econ.*, 57, 75, 2006.

46. Malhi, Y. and Wright, J., Spatial patterns and recent trends in the climate of tropical rainforest regions, *Phil. Trans. R. Soc. Lond. B, Bio. Sci.*, 359, 1443, 2004.

47. USGS, GTOPO30, Global 30-arc second digital elevation model, Sioux Falls, 1996.

48. Batjes, N., Bernoux, M., and Cerri, C. E., Soil data derived from SOTER for studies of carbon stocks and change in Brazil, version 1.0, GEFSOC Project, Technical Report 2004/03, ISRIC-World Soil Information, Wageningen, 2004.

49. New, M., Hulme, M., and Jones, P., Representing twentieth-century space-time climate variability. Part I: Development of a 1961–90 mean monthly terrestrial climatology, *J. Clim.*, 12, 829, 1999.

50. New, M., Hulme, M., and Jones, P., Representing twentieth-century space-time climate variability. Part II: Development of 1901–96 monthly grids of terrestrial surface climate, *J. Clim.*, 13, 2217, 2000.
51. WDPA Consortium, World database on protected areas, IUCN and UNEP-WCMC, 2005.
52. CIESIN-CIAT, Gridded Population of the World, version 3.0 beta, Columbia University, Palisades, NY, 2004.
53. CIESIN, Brazil's localities vector data, Socioeconomic data and applications center, Center for International Earth Science Information Network, 2005, available at: ftp:// ftp.ciesin.org/, accessed January 2006.
54. FAO, Cattle density for Latin America. Edition 1.0. United Nations Food and Agriculture Organization's Animal Production and Health Division, FAO-AGA, Rome, 2003.
55. Sutton, P. and Costanza, R., Global estimates of market and non-market values derived from nighttime satellite imagery, land cover, and ecosystem service valuation, *Ecol. Econ.*, 41, 509, 2002.
56. IBGE, Base Cartográfica Integrada do Brasil ao Milionésimo Digital, Instituto Brasileiro de Geografia e Estatística, 1999.
57. Peres, C., personal communication, 2006.
58. ANA, Agência Nacional de Águas. HidroWeb Sistema de Informações Hidrológicas, 2005, available at: http://hidroweb.ana.gov.br/, accessed January 2006.
59. Andrade, L. A., Felfili, J. M., and Violatti, L., Fitossociologia de uma área de cerrado denso na recor-Ibge, Brasília-DF, *Acta Bot. Bras.*, 16, 2002.
60. Prance, G. T., Islands in Amazonia, *Phil. Trans. R. Soc. Lond. B, Bio.* Sci., 351, 823, 1996.
61. Prance, G. T. and Schaller, G. B., Preliminary study of some vegetation types of the Pantanal, Mato Grosso, *Brazil, Brittonia*, 34, 228, 1982.
62. Anderson, A., Overal, W., and Henderson, A., Pollination ecology of a forest-dominant palm (Orbignya phalerata Mart.) in northern Brazil, *Biotrop.*, 20, 192, 1988.
63. Skole, D. L. and Tucker, C. J., Tropical deforestation and habitat fragmentation in the Amazon: Satellite data from 1978 to 1988, *Science*, 260, 1905, 1993.
64. Odland, J., *Spatial autocorrelation*, Sage, Newbury Park, CA, 1988.
65. Wrigley, N., Holt, D., Steel, D. G., and Tranmer, M., Spatial modelling and the ecological fallacy, in *Spatial analysis: Modelling in a GIS environment*, Longley, P. A. and Batty, M., Eds., GeoInformation International, Cambridge, 1996.
66. Fotheringham, A. S., Brunsdon, C., and Charlton, M., *Geographically weighted regression: The analysis of spatially varying relationships*, John Wiley & Sons, Chichester, 2002.
67. Lesch, S. M., Strauss, D. J., and Rhoades, J. D., Spatial prediction of soil salinity using electromagnetic induction techniques 1. Statistical prediction models: A comparison of multiple linear regression and cokriging, *Water Resour. Res.*, 31, 373, 1995.
68. LeSage, J., Pace, R., and Tiefelsdorf, M., Methodological developments in spatial econometrics and statistics, *Geo. Anal.*, 36, 87, 2004.
69. Verburg, P. H. and Veldkamp, A., Introduction to the special issue on spatial modeling to explore land use dynamics, *Int. J. Geo. Inf. Sci.*, 19, 99, 2005.
70. Veldkamp, A. and Verburg, P. H., Modelling land use change and environmental impact, *J. Env. Manag.*, 72, 1, 2004.
71. Fotheringham, A. S., GWR Workshop material, Nottingham, 2006.
72. Charlton, M., GWR Workshop material, Nottingham, 2006.
73. Brunsdon, C., GWR Workshop material, Nottingham, 2006.
74. de las Heras, A. and Lake, I. R., Modelling tree-cover change in the Brazilian Amazon and beyond, in *Proceedings of the GIS Research UK 14th Annual Conference, Nottingham*, Priestnall, G., Aplin, P., Eds., 2006.
75. Nakagawa, S., A farewell to Bonferroni: The problems of low statistical power and publication bias, *Behav. Ecol.*, 15(6), 1044, 2004.

76. Moran, M. D., Arguments for rejecting the sequential Bonferroni in ecological studies, *Oikos*, 100(2), 403, 2003.

77. Coppin, P., Jonckheere, I., Nackaerts, K., Muys, B., and Lambin, E., Digital change detection methods in ecosystem monitoring: A review, *Int. J. Rem. Sens.*, 25, 1565, 2004.

78. Lu, D., Mausel, P., Brondízio, E., and Moran, E., Relationships between forest stand parameters and Landsat TM spectral responses in the Brazilian Amazon Basin, *Forest Ecol. Manag.*, 198, 149, 2004.

79. Lu, D., Mausel, P., Batistella, M., and Moran, E., Land-cover binary change detection methods for use in the moist tropical region of the Amazon: A comparative study, *Int. J. Rem. Sens.*, 26, 101, 2005.

80. Mas, J. F., Monitoring land-cover changes: A comparison of change detection techniques, *Int. J. Rem. Sens.*, 20, 139, 1999.

81. Wu, F., SimLand: A prototype to simulate land conversion through the integrated GIS and CA with AHP-derived transition rules, *Int. J. Geo. Inf. Sci.*, 12, 63, 1998.

82. Pearce, D. and Brown, K., Saving the tropical forests, in *The causes of tropical deforestation: The economic and statistical analysis of factors giving rise to the loss of the tropical forests*, Brown, K. and Pearce, D., Eds., UCL Press, London, 1994.

83. Barbier, E. and Burgess, J., The economics of tropical deforestation, *J. Econ. Surv.*, 15, 413, 2001.

84. Schreuder, H. T., Gregoire, T. G., and Wood, G. B., *Sampling methods for multiresource forest inventory*, John Wiley & Sons, New York , 1993.

85. Husch, B., Beers, T. W., and Kershaw, J. A., *Forest mensuration*, John Wiley & Sons, Hoboken, NJ, 2003.

86. Tucker, C. J. and Townshend, J. R. G., Strategies for monitoring tropical deforestation using satellite data, *Int. J. Rem. Sens.*, 21, 1461, 2000.

87. FAO, Global Information and Early Warning System on Food and Agriculture, 2006, Available at: http://www.fao.org/giews/english/spot4/sam/index.htm, accessed December 2006 and February 2007.

88. Obregón, G., personal communication, June 2007.

89. Clevelario Junior, J., personal communication, June 2007.

90. Tobler, W. R., A computer movie simulating urban growth in the Detroit region, *Econ. Geo. Supplement: Proceedings of the International Geographical Union Commission on Quantitative Methods*, 46, 234, 1970.

15 GM(1,1)-Kriging Prediction of Soil Dioxin Patterns

Danni Guo, Renkuan Guo,
Christien Thiart, and Tonny Oyana

CONTENTS

OVERVIEW

Data to establish patterns of soil dioxin are not readily available. The costs of collection and analysis are high, and what becomes available is often insufficient for further analysis. Methods commonly used to construct contour maps from noisy data with spatial covariance include kriging, but feasibility depends on data quality and quantity, which may be too limited in practice. The number of observation points may be too small, and spatial covariance may be weak and difficult to model. In situations of this type, this chapter proposes that the mixed approach of combining gray differential equation models such as the GM(1,1) model with ordinary kriging can produce a GM(1,1)-kriging map that cannot be constructed by other contouring methods. The new approach can provide meaningful patterns even if the original spatial data sampling was poorly designed and resampling is not feasible because of time and cost constraints.

15.1 INTRODUCTION

Environmental data are costly and difficult to collect, and quite often the sampled data are insufficient for further analysis. Kriging is a commonly accepted spatial interpolation method. The feasibility of kriging analysis depends on data quality and quantity. The quality of data refers to the adequacy of the data spread for the assumed spatial prediction task. The quantity refers to the size of the sample and whether it is large enough. Today, we often face circumstances where a set of data is already collected, although from the viewpoint of kriging analysis the data are insufficient, and resampling is impossible because of cost and time limits. Therefore, a solution must be found. In this chapter, a mixed approach is achieved by combining gray differential equation models, particularly the GM(1,1) model, with an ordinary kriging approach. The combined approach is named GM(1,1)-kriging. The existing limited sample data available are expanded to produce a GM(1,1)-kriging map of a larger geographical area. The new approach addresses the issue of spatial data sampling design and provides improved spatial analysis results. This approach is illustrated using soil dioxin data collected from Midland County, Michigan.

15.2 EXAMPLE: SOIL DIOXIN IN MIDLAND COUNTY

Dioxins are a complicated family of chemicals that includes dioxins, furans, and poly-chlorinated biphenyls (PCBs) that have related properties and toxicity [1]. Dioxins are man-made chemical compounds that enter the air through fuel and waste emissions, including motor vehicle exhaust fumes and garbage incineration [2]. They are not deliberately manufactured, but are rather an unintended by-product of industrial processes that use or burn chlorine [1]. Dioxins are one of the most studied chemicals on the planet and are found throughout the environment and in our food supply. They cause a wide range of adverse health effects including cancer, birth defects, diabetes, learning and developmental delays, endometriosis, and immune system abnormalities, and represent the most potent animal carcinogens ever tested [1,3]. Most human exposure to dioxins occurs through the consumption of contaminated foods, especially animal fats [3].

Dioxin emissions from incinerators reach people. Dioxins enter the air and people breathe in the particles. But a bigger problem is that the particles settle on grazing land where cows eat the grass, and become concentrated in the fat of their meat and milk. Particles may also become concentrated in cattle and hogs that are fed dioxin-tainted grain and may be transferred through the hydrological cycle. The dioxin particles can fall directly into rivers, streams, and other bodies of water, or reach these waterways in surface water runoff. Particles settle on the bottom where fish and shellfish ingest small particles of sediment. Dioxin then builds up in their fat or organs [1,3].

The chapter uses soil dioxin data from Midland County, Michigan, to illustrate the proposed combined GM(1,1)-kriging method. The reported dioxin concentration is provided in toxic equivalents (TEQs). TEQs are used to report the toxicity-weighted masses of mixtures of polychlorinated dibenzo-p-dioxins and furans (PCDD/Fs) [4]. The measurement units are nanograms per kilogram (ng/kg), which is parts per trillion (ppt). Soil dioxin samples were collected in Midland City and along the Tittabawassee River (Figure 15.1). These soil dioxin samples are concentrated in a small area of Midland County. In this particular case, due to the spread of the sample data, a kriging map can be produced; however, it can only cover a small area around the samples (Figure 15.2). In order to predict dioxin levels for the entire Midland County, more samples are needed. Since sampling is a lengthy and expensive as well as time-consuming process, other options can be considered. In this case, gray predictions can be used to add to the sample data.

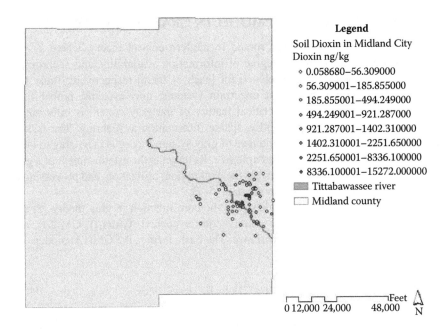

Legend

Soil Dioxin in Midland City
Dioxin ng/kg

- 0.058680–56.309000
- 56.309001–185.855000
- 185.855001–494.249000
- 494.249001–921.287000
- 921.287001–1402.310000
- 1402.310001–2251.650000
- 2251.650001–8336.100000
- 8336.100001–15272.000000

▨ Tittabawassee river
☐ Midland county

0 12,000 24,000   48,000 Feet N

FIGURE 15.1   Soil dioxin samples in Midland County.

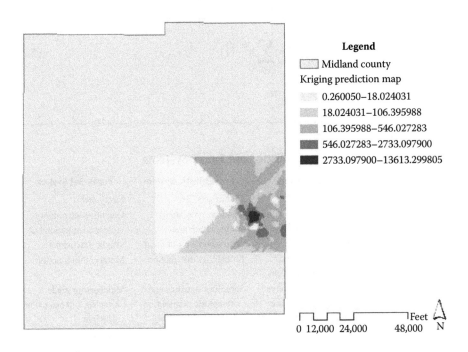

Legend

☐ Midland county

Kriging prediction map

- 0.260050–18.024031
- 18.024031–106.395988
- 106.395988–546.027283
- 546.027283–2733.097900
- 2733.097900–13613.299805

0 12,000 24,000   48,000 Feet N

FIGURE 15.2   Kriging map of soil dioxin in Midland County.

15.3 A REVIEW OF GM(1,1) MODEL IN GRAY SYSTEM THEORY

The gray system theory [5] was rooted in modern control theory, where system dynamics are classified by the degree of information availability, and accordingly different methodologies are developed for [each of them] respectively. Table 15.1 compares three commonly faced uncertain systems: gray systems, probabilistic systems, and fuzzy systems. A critical feature of the gray system is information incompleteness or, more specifically, sparse information availability. The task of establishing a model under the guidance of gray system theory is inevitably to build a model based on data of a small sample size. Its target is the establishment of a gray differential equation and it emphasizes the exploration, utilization, and processing of dynamic information contained in the data [5–7].

Gray differential equation models play the core function in gray theory [5] and its modeling developments. In a quality control context, GM(1,1), GM(2,1), and GM(1,N) are of fundamental importance. The basic form is the GM(1,1) model.

Definition 1:

$$x^{(0)}(k)+\alpha z^{(1)}(k)=\beta, \quad k=2,\cdots,n \tag{15.1}$$

is called a one-variable first order gray differential equation (GM(1,1)) with respect to the time series sequence $X^{(0)}=(x^{(0)}(1), x^{(0)}(2), ..., x^{(0)}(n))$, where

$$z^{(1)}(k)=\frac{1}{2}\left[x^{(1)}(k)+x^{(1)}(k-1)\right]$$

$$x^{(1)}(k)=\sum_{i=1}^{k}x^{(0)}(i) \tag{15.2}$$

$$k=2,\cdots,n$$

TABLE 15.1
Comparisons of Gray, Probabilistic, and Fuzzy Systems

Aspect	Gray System	Probabilistic System	Fuzzy Set System
Set foundation	Haze sets	Cantor sets	Fuzzy sets
Connotation and extension	Connotation haze with clear boundary and extension	Random event with connotation and extension well-defined	Cognitive uncertainty (clear connotation but vague extension)
Core concept	Gray derivative and differential equations	Probability distribution	Membership function
Data treatment	(Inverse) accumulative generating operation	Sampling statistics and asymptotic distribution	Membership grade, λ-cut set and extension principle
Data requirements	Small sample size	Large sample size	Empirical (plus sampling data)

and α is called the developing coefficient, β is the gray input, and $x^{(0)}$ is a *gray derivative* that maximizes the information density for a given series to be modeled. This model is called the GM(1,1) model with equal-gap.

Furthermore, the differential equation $dx^{(1)}/dt + \beta x^{(1)} = \alpha$ is called the whitenization differential equation or the *shadow* equation of the gray differential equation 1. The unknown parameter values (α, β) can be determined in terms of the classical least-square approach. Writing Equation 15.1 as:

$$\beta + \alpha\left(-z^{(1)}(k)\right) = x^{(0)}(k), \quad k = 2, 3, \cdots, n \tag{15.3}$$

a standard matrix form of the equation can be formed in terms of least-square theory,

$$X\begin{bmatrix} \alpha \\ \beta \end{bmatrix} = y \tag{15.4}$$

where

$$X = \begin{bmatrix} -z^{(1)}(2) & 1 \\ -z^{(1)}(3) & 1 \\ \vdots & \vdots \\ -z^{(1)}(n) & 1 \end{bmatrix} \quad \text{and } y = \begin{bmatrix} x^{(0)}(2) \\ x^{(0)}(3) \\ \vdots \\ x^{(0)}(n) \end{bmatrix} \tag{15.5}$$

which leads to the estimate for parameter (α, β)

$$\begin{bmatrix} \alpha \\ \beta \end{bmatrix} = \left(X^T X\right)^{-1} X^T y \tag{15.6}$$

Based on the estimates parameter (α, β) and differential equation theory, the predicted equation (i.e., the response or the filtering function) is

$$\hat{x}^{(1)}(k+1) = \left[x^{(0)}(1) - \frac{\hat{\beta}}{\hat{\alpha}} \right] e^{-\hat{\alpha}k} + \frac{\hat{\beta}}{\hat{\alpha}} \tag{15.7}$$

which corresponds to the GM(1,1) differential equation

$$\frac{dx^{(1)}}{dt} + \alpha x^{(1)} = \beta \tag{15.8}$$

As to the gray derivative sequence $X^{(0)} = \{x^{(0)}(i),\ i = 1,2,\ldots,n\}$, it can be obtained in terms of the inverse accumulative generating operation:

$$\hat{x}^{(0)}\left(k\right) = \hat{x}^{(1)}\left(k\right) - \hat{x}^{(1)}\left(k-1\right),\ k = 2,3,\cdots,n \tag{15.9}$$

For nonmonotone data patterns, GM(2,1), the second order one variable gray differential equation model may offer a better model fitting. Therefore, let us briefly review the related developments [9].

Definition 2:

Given the original discrete data sequence $x^{(0)} = (x^{(0)}(1), x^{(0)}(2), \cdots, x^{(0)}(n))$, the second order one variable gray differential equation, abbreviated as GM(2,1), possesses the following form:

$$a^{(1)} x^{(0)}\left(k\right) + a_1 x^{(0)}\left(k\right) + a_2 z^{(1)}\left(k\right) = b,\ k = 2,3,\cdots,n \tag{15.10}$$

with the corresponding whitenization equation

$$\frac{d^2 x^{(1)}}{dt^2} + a_1 \frac{dx^{(1)}}{dt} + a_2 x^{(1)}\left(k\right) = b \tag{15.11}$$

where $a^{(1)} x^{(0)}(k) = x^{(0)}(k) - x^{(0)}(k-1)$, $k = 2,3,\ldots,n$ is the first order inverse AGO (i.e., 1-IAGO), and $Z^{(1)} = \{z^{(1)}(k),\ k = 2,3,\ldots,n\}$ are defined by equation 2.

The solution of the GM(2,1) model is typically obtained by a two-step computation. The first step is to estimate the parameter vector $P = (a_1, a_2, b)^T$. Let

$$B = \begin{bmatrix} -x^{(0)}\left(2\right) & -z^{(1)}\left(2\right) & 1 \\ -x^{(0)}\left(3\right) & -z^{(1)}\left(3\right) & 1 \\ -x^{(0)}\left(4\right) & -z^{(1)}\left(4\right) & 1 \\ \vdots & \vdots & \vdots \\ -x^{(0)}\left(n\right) & -z^{(1)}\left(n\right) & 1 \end{bmatrix} \quad Y = \begin{bmatrix} a^{(1)} x^{(0)}\left(2\right) \\ a^{(1)} x^{(0)}\left(3\right) \\ a^{(1)} x^{(0)}\left(4\right) \\ \vdots \\ a^{(1)} x^{(0)}\left(n\right) \end{bmatrix} \tag{15.12}$$

Then the least-square estimator for the parameter vector $P = (a_1, a_2, b)^T$ is

$$P = \left(B^T B\right)^{-1} B^T Y \tag{15.13}$$

The solution to the whitenization equation 14, which is a second order ordinary differential equation with constant coefficients, $x^{(1)} = x_P^{(1)} + x_H^{(1)}$, where $x_H^{(1)}$ is the general solution to the corresponding homogeneous second order ordinary differential equation

$$\frac{d^2 x^{(1)}}{dt^2} + a_1 \frac{dx^{(1)}}{dt} + a_2 x^{(1)}(k) = 0 \tag{15.14}$$

while $x_P^{(1)}$ is a particular solution to Equation 15.11.

$$x_H^{(1)}(s) = \begin{cases} c_1 e^{r_1 s} + c_2 e^{r_2 s} \\ e^{\alpha s}\left(c_1 \cos(\beta s) + c_2 \sin(\beta s)\right) \end{cases} \tag{15.15}$$

which depends on the form of the solutions r_1 and r_2, respectively. The particular solution is $x_P^{(1)} = b/a_2$.

15.4 A REVIEW OF ORDINARY KRIGING METHOD

Ordinary kriging is a widely used geostatistical method for modeling spatial data [10]. It assumes that the local means are not necessarily closely related to the population mean, and therefore uses only the samples in the local neighborhood to produce estimates. Ordinary kriging relies on the spatial correlation structure of the data to determine the weighting values, and correlation between data points determines the estimated value at an unsampled point for inference, assuming normality among the data points [11].

The objective of the approach is to apply ordinary kriging to soil dioxin sample results generated from the GM(1,1) model. Once the GM(1,1) soil dioxin values are produced, additional values for the entire study region can be estimated. The basic mathematical idea behind ordinary kriging is to take N measurements $Z(r_1), \dots, Z(r_N)$ of soil dioxin derived from the GM(1,1) model at known locations $r_1, \dots r_N$ to obtain an estimate $\hat{Z}$ of Z at unsampled location r_0. The following linear equation can be used to estimate neighboring observations of soil dioxin:

$$\hat{Z}(r_0) = \sum_{i=1}^{N} \lambda_i Z(r_i) \tag{15.16}$$

where λ_i is the weight assigned to each observation. Only the closest observations within the searched neighborhood are used to compute average weight and in the production of the estimate for soil dioxin values.

15.5 NEW APPROACH: COMBINING GM(1,1) PREDICTION AT EQUAL-SPACED GRID WITH ORDINARY KRIGING

From Figure 15.1, it is obvious that within Midland City soil dioxin data are rich along the Tittabawassee River. However, the samples are limited to a small area making it difficult to perform an ordinary kriging analysis for the whole of Midland County. First, Midland County was divided into six areas (Figure 15.3). An ordinary

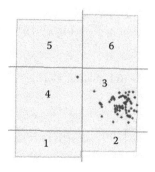

FIGURE 15.3   Area division in Midland County.

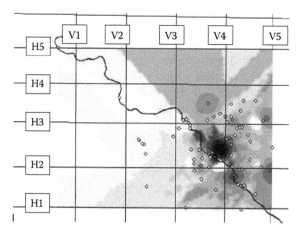

FIGURE 15.4   Grid creation in Area 3.

kriging map could only be produced for Area 3 (Figure 15.2). Area 3 was therefore divided by equally spaced vertical lines V1 to V5 and equally spaced horizontal lines H1 to H5 so that 25 intersection points can be recorded (Figure 15.4).

Eastings, northings, and soil dioxin values were recorded for each intersection point (Table 15.2). Once the equally spaced grids over Area 3 were created, GM(1,1) modeling was performed along vertical lines V1 to V5 upward with division points for covering Area 6. Similarly, GM(1,1) modeling was performed along vertical lines V1 to V5 downward with same-spaced points for covering Area 2. Area 4 can be GM(1,1)-predicted using horizontal lines H1 to H5 leftward. Finally, Area 2 GM(1,1)-predicted points are used to GM(1,1)-predict Area 1. Following GM(1,1) prediction of Areas 1, 2, and 6, well-spread, equally spaced grids were generated over the unsampled areas. These GM(1,1)-predicted data are combined with originally sampled observations in Area 3 and form an enlarged data set (Figure 15.5).

The gray sample predictions appear as a grid of sample points (Figure 15.5), expanding to cover much more of Midland County. With the newly enlarged gray prediction data set, an ordinary kriging map, the GM(1,1)-kriging map, can now be produced for the entirety of Midland County (Figure 15.6). The predictions from the

TABLE 15.2

Equal-Spaced Coordinated Points with Kriging Dioxin Values

Point ID	Easting	Northing	Kriging Dioxin Values
1	13123134.239112	749177.692346	12.296460
2	13136803.482106	749313.031385	21.675454
3	13150202.047022	749177.692346	51.224597
4	13163600.611937	749177.692346	66.166684
5	13176051.803575	749177.692346	416.056757
6	13123269.578152	759598.798391	3.7297130
7	13136803.482106	759598.798391	13.725505
8	13150377.386061	759598.798391	36.230734
9	13163600.611937	759598.798391	0.0000000
10	13176187.142615	759734.137431	968.863086
11	13123269.578152	771237.955792	4.0243420
12	13136803.482106	771373.294832	5.0239170
13	13150202.047022	771237.955792	5.4372570
14	13163735.950976	771237.955792	79.625813
15	13176051.803575	771102.616753	398.105374
16	13123134.239112	781388.383758	16.7457250
17	13136938.821146	781388.383758	5.8539130
18	13150202.047022	781523.722798	161.311452
19	13163600.611937	781388.383758	176.095226
20	13176051.803575	781388.383758	265.932684
21	13123269.578152	790320.760368	5.8600940
22	13136803.482106	790320.760368	167.724327
23	13150337.386061	790320.760368	136.376999
24	13163735.950976	790320.760368	18.8408860
25	13176251.803575	790050.082289	205.140055

GM(1,1)-kriging map cover the entire county and are clear, detailed, and consistent with the original soil dioxin samples.

15.6 CONCLUSION

The purpose of this research was twofold: (1) to combine the GM(1,1) model with ordinary kriging to account for small samples of soil dioxin data; and (2) to solve the spatial prediction and analysis problem. The GM(1,1)-kriging method can make better use of any kind of data, by taking advantage of the GM(1,1) model that can deal with small sample sizes (as small as $n = 4$). This method demonstrates strong predictive ability and is founded on very simple numerical computations [7], which can be carried out in Microsoft Excel. This chapter provides an illustration of this new method and further refinements will be carried out through Visual Basic for Applications (VBA) programming. For fluctuated data, it is suggested that the GM(1,1) model may be less powerful, and the GM(2,1) model may be used instead.

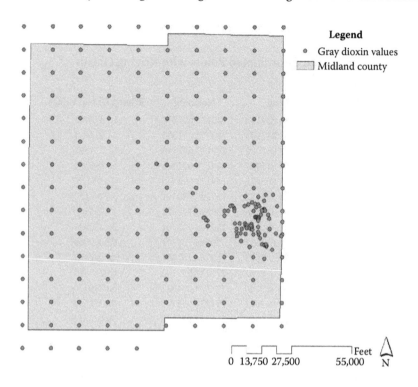

FIGURE 15.5   Gray dioxin predictions for Midland County.

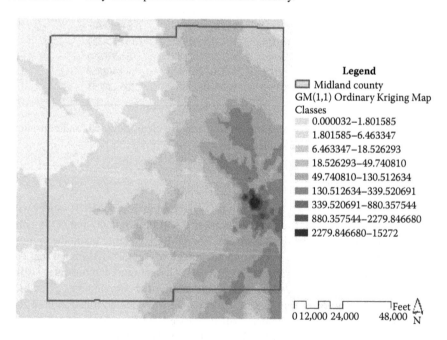

FIGURE 15.6   GM(1,1) Ordinary Kriging Map for Midland County.

REFERENCES

1. Alliance for Safe Alternatives, Dioxin—Common Questions and Answers, Alliance for Safe Alternatives, 2005, available at: http://www.safealternatives.org/faq.html.
2. Environmental Protection Agency, Dioxins, U.S. Environmental Protection Agency, http://www.epa.gov/, 2005.
3. World Health Organization (WHO), WHO Revises the Tolerable Daily Intake (TDI) for Dioxins, World Health Organization European Centre for Environment and Health; International Programme on Chemical Safety, Organohalogen Compounds, 38, 295, 1998.
4. TRIfacts, TEQ vs. TM-17 Method, TRIfacts.org, http://www.trifacts.org/teq_tm17/index.php, 2005.
5. Deng, J. L., Control problems of grey systems, *Systems and Control Letters*, 1, 6, 1982.
6. Guo, R., A fuzzy system analysis on repairable system via interval-valued fuzzy set approach, in *Proceedings of the 9th Annual International Conference on Industrial Engineering—Theory, Applications and Practice*, Auckland, 2004, 146.
7. Guo, D., Guo, R., and Thiart, C., Grey GIS, in *Proceedings of the 13th Annual Conference on Geographical Information Systems Research U.K.*, Glasgow, 2005, 379.
8. Guo, R., and Love, C. E., Grey repairable system analysis, plenary lecture, in *Proceedings of the European Safety and Reliability Conference: Advances in Safety and Reliability*, Tri City, 2005, 753.
9. Liu, S. F., Dang, Y. G., and Fang, Z. G., *Grey System Theory and Applications* [in Chinese], Scientific Publishing House, Beijing, 2004.
10. Cressie, N. A. C., *Statistics for Spatial Data*, John Wiley & Sons, New York, 1993.
11. Spatial Analysis and Decision Assistance (SADA), *Geospatial Methods*, University of Tennessee Corporation, http://www.tiem.utk.edu/~sada/help/, 2003.

Section 3

Visualizing the Natural Environment

16 Keynote Paper
Information Access, Depicting Geography, and Geographical Visualization Tools

William E. Cartwright

CONTENTS

OVERVIEW

Users looking for geographical information would typically access resources from hard-copy materials, discrete media, and via intranets and the Internet. However, this is no longer a simple task, as the amount of information at hand, physically and virtually, has increased enormously. As a result, users have difficulty in finding this information and, once found, have even greater difficulty consuming that information. Visualizing enormous amounts of geographical information becomes impossible without tools that interrogate databases, synthesize the essential elements of that information, and then display the results in the most appropriate manner for the user and use. Efforts within the international geographical visualization community have resulted in many innovative techniques that build visualizations, contemporary and traditional, upon the opportunities afforded by the creation of detailed, global coverage databases. Users have improved understanding of the geography of the world through the automating of the process of database interrogation, information analysis, and visualization output. This chapter looks at some of the realities of current data availability via contemporary Internet-accessible repositories and the usefulness of such databases. It then provides a snapshot of the chapters in this part of the book and, finally, proposes some areas of research that could be considered for evaluating the effectiveness of such visualizations.

16.1 INTRODUCTION

Relative to the slow evolution of maps, two things have occurred quite recently: the field of geographical visualization has matured; and large geographic data sets and systems have been established and made available online. Users have begun to concentrate more on the application and usage of their geographical information, rather than the process of capturing the data itself. Although geographical visualization tools already have the ability to transform data into information, users are now demanding that their systems become smarter and migrate to the next generation enabling them to turn information into knowledge.

This is not a new phenomenon; almost two decades ago Taylor [1] acknowledged the importance of scientific technologies in cartography (and geographical visualization) but stressed that future directions were not primarily governed by technical issues. He argued that although maps have always asked the question where, in the information era they must also answer new questions such as why, when, by whom, and for what purpose; and they must convey to the user an understanding of a much wider variety of topics than was previously the case. Looking at the future of geographical information systems (GIS), it was stated that to meet this challenge GIS will have to be easier to use, inexpensive, available to a multitude of users, and offer "the right stuff at the right place at the right time" [2]. There exists a need to exploit the potential that geographical visualization [offers geography] by leveraging the existence of already-established comprehensive data sets to facilitate not just large repositories of geographical information, but ways to exploit this information to support decision making where geolocation is important. The need for the provision of geographical information in an integrated and holistic manner has already been identified by researchers in Europe [3] and North America [4].

Geographical visualization offers the potential to provide an important conceptual development in an area of critical interest to the geographical information industry, academia, and a wide community of consumers. It provides the means to enhance access to, and the communication of, geographical information. In doing so, resources can be better utilized in developing new applications, which, to date, have proven to be very labor intensive. Developing better techniques to visualize geographical information has widespread significance, locally and nationally, to public and private sectors of the geographical community alike, in areas as diverse as education, planning, and decision making, as well as natural resource monitoring and assessment.

16.2 ACCESSING AND CONSUMING GEOGRAPHICAL INFORMATION

A number of information access and usage realities now exist: the sheer amount of information now available via the Internet; how we access information (including geographical information); the importance of location in information use; and the impact of understanding the importance of location during the decision-making process. These are expanded upon in the following sections.

16.2.1 THE EXPLOSION OF DATA AVAILABLE VIA THE INTERNET

The amount of information on the Internet is increasing each year. Bergman [5] conducted a study that quantified the size of the Web and results were reported in two categories, under the headings of "the surface Web" and "the deep Web." Surface Web pages are those that are static. Web pages in the deep Web are those that are produced dynamically as a result of an enquiry to a searchable database (e.g., ProQuest Direct is able to provide access to many online periodicals [6]). In a study conducted in early 2000, it was found that the deep Web contained approximately 7,500 terabytes of information compared to the surface Web, which contained 19 terabytes. The top 20 public deep Web sites contained 656 terabytes of information and the largest deep Web site contained 366,000 gigabytes (or 357 terabytes) of information [5]. In a more recent study, a group of researchers at the University of Berkeley conducted a study that estimated the amount of information available in the world, including both analog and digital information. Their findings found that the surface Web contained 167 terabytes and that the deep Web contained 91,850 terabytes of information [7].

In 1998, a report of the Commission on Geosciences, Environment and Resources of the US National Research Council focusing on spatial information and its use forecast that "there is certainty that the society of 2010 will require increased use of spatial data and spatial thinking in problem solving, at scales from the human genome to the human body to the environment to galaxies" [8]. The truth of this statement is reflected in more recent figures. In 2000 the largest deep Web site was The National Climatic Data Center (NOAA) site, which was found to contain 366,000 gigabytes (or 357 terabytes) of information [5]. This site contains predominantly spatial information. In fact, half of the top 20 deep Web sites provide access to approximately 642,283 gigabytes (or 627 terabytes) of spatial information. Although, there are no studies that explicitly estimate the volume of spatial information on the Web, it is clear that it accounts for a substantial amount of the information we store. Lesk [9] estimated that textual documents on the Web have the potential to increase to the size of about 800 terabytes. Over 70% of all textual documents contain spatial references [10]. A substantial quantity of textual documents on the Web can be linked to spatial locations. From Australian Spatial Data Directory (ASDD) searches of 36,978 metadata records, it was revealed that large environmental data-provision organizations make available massive data sets. For example, the US National Climatic Data Center (NCDC) has 366,000 gigabytes online and NASA EOSDIS has 219,600 gigabytes online [11].

16.2.2 INCREASED ACCESS TO SPATIAL INFORMATION

The Internet has produced a new means to access spatial information [12,13] and this has been tapped into by national mapping organizations such as Ordnance Survey in the United Kingdom, the United States Geological Survey, and Geoscience Australia. New companies are now competing to provide spatial data and services. These companies are changing the profile of spatial data from a base map data set collected and used by government agencies to a consumer driven service. MapQuest, launched in February 1996, was the first consumer-based interactive mapping site on the Web [14]. After nearly three months, MapQuest was generating 200,000 maps a day; after one year, 700,000 maps a day [13]; 400 million maps requested by 60 million user sessions was its average monthly rate for 2002 [15]; and in 2004 over 20 million maps a day [14]. An analysis of popular mapping sites reveals that the total audience of Internet-delivered maps increased from 71,387,000 unique visitors in 2006 to 74,563,000 in 2007. MapQuest increased by 3%, and Google Maps [16] increased by 26% in the corresponding period.

This early interest by consumers of accessing geographical information and artifacts differently via MapQuest has been dwarfed recently by Google Maps [16] and Google Earth [17]. Both products have generated an enormous interest in Web-delivered geographical information. This is further driven by Google's search engine that directs users searching for maps to its mapping and imagery sites. Google Maps grew by 51.57% during 2007 and the site had almost 90,000,000 unique visitors in that year, up from around 60,000,000 in 2006 [18]. In 2007, MapQuest had 5 times the number of American visits than Google Maps. In January 2008, this lead over Google Maps dropped to 1.25 times [19].

As a consequence of the growth in this emerging marketplace, it is no longer the case that it is just the research community that is now interested in better methods for visualizing geographical information. The general public has a high expectation of free, high-quality data with associated easy-to-use tools for exploring that data.

16.2.3 THE IMPORTANCE OF LOCATION

It is widely acknowledged that 80% of all decisions made by humans have a spatial element [12]. Frank, Raubal, and van der Vlugt [20], the editors of *PANEL-GI Compendium—A Guide of GI and GIS*, stated that geographical information at a range of scale is critical to dynamic and precise decision making. Frank and Raubal [21] suggested that geographical information could improve economic efficiency by about 20%, which means geographical information has the potential to contribute up to 16% of gross domestic product.

Location plays an important role in wireless e-commerce. Wireless e-commerce provides services and information that can be accessed at any time and at any place. There are a number of wireless e-commerce services, more commonly known as location-based services (LBS), where location is of central importance, including those services that provide navigational assistance, directory search, and maps [22]. But in *The Economist* article "Technology Review: The Revenge of Geography," Manasian [23] reported that "finding information relevant to a particular place, or the location associated with a specific piece of information, is not always easy." Manasian went on to present interesting and innovative applications able to link virtual information and physical location, suggesting that it is critical that we address the issue of ease of use.

16.3 MAKING IT WORK

It is not sufficient to offer a toolbox of visualization tools. We need to consider issues of interoperability; we need to fully evaluate these tools to ensure their utility among a broad community of users. When geographical visualization tools are embedded within or appended to a contemporary mapping package it is essential that effective evaluation procedures be developed and tests conducted. Proper evaluation of geographical visualization tools needs to be conducted in a manner that is sympathetic to both the user and the environment in which it will be used.

Chapters in Part 3 of this book provide examples of efforts being made to develop and evaluate visualization tools. Rider and Reitsma (Chapter 22) outline the development of an agent-based application for modeling sheep grazing patterns. In their work they explored the effects of flocking, whereby four of the most essential animal state variables were modeled and visualized: location, direction, hunger, and thirst. Modeling explored the effects of siting water troughs in different patterns in the fields and how the shape of fields, generally square or rectangular, influenced where sheep flocks grazed. The resulting visualization generated was a grazing pattern overlaid atop an air photograph. An accuracy of approximately 1 square meter was achieved. The authors identified its strength as an interactive learning tool, allowing users to see the results of alternative pasture management methods.

In Chapter 21, Bleisch and Dykes describe how they tested the usefulness of Web-based realistic visualizations. Here, they generated interactive 3-D visualizations by draping orthoimagery over digital terrain models. Comparisons were made between the use of Internet-delivered 3-D visualizations and the paper maps most popular for hike planning.

Jardine and Mackaness write about the generation of paper-based map products for visualizing risk for hill walkers in Chapter 20. The focus of the research was on the exploration of ways in which risk can be visualized when planning and executing routes. They sought to develop better presentation methods to illustrate risk using static maps at 1:25,000 scale. Walking risk was modeled using a subjective risk ranking assessment method to generate a quantitative model. Using the Ordnance Survey (OS) MasterMap, attribute tables risk values were calculated and visualized by superimposing color-coded information on to the contour lines of the map. The resulting maps were evaluated and the improvement in map users' understanding of risk using annotated OS MasterMaps was determined.

Chapter 19 looks at the use of geostatistics and GIS for examining spatial patterns, in a conservation context. In this chapter, Lieske and Bender describe their development of a method for visualizing broad-scale species distributions. They used data relating to the distribution and abundance of four bird species. Using ArcGIS, the authors generated 3-D plots of simulations of abundance patterns.

16.4 QUESTIONS ARISING FROM THE GEOGRAPHICAL VISUALIZATION RESEARCH REPORTED IN THIS BOOK PART

The chapters summarized in the previous section illustrate some of the endeavors related to developing and evaluating visualization. Information is delivered as 2-D or 3-D visualizations, as stand-alone products (paper or screen), or Internet-delivered interactive tools. One challenge is to assess the efficacy of 2-D and 3-D maps, perhaps to assess which technique is best, and what additional elements should future research in geographical visualization address. Some areas that might be topics for further research are discussed next.

16.4.1 ARE PAPER PRODUCTS BETTER THAN COMPUTER-GENERATED VISUALIZATIONS?

Ptolemy believed maps to be a means to "exhibit to human understanding ... the earth through a portrait" [24]. Historically, maps have been used to provide information to users about places recently discovered, voyages completed to unknown worlds, or hitherto seemingly impossible journeys. In addition, used less formally than a navigator might, maps allow armchair travelers to "go" to places from the comfort of their lounge or study. Maps, like that described by Lewis Carroll [25], employ information graphics to inform about the unknown:

> He had brought a large map representing the sea,
> Without the least vestige of land:
> And the crew were much pleased when they found it to be
> A map they could all understand.
> "What's the good of Mercator's North Poles and Equators,
> Tropics, Zones, and Meridian Lines?"
> So the Bellman would cry: and the crew would reply
> "They are merely conventional signs!
> Other maps are such shapes, with their islands and capes!
> But we've got our brave Captain to thank"
> (So the crew would protest) "that he's bought us the best—
> A perfect and absolute blank!"

In the context of visualization, it can be argued that the basic premise of using maps and other (geo)visualization artifacts is to build mental models of systems, processes, associations, or geographical reality. The rules that govern their design, production, and consumption have evolved over centuries, and the methods of producing maps via the printing press have been established by 500 years of experiment and development [26].

In the context of the developments presented in this book, we should ask whether paper maps are still appropriate for depicting the visualizations generated by contemporary geographical visualization systems. Paper maps work, and frequently are still the preferred visualization/interpretation tools of many users, as noted in Chapter 21. Paper maps are still needed; the question becomes how best do we design and produce them as adjuncts to computer depiction?

16.4.2 CAN OFF-THE-SHELF GIS SOFTWARE PRODUCE INNOVATIVE VISUALIZATIONS?

Developers of geographical visualization tools use integrated media devices to provide a more innovative approach to the portrayal of geographical information. Their products invite users to explore large databases and to view impressive displays of digital excellence. This ranges from simple graphic depictions to animations and fly-throughs. But, are the means of data presentation and access really any different from that used in conventional maps or any other digital map-related tools? How does the user interact with the data? Is it a simple view, or are users allowed to explore the information (in the many forms available with contemporary packages) and discover the location of geospatial phenomena of interest, thus gaining an understanding of why that particular element of the landscape occurs at that particular location?

Certain conventions lead the way in which data are collected and presented; they also govern the way in which geographical information packages are used. However, individual users may prefer to use geographical visualization tools in different, personalized ways, rather than accepting a standard product. This can range from being able to specify the look and feel of the paper maps that are generated from an automated system (as being researched by IGN France [27]), or modifying the interface or operating methods of an automated system. They might demand a system that allows them to construct a tailored package for their unique use needs. Therefore, research can address whether tailored packages built to suit individual users' access requirements, respond with more active use of packages and thus allow the viewing and interpretation of geographical information products from various perspectives.

16.4.3 DO INTEGRATED INTERACTIVE RICH-MEDIA VISUALIZATION TOOLS ASSIST OR MERELY CONFUSE?

There is interest in using the expertise of many professions to understand how geographical visualization tools work. Fabrikant and Buttenfield [28] noted: "Research questions such as 'How do people learn about geographical information?' or 'How do people develop concepts and reason about geographical space?' beg for an interdisciplinary approach, drawing upon expertise in cognitive psychology, geographic information science, cartography, urban and environmental planning and cognitive science." Rich media can be used to provide many views of geography from one information resource. [By assembling a suite of artifacts, maps output to screen or are printed on demand (from computer mapping systems and GIS or from linked pdf files), as computer images, animations, fly-throughs; interpretations of a geographical space made from different disciplines can be made available.] Multiple (disciplinary) windows into the world can be provided and not just one view becomes paramount.

We need to understand how new media enhance the communication effectiveness of geographical information portrayal through the use of geographical visualization tools. Problems encountered when using these tools relate to how geography is portrayed, viewed, and perceived. Research regarding these elements could provide much useful information about the limitations of geographical visualization tools.

16.4.4 WHAT IS THE AGREEMENT BETWEEN PRODUCER AND USER REGARDING THE LOGIC UNDERPINNING THE ACCEPTANCE OR REJECTION OF GEOGRAPHICAL VISUALIZATION TOOLS?

Compared to paper maps, computer-generated geographical visualization tools are relatively new, and their use as a tool for visualizing geography is still largely unproven. There is little real information about best use or what users see with contemporary geographical visualization tools, and whether they see geography differently with these tools. Understanding how paper-map counterparts work best is no easy task, as the real reasons that certain artifacts are accepted and others rejected can be hard to unearth. It is somewhat hard to determine whether the artifact itself, or the users knowledge about the area/topic being mapped contributes to the success of the communication device.

Take for example what is considered to be one of the most successful and widely used maps to support navigation and decision making in an urban space, albeit underground: the London Underground map. By distorting geography, the designer, Beck, made the map more usable and an effective communicator about how to move about London. His original design moved away from the concept that the maps had to follow the actual geographical route of the lines. According to Hadlaw [29], what Beck did was to set aside geographic space in favor of graphic space. When it was introduced in 1933 it was an outstanding success, with 850,000 copies of the map in circulation within two months of its introduction [30]. However, Hadlaw [29] says that the actual success of the map lies in the fact that both the map and the underground users shared the same sensibility, and it was comprehensive because of the logic that underpinned the design, which was "coherent with their experience, as modern individuals, of a historically particular time and space" [29]. Here, neither good map design nor an intimate knowledge of the place of the underground was responsible for better understanding of how the map worked, but an agreement about the underpinning logic. How can this agreement about the underpinning logic be discovered? And, does the underpinning logic differ from visualization artifact to visualization tool?

16.4.5 How Best to Ensure That the "New" Is Not a Reason for Rejecting Alternative Views of the World?

There is no guarantee that the new will always be accepted. Just because an innovative contemporary visualization tool has been developed, the assumption that users will just like it and exploit what is offers cannot be made. Take, for example, the weather maps provided on the British Broadcasting Service (BBC) news. In 2005 the BBC changed its traditional computer-generated weather maps to virtual reality maps and 3-D landscapes. These new products were seen as "the biggest change to the way the weather forecasts are presented since computer generated maps replaced magnetic symbols in 1985" [31]. Once the new-look news graphics were introduced, the BBC was inundated with complaints like those from Scottish viewers, who were upset that Scotland was distorted (this was also a view of the Scottish National Party), that Britain no longer looked green, but it resembled a desert country; that the new maps were harder to visualize; and so forth. This generated so much interest that the BBC, which drew comments from both sides of the "new maps" argument, established an online debate [32]. Comments from detractors included: "I prefer the old map as it was easier to visualize"; "Sad, very sad and dull"; "The new format is a complete disaster"; and "I'm afraid that I'm not happy about England appearing 10 times bigger than Scotland" (from an Edinburgh contributor). From the supporters of the new maps: "At least the Beeb are trying to drag their systems into the 21st Century"; "I actually like the map, it did not take long to figure it out"; "It was about time to have an upgrade to the outdated model and I welcome the change"; and "The new technology is excellent." And the noncommittal: "I'm sure that we'll all get used to it in time"; and "The old graphics were probably badly received when they first came out. We'll get used to it."

When assembling the criteria for improving geographical visualization tools, consideration must be given to the individual preferences of potential users. Once their particular likes and dislikes of information visualization tools are ascertained, then products can be designed and delivered accordingly. Although the rules for design thus developed may not afford what the designer/producer considers to be good design rules, they are nevertheless the rules that must be applied to contribute to the eventual acceptance of new products. Bad design rules according to theoretical considerations become good rules in practice. Just producing new geographical visualization tools goes only part of the way to ensuring actual acceptance of tools. Tools need to be developed in concert with users' real needs and likes. The BBC example illustrates how difficult this is to do.

16.4.6 Do Different Geographical Visualization Tools Provide Different Views of Reality When Visualizations Are Generated from the Same Database?

Providing different ways of seeing geographical information can be achieved by providing different views of information. Delivering many computer-generated views from one database could, hopefully, ensure that any voids when viewing visualizations of a particular geography are eliminated. Voids could be filled with information seen from other perspectives and used to assemble a more complete picture of reality.

But, because each user of a visualization package might choose to view geography using different views or combinations of views, each individual mental map will be different. Therefore, this begs the question: Does the way of seeing influence the way of knowing? Research is needed to determine how different visualizations generate different mental images of an area being studied. As well, other questions that could be addressed are: What is the most appropriate pedagogy for using visualization tools to assist users in understanding geography. How do humans learn about geographical information, and how does this learning vary as a function of the medium through which it occurs (direct experience, maps, descriptions, virtual systems)? The use of many media types, conventional and digital, might provide different viewpoints from which to view geography. Traditional map products only provide one viewpoint of geography—through the map—and geographical visualization tools can offer many media views of geography.

16.4.7 How Can Multimedia, Delivered as Interactive Integrated Media Tools, Be Assembled in the Most Effective Manner to Ensure That a Cohesive Tool Suite Is Provided?

When assembling integrated media there can be a tendency to just throw many different media tools into one toolbox without much thought about how they will work together. Each individual tool might operate quite efficiently as individually presented elements, but they cannot be guaranteed to work well together, if at all, when provided and used as a conglomerate media toolbox. Concepts need to be developed that can be used to design a framework around which integrated media tools can be inserted.

These concepts I call viscosity, friction, and gravity. As the toolbox would be composed of many integrated media elements, it is essential to understand how these elements coexist and interact with one another. And this understanding can be applied to designing better-integrated installations. In this context, *viscosity* refers to the speed at which technology changes. Technologies with a high viscosity, or which are semifluid, would change more quickly than those with a low viscous value. How do we marry many information resources, each with different viscous values? *Friction* refers to how close different application types need to be so as to ensure smooth information flow. One material with a high friction value would essentially stick to other applications, indicating that this application was always needed in any part of an entire application or product. *Gravity* is the pulling component of integrated elements, where due to their very nature they attract other media types. Here, strategies are needed to separate unwanted media types away from the attractive element.

16.5 CONCLUSION

As noted earlier in this chapter, geographical visualization is still relatively unexplored, compared to paper maps, even though users have embraced it since the mid-1980s. Geographical visualization has been adopted, adapted, and employed as a tool to facilitate the provision of geographical information in an exciting and innovative manner, but it has not been adequately evaluated. There exists much interest in exploring some ideas about the best use of geographical visualization tools and how they might be different from conventional methods.

The chapters in this book part illustrate the endeavors to develop and test diverse geographical visualization tools. Tools have been developed using methodologies that range from GIS to specialist software. Applications are output as screen imagery, tailored presentations illustrating the results of extensive analysis, and as paper maps, with specialist information overlays. Endeavors to develop innovative applications of existing technology and explorations of potential new tools provide challenges for those undertaking research in this field.

REFERENCES

1. Taylor, D. R. F., A Conceptual Basis for Cartography: New Directions for the Information Era, *The Cartographic Journal*, 28(2) 213, 1991.
2. Crane, D. E., Future Directions for GIS, *Proceedings of the 21st Annual AURISA Conference, Adelaide* (Vol. 1), 466, 1993.
3. Niederer, S., Kriz, K. and Pucher, A., Open Source Spatial Decision Support System for Sustainable Water Management, *Proceedings of the 21st ICC, A Coruña, Spain* CD-ROM], ICA, July, 2005.
4. Taylor, D. R. F., Eddy, B., Pulsifer, P., and Lauriault, T., Cybercartography: Theory and Practice, *Proceedings of the 21st International Cartographic Conference, A Coruña, Spain* [CD-ROM], ICA, July 2005.
5. Bergman, M., *Deep Content—The Deep Web: Surfacing Hidden Value*, BrightPlanet Corps, 2001, available at: http://brightplanet.com/pdf/deepwebwhitepaper.pdf, accessed February 2008.
6. ProQuest Direct, http://www.proquest.com, accessed February 2008.
7. Lyman, P. and Varian, H. R., How Much Information?, University of California, Berkeley, 2003, available at: http://www.sims.berkeley.edu/how-much-info-2003, accessed February 2008.
8. US National Research Council, A Forecast, Commission on Geosciences, Environment and Resources Report, 3, 1998.
9. Lesk, M., How Much Information is There in the World?, 1997, available at: http://www.lesk.com/mlesk/ksg97.html, accessed February 2008.
10. MetaCarta, The Geography of Knowledge, http://www.metacarta.com/, 2004, accessed February 2008.
11. Jamieson, P., Customer Support Unit, Sales and Distribution, Geoscience Australia, personal correspondence, 2005.
12. Albaredes, G., Network-Centric GIS: New Means to Access Spatial Information, *Proceedings of GIS PLANET, Lisbon, Portugal* [CD-ROM], 1998.
13. Peterson, M., Trends in Internet Map Use, *Proceedings of the 18th ICC, Stockholm, Sweden*, 1997.
14. MapQuest, http://www.mapquest.com/about/main.adp, accessed February 2008.
15. Tang, W. S. M. and Selwood, J. R., The Development and Impact of Web-Based Geographic Information Services, *GISdevelopment.net*, 2002, available at: http://www.gisdevelopment.net/technology/gis/mi03002pf.htm, accessed February 2008.
16. Google Maps, http://maps.google.com/, accessed February 2008.
17. Google Earth, http://earth.google.com, accessed February 2008.
18. Techcrunch, 2007 in Numbers: iGoogle Google's Homegrown Star Performer This Year, 2007, available at: http://www.techcrunch.com/2007/12/22/2007-in-numbers-igoogle-googles-homegrown-star-performer-this-year/, accessed February 2008.

19. PC World, Google Maps Gaining on MapQuest, http://www.pcworld.com/article/id,141391-pg,1/article.html, accessed January 2008.
20. Frank, A. U., Raubal, M., and van der Vlugt, M., *PANEL-GI Compendium—A Guide to GI and GIS*, Geographical Information Systems International Group (GISIG) and European Commission, Genoa, Italy, 2000.
21. Frank, A., and Raubal, M., GIS Education Today: From GI Science to GI Engineering, *Journal of the Urban and Regional Information Systems Association (URISA)*, 13(2), 5, 2001.
22. Steinfield, C., The Development of Location Based Services in Mobile Commerce, in *Elife After the Dot Com Burst*, Preissl, B., Bouwman, H., and Steinfield, C., Eds., Springer, Berlin, 2004, 177.
23. Manasian, D., Technology Review: The Revenge of Geography, *The Economist*, 13 March 2003.
24. Crane, N., Changing Our View of the World, *Geographical, The Royal Geographical Society*, 33, 2003.
25. Carroll, L., *The Hunting of the Snark: An Agony in Eight Fits*, Macmillan, 1876, chap 2.
26. Cartwright, W. E., 500 Years of Flatness, 50 Years of Compromise, 5 Years of Innovation: The Emergence of Multimedia Cartography as a Unique Visualization Form, Paper presented at the FELCS Postgraduate Research Conference, Melbourne, 1999.
27. Christophe, S., Making Legends by Means of Painters' Palettes, *Proceedings of the Art and Cartography: Cartography and Art Symposium*, Cartwright, W. E., Gartner, G., and Lehn, A., Eds., International Cartographic Association Working Group on Art and Cartography, Vienna, 2008.
28. Fabrikant, S. I. and Buttenfield, B. P., Formalizing Semantic Spaces for Information Access, *Annals of the Association of American Geographers*, 91(2), 280, 2001.
29. Hadlaw, J., The London Underground Map: Imagining Modern Time and Space, *Design Issues*, 19(1), 35, 2003.
30. Garland, K., *Mr. Beck's Underground Map*, Capital Transport, London, 1994.
31. BBC News, Winds of Change for BBC Weather, 13 May 2005, available at: http://news.bbc.co.uk/1/hi/entertainment/tv_and_radio/4546141.stm, accessed February 2008.
32. BBC News, Have Your Say: What Do You Think of the New Look Weather?, 20 May 2005, available at: http://news.bbc.co.uk/1/hi/talking_point/4551051.stm, accessed February 2008.

17 Keynote Paper
Wiki Cartography and the Visualization of the Natural Environment

Daniel Z. Sui

CONTENTS

OVERVIEW

Following the success of earlier open-source software development such as Linux, consumer-driven business development such as eBay, and, most recently, user-led knowledge production such as Wikipedia, the past five years have witnessed the emergence of user-created Web content using Web 2.0 technologies as evidenced by the growing popularity of MySpace, Facebook, YouTube, and more broadly the reality TV or game/competition programs with increasing viewer involvement. The wind of this general societal trend of wikification has started blowing in the cartographic

269

community during the past two years. The wikification of cartography includes the growing effort in open-source and free mapping software, increasing availability of volunteered geographic information (VGI), and the growing deployment of grid and ubiquitous computing in mapping practices. This chapter reviews the recent development of wiki cartography and discusses the implications of this new trend for visualizing the natural environment and future cartographic practices. It is argued that emerging Map 2.0 as a result of wiki cartography must be understood as an integral part of the spatial media.

17.1 MAPPING BY THE CREATIVE COMMONS: THE RESURRECTION OF CARTOGRAPHY

The development of a discipline, in many ways similar to an individual's life, is often full of ups and downs. Despite its long and rich history, cartography had certainly reached its low points during the past two decades due to the phenomenal growth of geographic information systems (GIS) as a discipline (geographic information science [GIScience]) and an industry. I recall that 12 years ago, a plenary session during the 1996 Association of American Geographers (AAG) annual meeting was devoted to the topic of "Has GIS killed cartography?" Although both geographic information scientists and cartographers answered no to this question, the justification for their answers was quite different. Most cartographers argued instead that cartography had actually committed suicide [1], as evidenced by the changes the discipline had to make to accommodate the impacts of GIS. GIS followers also did not believe it; they argued what had happened was that GIS and cartography had actually gotten married.

Regardless of the answer one gives to the question, one thing for sure is that the development of GIS in the 1980s and 1990s has transformed cartography in some very fundamental ways. With the focus on spatial analysis and modeling, and data representation issues dominating the research and education agenda, traditional concerns for mapping and cartography have been pushed to the sideline, and for quite sometime they no longer seemed to be a primary concern within the broader geospatial community. Many geography departments/programs even questioned the relevance (or even the need) of cartography in the age of GIS. In the late 20th century, the shift of dominant metaphors in geographic thought from visual to aural ones, and the postmodern critiques on ocularcentrism certainly did not help in advancing the development of cartography [2].

As we move into the 21st century, GIS has increasingly become an integral part of the media [3], and maps and visualization have gained ascendance among the geospatial research community in recent years. A watershed year in the development of GIS was 2005 due to the release of Google Maps/Earth, after which GIS users worldwide increased from 1 million to 200 million in less than four years. Recent innovations in technologies have drastically flattened the learning curve for geospatial and mapping technologies. The development of Google Maps/Earth and Microsoft's Virtual Earth, along with precursors like MapQuest, Yahoo Maps, Multimap, and StreetMap have widened the use and production of massive online geospatial information. The accelerating democratization of mapping and cartography has put powerful mapping tools in the hands of the masses, who can now perform relatively complex mapping tasks (once requiring years of training) with little or no training at all. Google Earth seems to have mobilized millions of people around the world to engage in various kinds of mapping activities—something cartographers never dreamed before. With all these new developments in recent years, cartography and mapping seem to have moved back to the center stage. Indeed, the rapid growth of wiki mapping sites in recent months has given a new life to resurrected cartography.

The goal of this chapter is to review the recent development of wiki cartography—the online mapping by a mass of users using Web 2.0 technology [4]—and discuss the implications of this new trend for visualizing the natural environment. The chapter is organized into five sections: After a brief introduction, the recent development of wiki cartography is reviewed in Section 17.2, followed by examples of its use for visualizing the natural environment in Section 17.3. Section 17.4 discusses the problems and challenges for wiki cartography, and the final section contains a summary and conclusions situating wiki cartography in the broader context of recent developments in the geospatial Web and spatial media.

17.2 WIKI CARTOGRAPHY: CARTOGRAPHY WITHOUT CARTOGRAPHERS

Wiki cartography refers to the bottom-up, grassroots approach for collaborative mapping without any government, corporate, or academic oversight. For example, two Russian Internet entrepreneurs, Alexandre Koriakine and Evgeniy Saveliev, were inspired by the success of Google Maps and Wikipedia and launched the WikiMapia Web site [5] on May 24, 2006. Its goal is to enable users to interactively edit online geospatial information for areas of their interest. During the past 15 months, geospatial information for more than 4 million places has been added via WikiMapia.

Online mapping using tools available on the Internet was not new to the cartographic community [6]. There has been substantive literature related to multimedia cartography [7], Web cartography [8], and cybercartography [9]. However, all the previous work on Web or cybercartography had focused predominantly on how professional cartographers used Web-based tools for mapmaking. Very few had anticipated the development of wiki mapping and cartography involving the general public at such an unprecedented magnitude, perhaps with the exception of some technical discussions on peer-to-peer data sharing. The wikification of cartography has been manifested in three major aspects related to data, software, and new computing infrastructure/hardware.

17.2.1 Volunteered Geographic Information and Geotagging

The phenomenal growth of wiki mapping and cartography has been driven primarily by the massive and voluntary collaboration among users, both amateurs and experts, of Web 2.0 technology. One of the defining characteristics of Web 2.0 technologies is its capability to allow users to create and broadcast their work with great ease through user-oriented social networks, wikis, blogs, and information-tagging devices, thus turning the Web from a one-way path to access information into a two-way, peer-to-peer exchange. Obviously, wiki cartography is part of the much wider trend of user-created content on the Web, evidenced by the growing popularity of MySpace,

Facebook, YouTube, and even of reality TV or game/competition programs with increasing user/viewer participation. More broadly speaking, I see wiki cartography as a continuation of the earlier success of open-source software development such as the Linux operating system, consumer-driven business development such as eBay, and user-led knowledge production such as Wikipedia. The core of this new trend lies in Web-based mass collaboration, which relies on free individual agents coming together and cooperating to improve a given operation or solve a problem. The business community has regarded mass collaboration as a special type of outsourcing, often referred to as crowd-sourcing within the business community [10]. The cult of amateurism has been described as a defining characteristic of this new societal trend [11].

One of the core enabling technologies for wiki cartography is a *wiki*, a piece of server software that allows users to freely create and edit Web page content using any Web browser. Generally speaking, wikis support hyperlinks and have a simple text syntax for creating new pages and crosslinks between internal pages on the fly [12]. The wind of wikification has also reached the cartographic community during the past two years. More commonly known as collaborative mapping, wiki cartography has been achieved through the combination of Web-based maps and user-generated content, or what Michael Goodchild [13] called volunteered geographic information. Collaborative mapping requires the involvement of individual users to contribute either spatial or attribute information to an online geospatial database, but it does not include generic applications such as wayfinding or navigation where the maps are not meant for user-based modification or editing.

The development of wiki cartography has quietly transformed the masses from being passive consumers to becoming active producers of geospatial information. Although still rather primitive, wiki cartography currently provides users with two general capabilities: map generation and map annotation [14]. In the case of map generation, some Web sites allow users to create maps collaboratively, using GPS devices like OpenStreetMap [15]. For instance, a substantial number of street maps for Dublin, Ireland—a city known for its lack of detailed digital street databases—was created by users of OpenStreetMap. For map annotation, maps from a third-party (such as digital globe or Google Maps) are usually deployed, and users can add their own edited overlays in a wiki fashion; for example, WikiMapia adds user-generated place names and descriptions to locations, and Google Maps/Earth can also allow users to annotate maps through Keyhole Markup Language (KML).

Unlike most traditional cartographic practices, wiki cartography has enabled users to map and visualize many nonconventional forms of information using a cartographic approach [16]. Maps have often been used as metaphors to understand aspects of nonspatial or imaginary worlds [17,18]. Some of these initiatives tie in very closely with the new blogging community [19,20]. GeoURL and Blogmapper allow us to locate blogs all over the world, based on coordinates expressed in latitude and longitude. Others use tube stops as reference points, like the NYCBlogger and Londontubeblogger.

Another interesting development closely related to wiki mapping has been geotagging, which implies georeferencing/coding materials that may have already existed in the Web, ranging from photos [21,22] to the location of wikipedia articles [23–25] to geosocial networking. A good example of geotagging is Placeopedia [26], which is an online gazetteer that integrates Google Maps images (including satellite photos) and Wikipedia encyclopedia articles using user-generated content. Placeopedia was constructed by UK-based mySociety and started in September 2005.

17.2.2 OPEN-SOURCE AND FREE SOFTWARE

In addition to the wikification of cartographic data resulting from volunteered geographic information, both software and hardware used for mapping are also going through a wikification process. In line with the open source/free paradigm, the wikification of cartographic software is expected to continue in the years to come. Several open-source, free mapping software have been produced recently by the volunteer user community [27]. For example, two particle physicists made headlines in 2004 and appeared on the cover of the *Proceedings of the U.S. National Academy of Science*, not due to breakthroughs in particle physics, but for the new algorithm and software for making cartograms they had developed [28]. Perhaps more than professional cartographers, these physicists have popularized the use of cartograms in a widening range of fields from politics to epidemiology and economics [29,30].

As for visualization of textual information, wiki Web sites like touchgraph.com have made the process so easy, just like a simple Google search (www.wikiviz.com [31] can show more wiki tools for visualizing textual information). Although the mapping software dominant in the market is still not free, major software developers like ESRI have started to release some modules of their software in open source. And perhaps more important, the open-source movement has led more software vendors to design their products in a way that encourages and facilitates user-led secondary development and customization for specific applications. In recent years, we have witnessed the rapid growth of "free" GIS software created within the user community for a variety of applications [32–35]. In this spirit, free, open-source software has become one of the driving forces in the wikification of mapping software development [36–38]. Google's KML obviously brought the wikification of GIS software to an even broader community.

Undoubtedly, the open source community will be movers and shakers for the development of Map 2.0. However, it would be naïve to discount the importance of the commercial sector that provides high quality and innovative, standard-based products, which allow users and developers alike to further expand geospatial functionalities and data access without completely relying on open-source tools such as Google's API. Many recent commercial developments have relied on stable and powerful open-source server stacks, including PostGIS, GeoServer, TileCache, and OpenLayers. This armory of free, accessible, stable, and powerful tools, themselves built upon open source libraries such as GeoTools, Proj4, and GDAL within a standards framework, may play a major part in future interoperability of Map 2.0.

17.2.3 GRID COMPUTING AND UBICOMP

In the age of ubiquitous computing, computing infrastructure is increasingly moving toward the thin client and bigger (fat) server model [39]. At the client end, the devices (RFID, smart dust, etc.) have increasingly become smaller and smaller, and can even be implanted inside human bodies. Everything and everybody can be theoretically tracked anywhere and anytime, which demands faster and bigger servers to process the data. The recent development of grid computing [40] has provided the next wave of infrastructure for mapping and geocomputation. In fact, grid computing serves as the perfect, enabling metaphor for wiki cartography from a hardware perspective. Until recently, computer networks in general and the Internet in particular have enabled us to share, exchange, and access information. Little progress has been made toward sharing computing power however, despite the fact that most computers only use about 30% to 40% of their computing resources at any given time. The goal of grid computing is to make networked computers share computing power so that they can work collaboratively to process an increasingly large amount of information requested by clients [41]. The framework proposed by Keith Clarke and his colleagues for grid-based GeoComputation can be considered as a blueprint for the future of geocomputation itself [42,43]. Together with Goodchild's [13] concept of "citizens as sensors," we are gaining a glimpse of future computing infrastructure for wiki cartography—a hybrid of vast numbers of computers (linked in the computing grid) and human sensors (linked by Web 2.0). One thing I should emphasize is that GRID computing highlights the potential battle between top-down standards (required for all the parts to work together) and bottom-up or ad hoc "business"-led standards (e.g., KML, SenseWeb, etc.). The lightweight implementations of standards used by Google and Microsoft are quite different from the Open Geospatial Consortium's (OGC) Sensor Web Enablement (SWE) framework or Geography Markup Language (GML). How and where these standards meet may define how much of the voluminous Map 2.0 data are reusable. It will also be interesting to see how these different approaches can be combined, which may set the stage to test the limits of interoperability.

Mobile mapping technologies will play increasingly important roles in this hybrid network of humans and machines [44,45], especially in the context of the rapidly expanding location-based services (LBS) and the concomitant growth of so-called telecartography [46]. The recent proposed acquisitions of TeleAtlas by TomTom or Navteq by Nokia further testifies to the increasing encroachment of digital geospatial data into the mobile phone user market. Indeed, the cartographers' challenge might be greatly increased when data collection and dissemination are largely carried out through screens no bigger than 6 cm × 4 cm with thumb-controlled joysticks.

17.3 MAP 2.0 AND VISUALIZATION OF THE NATURAL ENVIRONMENT: G-VIS BY THE CROWDS

The integration of traditional cartography with the recent advances in scientific visualization has promoted the growth of geovisualization (G-Vis) [47], and the recent development of wiki cartography and visualization [31] has pushed G-Vis to a new level, practiced by millions of people around the world.

17.3.1 MAP 2.0

The phenomenal growth of wiki cartography has not only resurrected a field that has been pushed aside by GIS, but also forced the rethinking of the definition of the map. Historically, each major technological innovation in cartography has promoted debates about what a map is (or should be). Although Mollering's [48] differentiation between real and virtual maps is still useful, neither his real or virtual maps can capture the fundamental characteristics of maps produced by wiki cartography. Following the general "2.0" neologism, it may not be a bad idea to loosely refer to the typical end product of wiki cartography as *Map 2.0.*

Map 2.0 has been used by Crampton [49] and is listed as one of the categories on the All Things Web 2.0 Web site (allthingsweb2.com), but neither Crampton nor allthingsweb2.com give a precise definition of Map 2.0. Here I'd like to venture to elaborate on the differences between Map 1.0 and 2.0 (Table 17.1). Map 1.0 refers to all maps that are produced without the use of a wiki. Map 2.0 generally refers to maps produced using an online wiki-type Web site. The driving metaphor for Map 1.0 is an image [50] or a model [51], whereas for Map 2.0 the driving metaphor is a wiki, which can be updated and broadcast quickly by large groups of professional and nonprofessional people (Table 17.1). In many interesting ways, Map 2.0 can be compared to what Kraak [52] calls the map plus. The main purpose of Map 1.0 is presentation and its emphasis is on map design; whereas the main goal of Map 2.0 is exploration and its main focus is on map use. In terms of mapmaking methods, Map 1.0 is produced using conventional mapping techniques, whereas Map 2.0 is often made through hacking or mashing-up processes and general understanding/visualization [53–56].

17.3.2 VISUALIZATION OF THE NATURAL ENVIRONMENT: G-VIS BY THE CROWDS

The practice of wiki cartography in the emerging tradition of Map 2.0 has been gaining momentum as evidenced by the growing Google Earth community and numerous other Web sites devoted to wiki cartography. Among the diverse maps posted on the Web so far, quite a few are devoted to the mapping and visualization of the natural environment, and techniques used are quite comparable to those as described in Mach and Petschek [57].

TABLE 17.1
Comparison of Characteristics of Map 1.0 and Map 2.0

Defining Characteristics	Map 1.0	Map 2.0
Medium	Analog/digital	Digital
Data used	Authority	Volunteered
Author	Trained cartographers	Citizens with little training
Dissemination	Atlas/maps, static, local, takes months or years	Wikis/geotagged, mashups, interactive, global, instantaneous
Metaphor	Image/model	Evolving wiki
Method	Mapping, GIS, emphasis on map design following accepted rules	Hacking, mashup, emphasis on map use and fun via continuous experimentation
End product	Maps for presentation	Maps for exploration
Map use	Visualizing data, analytical modeling	Geonarrative, geostorytelling

---

Even a cursory look at the Google Earth KML gallery would quickly reveal that some G-Vis (geovisualization) products for the natural environment are breathtaking. Some of these maps cover almost every aspect of the natural environment—atmosphere, lithosphere, biosphere, and human activities [58]. The maps are made at variable scales, ranging from the local all the way to the global (Figure 17.1; [59]). With the release of Google Sky [60], users now even have the capability to map and mash up at the extraplanetary level. If we also take into account the vast amount of information that can be obtained by biometric and genetic technologies for humans and other species, we now have, for the first time in human history, a remarkable mapping capability that can unify scales all the way from genetic to global/planetary (Figure 17.2). Figure 17.3 show some of the preliminary results of the Harvard AstroMed project—the visualization of star formation during very early stages of the universe using medical imaging technologies. Indeed, as a general process of science 2.0 [61], the Google Earth mashups by amateurs have produced extraordinary results that caught the cartographic establishment by surprise.

17.3.3 Chapters in This Book Part

The chapters included in this book part cover three different topics related to the visualization of the environment: planning hikes, mapping species distribution, and precision agriculture. The chapter by David Lieske and Darren Bender (Chapter 19) examines the role of geostatistics and GIS for visualizing species distributions and understanding large-scale spatial variation in bird breeding. In another study, Alastair Jardine and William Mackaness (Chapter 20) develop a multicriteria risk model and an interesting way to visualize the risks for hikers. Susanne Bleisch and Jason Dykes (Chapter 21) discuss the utility and usability of Web-based 3-D visualization tools for planning hikes. Conrad Rider and Femke Reitsma (Chapter 22) develop a visualization tool (PastureSim) for pasture management. All four chapters are related to wiki cartography in interesting ways, and they also demonstrate the necessity of the kind of work professional GIS/cartographic researchers can and should focus on in the age of wikification.

Recreational uses are one of the fastest growing areas of online geospatial information. In the last few years, online maps have become an integral part of life for millions of Web users around the world. From driving directions to planning a new vacation, more and more people seem to rely on Google Maps/Earth (or similar products) than on traditional paper maps. Several wiki sites have been developed for planning physical activities and recreational purposes; for example, TierraWiki [62] allows the sharing and visualization of routes by volunteers. The premises behind these sites are: (1) that each of us is an expert on something others would be interested in, and (2) that there are the hundreds of areas within our reach that we haven't explored because we don't know enough about them. Sites like TierraWiki aim to use the power of the Web to fix it. Any user can add a GPS track to visualize a route previously done and share it with the community, or link Google Earth to TierraWiki to find new routes in specific areas or visualize routes that were contributed (Figure 17.4). Results from the chapter by Bleisch and Dykes demonstrated

FIGURE 17.1 Google Earth mashup of the vertical dimension of air quality and groundwater pollution.

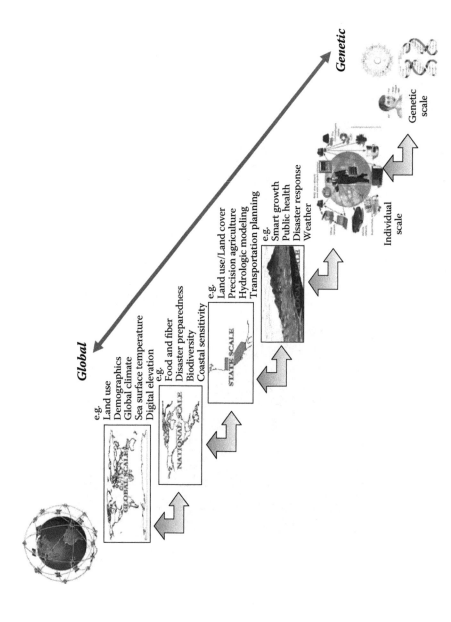

FIGURE 17.2   Mapping across the scale: from genetic to global levels.

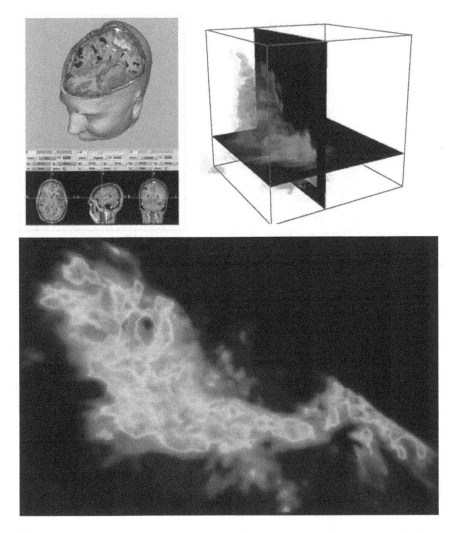

FIGURE 17.3   Harvard's AstroMed Project: linking astronomy with medical imaging.

the utility of Web-based 3-D visualization in planning hiking routes, and yet they also seem to indicate that at least for certain tasks, we cannot completely eliminate the need for paper maps. Most of the wiki sites for recreation could be improved if the methodology for visualizing risks developed by Jardine and Mackaness were incorporated.

In recent years, wiki sites devoted to nature conservation and biodiversity have also grown significantly. Web 2.0 technologies have enabled researchers from multiple disciplines to share, analyze, and visualize data collaboratively. For example, AmphibiaWeb currently (Feb 25, 2008) contains varied information for 6,308 species; the site has posted 1,167 distribution maps, 3,850 literature references, 156 sound files, and 11,479 photos of 2,161 different amphibian species. Figure 17.5 is an

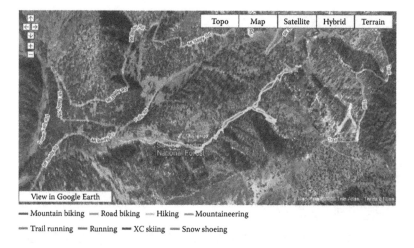

FIGURE 17.4   Hiking route planning by Tierrawiki (http://www.tierrawiki.org).

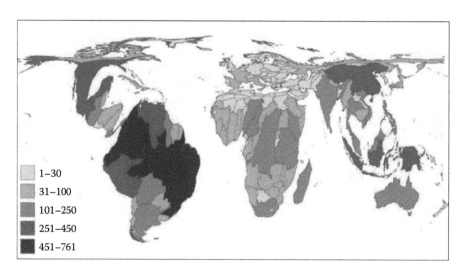

FIGURE 17.5   Cartogram for global amphibian diversity by country.

example of a global cartogram for amphibian diversity. The geostatistical approach developed by Lieske and Bender can be extended to examine the vast amount of species data posted on the Web.

The PastureSim developed by Rider and Reitsma is related to broader literature on precision agriculture. Rider actually has his own Web site [63] with wiki features built in. With a little bit more work, PastureSim can be implemented on the Web to reach a much wider audience. Many existing sites on precision agriculture, such as CropMaps.com, FarmGps.com, PrecisionFarming.com, and so forth have already demonstrated the potential and utility of Web 2.0 for new farming practices.

17.4 FROM G-VIS TO WIKIVIZ: PROBLEMS AND CHALLENGES

The development of wiki cartography has obviously further enhanced geocollaboration and the public participation in the mapping process. But unlike the public participation GIS (PPGIS) of an earlier era [64], the "public participating" in wiki cartography is much larger than that in PPGIS. Until recently, most people have been passive users of the vast geospatial information available online. Wiki cartography is quietly transforming users from being passive consumers to becoming active producers of geospatial information. The phenomenal growth of wiki cartography and Map 2.0 during recent years is both exciting and disturbing for the geospatial community in general, and the cartographic community in particular.

17.4.1 PROFESSIONAL CONCERNS: DEPROFESSIONALIZATION AND REPROFESSIONALIZATION

In January 2007, the Management Association for Private Photogrammetric Surveyors (MAPPS) filed a lawsuit in the United States seeking to limit competition for federal mapping contracts of nearly every type, including GIS services, to just firms of licensed architects, engineers, and surveyors. After months of legal maneuvering, the court ruled against MAPPS, arguing that MAPPS failed to "establish that an injury in fact was suffered by the individual surveyors or their firms" [65]. This is a significant legal victory for the continuing democratization of GIS, which will hopefully ensure all qualified individuals—not just licensed engineers and surveyors—can compete for government contracts related to mapping and GIS activities. Although the geospatial community has applauded the court's ruling in *MAPPS v. U.S.*, the fast-growing collaborative mapping by citizen cartographers obviously has deprofessionalized many of the conventional mapping tasks, and this raises new questions and concerns that deserve the attention of the cartographic community. The extent to which, and time it will take, to reprofessionalize mapping practices and conventions remains to be seen.

17.4.2 TECHNICAL CONCERNS: STANDARDS AND QUALITY ASSURANCES

The accelerating trend of wikification of mapping and GIS will undoubtedly intensify and enhance public participation in both geospatial data application and production at new levels. We can reasonably anticipate that more data will be created at the local and personal level with much improved spatial and temporal resolution, as has been already demonstrated by wiki mapping sites such as WikiMapia, OpenStreet, and Google Earth mashups. This bottom-up approach for geospatial data production initiated by citizens is in sharp contrast to the traditional top-down approach controlled by the government. Concerns have been voiced about standards and the need for quality assurances to ensure VGI's interoperability with the existing data and its contribution to the global geospatial data infrastructure. Without proper protocols and metadata standards, or enforcing metadata standards already established by the cartographic community, wiki cartography can quickly reverse itself into wacky cartography of limited value. In addition, unrestricted access to a wealth of information could pose serious threats to society at multiple scales, especially in areas related to public health and homeland security.

The wikification of geographic information data will very likely result in two parallel universes of those with access to "real" data and those with access only to Map 2.0 data. For many countries in Europe, the availability of high-quality base map data sets is still a contentious issue. The uptake of projects like OpenStreetMap or WikiMapia is in part related to the fact that the data can be reused without licensing restrictions. The power of the wikification of GI data is maximized when these two parallel universes of data are synergistically integrated, which demands more attention that should be given to semantic interoperability.

17.4.3 SOCIAL CONCERNS: PRIVACY AND THE DIGITAL DIVIDE

The development of wiki cartography and VGI is leading to a dramatic increase of geospatial data available at the individual level. The concept of "humans as sensors" may well run the risk of citizens spying on one another. In this age when almost everything in society can go through massive wikification, people are not only concerned about the growing resources of the usual Big Brothers (government and big corporations), but also of many "little brothers" (individuals and small citizen groups/organizations) who are equipped with a wealth of intrusive technologies to collect personal data without permission or even knowledge [66]. In addition to privacy issues, it is also too early to identify the future winners and losers of the wikification process. Will wiki cartography narrow the digital divide along geographic, socioeconomic, racial/ethnic, age, and gender lines, or will it further perpetuate (or worse, even enlarge) the digital divide from local to global levels?

Addressing these problems and concerns must be one of the starting points for setting a new cartographic research 2.0 agenda. The development of wiki cartography and Map 2.0 has been challenging traditional ontology and epistemology in cartography. Map 2.0 fits better with the dynamic, process-oriented cartographic ontology espoused by Kitchin and Dodge [67]. The elevation of hacking to one of the new mapping practices [68,69] resonates well with the geocollaboration cartographers have tried to promote via maps and geovisualization [70]. Wiki sites can certainly be designed in a way to serve as geocollaboratories. The further integration of wiki cartography with the geospatial Web and emerging new spatial media may lead future cartographic practices toward more analytical [71] as well as more artistic traditions [72–74].

17.5 WIKI CARTOGRAPHY AND MAP 2.0: THE MEDIUM AND THE MESSAGE

The recent development of wiki cartography and Map 2.0 has enabled the emergence of millions of citizen cartographers worldwide, now able to map and visualize the natural environment and their lives in many interesting and creative ways. As Fisher [75] articulated, maps and visualization are going to become more, not less, important for the wide adoption of geospatial technologies in our society. With the further development and maturity of the geospatial Web [76,77], mapping and GIS capabilities have increasingly become integral parts of the media [78]. With the emerging spatial media, cartographers are becoming a new type of journalist. Unlike the era of Map 1.0 when journalists became cartographers and used maps to provide additional information for their news stories [79], cartographers and citizens alike have become journalists themselves in the age of Map 2.0, as demonstrated by recent reporting of human rights abuses worldwide by the Science and Human Rights Group at the American Association for the Advancement of Science [80]. We can only hope that cartographers will be able to tell better stories by creatively deploying the new spatial media, thus sending more clear and powerful messages about the human and environmental condition through Map 2.0.

ACKNOWLEDGMENTS

Comments by Dr. Jose Gavinha on an earlier draft are gratefully acknowledged. Thanks are also due to Jeremy Crampton for sharing his paper on Map 2.0.

REFERENCES

1. MacEachren, A. M., Panel presentation during the Plenary Session on "Has GIS killed cartography?", AAG Annual Meeting, Charlotte, NC, 1996.
2. Sui, D. Z., Visuality, aurality, and the shifting metaphors in geographic thought in the late 20th century, *Annals of the Association of American Geographers*, 90(2), 322, 2000.
3. Sui, D. Z. and Goodchild, M. F., Are GIS becoming new media? *International Journal of Geographical Information Science*, 15(5), 387, 2001.
4. O'Reilly, T., What Is Web 2.0: Design Patterns and Business Models for the Next Generation of Software, San Francisco, CA, 2005, Available at: http://www.oreillynet.com/pub/a/oreilly/tim/news/2005/09/30/what-is-web-20.html, accessed September 2007.
5. WikiMapia, www.wikimapia.org, accessed March 2008.
6. Peterson, M. P., *Maps and the Internet*, Elsevier, Amsterdam, 2003.
7. Cartwright, W., Peterson, M. P., and Gartner, G., *Multimedia cartography* (2nd ed.), Springer, Berlin, 2007.
8. Kraak, M. J. and Brown, A., *Web cartography*, Taylor & Francis, London, 2001.
9. Taylor, D. R. F., *Cybercartography: Theory and practice*, Elsevier, Amsterdam, 2005.
10. Tapscott, D. and Williams, A. D., *Wikinomics: How mass collaboration changes everything*, Portofolio, New York, 2006.
11. Keen, A., *The Cult of the Amateur: How today's Internet is killing our culture*, Doubleday/Currency, New York, 2007.
12. Wiki.org, http://www.wiki.org, accessed March 2008.
13. Goodchild, M. F., Citizens as sensors: The world of volunteered geography, *GeoJournal*, 69(4), 211, 2007.
14. Fairhurst, R., *Geowiki: A map which you can annotate*, available at: http://www.geowiki.co.uk, accessed December 2007.
15. OpenStreetMap, http://www.openstreetmap.org, accessed March 2008.
16. Skupin, A. and Fabrikant, S. I., Spatialization, in *Handbook of Geographic Information Science,* Wilson, J. and Fotheringham, S., Eds., Blackwell Publishers, Malden, 2007.
17. Harmon, K., *You are here: Personal geographies and other maps of the imagination*, Architecture Press, Princeton, NJ, 2004.
18. Turchi, P., *Maps of the imagination: The writer as cartographer,* Trinity University Press, San Antonio, 2004.
19. Hansen, B., The GeoURL ICBM Address Server, http://geourl.org/, accessed October 2007.

20. Harlan, J., Blogmapper: Map your blog and blog your map, http:// blogmapper.com, accessed January 2008.
21. Flickr, http://www.flickr.com/, accessed March 2008.
22. Panoramio, http://www.panoramio.com/, accessed March 2008.
23. Alder-Digital, http://alder-digital.de/, accessed March 2008.
24. Pintomap, http://www.pintomap.com/site/en_GB/fs_top/map.html, accessed March 2008.
25. Geonames, http://www.geonames.org/, accessed March 2008.
26. Placeopedia, http://www.placeopedia.com/, accessed March 2008.
27. Teranishi, Y., Kamahara, J., and Shimojo, S., MapWiki: A ubiquitous collaboration environment on shared maps, *International Symposium on Applications and the Internet Workshops,* Phoenix, 2006, 146.
28. Gastner, M. T. and Newman, E. J., Diffusion based method for producing density-equalizing maps, *Proceedings of the National Academy of Sciences* 101, 7499, 2004.
29. Webb, R., Cartography: A popular perspective, *Nature,* 439, 800, 2006.
30. Worldmapper, http://www.worldmapper.org, accessed March 2008.
31. Wikiviz, http://www.wikiviz.com, accessed March 2008.
32. FreeGIS, http://www.FreeGIS.org, accessed March 2008.
33. Zonum Solutions, http://www.zonums.com, accessed March 2008.
34. Batch Geocode, http://www.batchgeocode.com, accessed March 2008.
35. Map-S, http://map-s.de/, accessed March 2008.
36. Kropla, B., *Beginning MapServer: Open source GIS development,* Apress, Berkeley, CA, 2005.
37. Quantum GIS, http://www.qgis.org, accessed March 2008.
38. GeoServer, http://www.GeoServer.org, accessed March 2008.
39. Botts, M., Robin, J., Davidson, A., and Simonis, I., *OpenGIS Sensor Web Enablement Architecture Document,* Open Geospatial Consortium Inc., 2006, Available at: http://www.opengeospatial.org/about/?page=ipr, accessed February 2008.
40. Liebhold, M., *Infrastructure for the new geography,* Institute for the Future, Menlo Park, California, 2004, available at: http://www.iftf.org/docs/SR-869_Infra_New_Geog_Intro.pdf, accessed November 2007.
41. Foster, I. and Kesselman, C., *The Grid 2: Blueprint for a new computing infrastructure,* Morgan-Kaufmann, San Francisco, CA, 2004.
42. Clarke, K. C., Geocomputation's future at the extremes: High performance computing and nanoclients, *Parallel Computing,* 29(10), 1281, 2003.
43. Guan, Q., Zhang, T., and Clarke, K. C., GeoComputation in the grid computing age, in *W2GIS 2006 (Lecture Notes in Computer Science 4295),* Carswell, J. D. and Tezuka, T., Eds., Springer-Verlag, Berlin, 2006, 237.
44. Meng, L., Zipf, A., and Winter, S., *Map-based mobile services: Interactivity and usability,* Springer, Berlin, 2007.
45. Tao, C. V. and Li, J., *Advances in mobile mapping technology,* Taylor & Francis, London, 2007.
46. Gartner, G., Cartwright, W., and Peterson, M. P., Eds., *Location based services and telecartography,* Springer, Berlin 2007.
47. Dykes, J., MacEachren, A. M., and Kraak, M. J., *Exploring geovisualization,* Elsevier, Amsterdam, 2005.
48. Moellering, H., Real maps, virtual maps and interactive cartography, in *Spatial statistics and models,* Gaile, G. L. and Willmott, C. J., Eds., Reidel, Dordrecht, 1980, 109.
49. Crampton, J. W., Map 2.0, *Progress in human geography* (in press).
50. Robinson, A. H. and Petchenik, B. B., *The nature of maps: Essays toward understanding maps and mapping,* University of Chicago Press, Chicago, 1976.
51. Board, C., Maps as models, in, *Models in geography,* Chorley, R. J. and Haggett, P., Eds., Methuen, London, 671, 1967.

52. Kraak, M.-J., *Is there a need for embedded cartographic knowledge in geovisualization software?* 2007, available at: http://www.itc.nl/personal/kraak, accessed February 2008.
53. Butler, D., Mashups mix data into global service, *Nature*, 439, 6, 2006.
54. Miller, C. C., A beast in the field: The Google maps mashup as GIS/2, *Cartographica*, 41, 187, 2006.
55. Erle, S., Gibson, R., and Walsh, J., *Mapping hacks: Tips & tools for electronic cartography*, O'Reilly Media, Sebastopol, CA, 2005.
56. Google Earth Hacks, http://www.gearthhacks.com, accessed March 2008.
57. Mach, R. and Petschek, P., *Visualization of digital terrain and landscape data*, Springer, Berlin, 2007.
58. Google Earth, http://earth.google.com, accessed March 2008.
59. Annotated Earth, http://www.annotatedearth.com/, accessed March 2008.
60. Google Sky, http://earth.google.com/sky, accessed March 2008.
61. Waldrop, M. M., *Science 2.0: Great new tool, or great risk?* 2008, Available at: http://www.sciam.com/article.cfm?id=science-2-point-0-great-new-tool-or-great-risk&print=true, accessed February 2008.
62. TierraWiki, http://www.tierrawiki.org/, accessed March 2008.
63. evilTree, http://www.eviltree.co.uk, accessed March 2008.
64. Sieber, R., Public participation geographic information systems: A literature review and framework, *Annals of the Association of American Geographers*, 96(3), 491, 2006.
65. University Consortium for Geographic Information Science, http://www.ucgis.org, accessed March 2008.
66. Sui, D. Z., *The Streisand law suit and your stolen geography*, GeoWorld, December issue, 26, 2006.
67. Kitchin, R. and Dodge, M., Rethinking maps, *Progress in Human Geography*, 31, 331, 2007.
68. Brown, M. C., *Hacking GoogleMaps and GoogleEarth.* New York: John Wiley & Sons, 2006.
69. Gibson, R. and Erle, S., *Google Maps hacks*, O'Reilly Media Inc., Sebastopol, CA, 2006.
70. MacEachren, A. M. and Brewer, I., Developing a conceptual framework for visually-enabled geocollaboration, *International Journal of Geographical Information Science*, 18, 1, 2004.
71. Thomas, J. J. and Cook, K. A., *Illuminating the path: Research and development of agenda for visual analytics*, National Visualization and Analytics Center, Richland, Washington, 2006.
72. Cosgrove, D., Maps, mapping, modernity: Art and cartography in the twentieth century, *Imago Mundi*, 57, 35, 2005.
73. Wood, D., Map art, *Cartographic Perspectives*, 53, 5, 2006.
74. Tufte, E. R., *Beautiful evidence*, Graphics Press LLC, Cheshire, 2006.
75. Fisher, P. F., Is GIS hidebound by the legacy of cartography? *The Cartographic Journal*, 35, 5, 1998.
76. Maguire, D., *GeoWeb 2.0*, 2006, available at: http://gismatters.blogspot.com/2006/06/¬geoweb-20.html, accessed December 2007.
77. Scharl, A. and Tochtermann, K., *The geospatial web: How geobrowsers, social software and the Web 2.0 are shaping the network society*, Springer, Berlin, 2007
78. Ball, M., Google verifies that GIS is media, *GeoWorld*, 5, 2005.
79. Monmonier, M. S., *Maps with the news: The development of American journalistic cartography*, University of Chicago Press, Chicago, 1989.
80. American Association for the Advancement of Science, http://shr.aaas.org/geotech/ge.shtml, accessed March 2008.

18 Invited Paper
GIS-Based Landscape Visualization—The State of the Art

Andrew Lovett, Katy Appleton, and Andy Jones

CONTENTS

OVERVIEW

This chapter reviews the current state of the art regarding geographic information systems (GIS)-based three-dimensional (3-D) visualization of rural environments. It begins by discussing the reasons for creating visualizations, followed by developments in the availability of data and necessary software tools. Subsequently, some of the practical issues involved are expanded upon in presentations of two examples of recent work. The first of these involved the creation of photorealistic still images to represent a long-term vision for the creation of ecological corridors in part of the Norfolk Broads. In the second example, several different software tools are compared in terms of their ability to generate a real-time visualization showing a rural landscape with increased planting of a biomass crop for energy generation. The chapter concludes with a discussion of issues associated with the use of landscape visualization techniques and identifies some challenges for future research.

18.1 INTRODUCTION

The generation of three-dimensional (3-D) landscape visualizations from geographic information systems (GIS) databases has become much more common during the past decade. Applications of such revisualization techniques have occurred in both urban and rural environments, with a particular focus on the communication of information and facilitating stakeholder engagement in decision making [1–3].

Three main factors have driven the increased use of these visualization tools. A key influence has been a growing recognition of the importance of a landscape-scale perspective in tackling environmental management and planning problems [4,5]. In Europe this has been reflected by measures such as the Landscape Convention, which is now ratified by 29 countries and introduces "landscape quality objectives" into the protection, management, and planning of geographical areas [6]. Another influential piece of legislation is the EU Water Framework Directive, which has set objectives for improving water quality through mechanisms such as river basin management plans [7]. There are also broader debates regarding future land use [8,9] and a common response has been to call for more integrated or "whole landscape" approaches [10,11].

Representing the appearance of alternative future landscapes is an effective way of conveying integrated policy options, reflecting the benefits that visualizations can provide in terms of increasing the accessibility of complex spatial and environmental information to nonexpert users [12–14]. Such activities also relate to a second driver, namely, initiatives that seek to enhance public access to information and participation in planning processes (e.g., the Aarhus Convention of 1998) [15–17]. Several government agencies, such as the Scottish Executive [18], have now identified landscape visualization as a useful means of facilitating community engagement in planning issues.

Third, there have been a series of technical developments that have made it much easier to generate 3-D landscape visualizations. These include advances in computer performance, better integration of software, and significant enhancements in the availability of spatial data [19–20].

Landscape visualization techniques can be applied in both urban and rural contexts, but the challenges and emphases involved are often slightly different (e.g., the representation of buildings is often a key aspect of urban visualizations, whereas vegetation becomes more significant in rural settings). To provide focus, this chapter concentrates primarily on reviewing and illustrating the current state of the art regarding GIS-based 3-D visualizations of rural environments. The following section examines developments in the availability of data and software tools for the creation of visualizations. Subsequently, some of the practical issues involved are expanded upon in presentations of two examples of recent work at the University of East Anglia (UEA), Norwich, United Kingdom. The final section discusses several issues associated with the use of these landscape visualization techniques and identifies some challenges for future research.

18.2 RESOURCES FOR VISUALIZATION

Landscape visualizations may be created for a variety of purposes. These include:

- Testing factors influencing landscape perceptions or preferences
- Illustrating possible scenarios or principles regarding directions of change
- Communication of specific plans (e.g., for a particular site)
- Engagement of stakeholders in discussions of policy options

The reasons for producing a visualization usually have important implications concerning data requirements and the appropriateness of different software techniques. For instance, they may influence the size of area that needs to be shown, the level of feature detail required, and the extent to which functions such as an ability to readily alter viewpoints need to be supported. Factors such as the type of audience involved and the presentation method envisaged (e.g., a printed poster or computer-based display) are also important considerations in the design of visualizations. Developments in data resources and software capabilities are discussed at greater length in the following.

18.2.1 DATA

The creation of landscape visualizations invariably requires information on terrain and land cover characteristics [21]. With respect to the former, a key development has been the supplementation of traditional map-derived digital elevation models with data from a variety of radar-based sensors on aircraft or space platforms. On an international scale this includes the release of 90 m resolution elevation data from the Shuttle Radar Topography Mission (SRTM) [22] in February 2000. At a more local level, an important innovation has been the growing use of interferometric synthetic aperture radar (IfSAR) or light detection and ranging (LiDAR) systems to provide height information at far better spatial resolutions (<10 m) and vertical accuracies than has previously been possible without extensive photogrammetric or surveying work. This additional detail has proved particularly valuable in applications where small variations in elevation can be very significant (e.g., coastal zone management [23]) and, with additional processing, allows the heights of surface features (such as buildings or trees) to be estimated. The use of such active sensing methods is now central to the construction of many 3-D city models [24] and has also been employed to assess vegetation characteristics [25].

Information on land cover is often obtained from aerial photography or satellite imagery. An important development in this context during the past 10 years has been the launch of civilian satellite systems such as IKONOS and QuickBird that have the potential to provide global coverage of 1 m panchromatic and <5 m multispectral imagery. New satellites are likely to improve this resolution to 50 cm panchromatic and <2 m multispectral imagery by the end of 2008 [26,27].

In many parts of the western world this satellite imagery has been supplemented by digital orthophotography at resolutions <1 m, and now increasingly <20 cm for major urban areas [28]. Imagery at such submeter resolutions permits reliable identification of features such as buildings, hedgerows, or trees that are often important for visualization purposes and, in addition to providing surface drapes or textures, can be used to enhance the attribute information included in large scale (better than 1:10,000) digital map databases such as the Ordnance Survey MasterMap® in Great Britain [29]. Such work, however, can be very time consuming [19] and substantial research efforts are currently occurring at the intersection of remote sensing, photogrammetry, and GIS to improve the efficiency with which visualization-ready databases can be created [30,31]. The success of these initiatives will be an important step toward making the generation of 3-D landscape visualization for large areas a more routine exercise.

Other data issues associated with landscape visualization are arising as a consequence of the release (since 2005) of free geobrowsers such as Google Earth [32] and Microsoft Virtual Earth [33]. These Internet-based software tools are having major impacts on public familiarity with geographical information (e.g., they typically use the sub-5 m resolution imagery mentioned earlier as a base layer) and are also starting to change expectations of the content provided in visualizations. Furthermore, it is relatively straightforward for users to create their own 3-D content (e.g., building models) using tools such as SketchUp [34] and then add this to the geobrowser display. Given time, there is the potential for this volunteered information to become an important resource for many GIS applications (including landscape visualization), but it also raises questions regarding data reliability and the scope for digital divides in terms of where such content is provided and by whom [35].

At present, it is possible to mash up (i.e., integrate) geobrowser displays with other spatial data to create various forms of georevisualization [36], but the restrictions of Internet bandwidth mean that this option has limitations as a platform for landscape visualization compared to the desktop software discussed next. Nevertheless, in 5 to 10 years this could change appreciably and might mean that landscape visualizations could be generated in a much more collaborative and distributed manner in the future.

18.2.2 Software and Display Facilities

The history of landscape visualization can be traced from early drawings and models through augmented photographs to computer-based methods [37]. Developments in the technology employed in computerized visualization have been driven by the needs of major users such as the military, entertainment industries, and some government departments (e.g., the US Forest Service). One consequence of this was that until at least the mid-1990s advances in this type of software were on a rather separate track from those in GIS and it was not straightforward to import data from the latter into the former. Following the introduction of desktop personal computers (PCs) and their rapid improvement in performance, the integration of GIS and 3-D visualization has gradually become much stronger; one early manifestation of this being the representation of GIS data as 3-D models using Virtual Reality Modeling Language (VRML) [38]. Today there are many examples of both GIS programs with 3-D display capabilities (e.g., ArcGIS 3D Analyst [39]), and more specialist visualization software that can directly import GIS data formats and utilize associated metadata, such as information on map projections (e.g., Visual Nature Studio [40]).

At present, the outputs from GIS-based landscape visualization tools vary across a range that includes photorealistic still images, animated sequences, and real-time models where the user can interactively alter their viewing position or the environmental content [19,41]. The latter includes models that can be shown on a laptop or desktop PC, as well as output generated through more sophisticated (and costly) software. Examples of the latter include Multigen Creator [42] or Terra Vista [43] where the output is displayed through a real-time viewer (e.g., Multigen Vega Prime [44] or Quantum 3-D Mantis [45]) and a multiprojector system onto a large screen (see Figure 18.1). Such a presentation method provides a greater sense of visual immersion than a PC screen and can also have the capacity to provide a stronger sense of depth through stereo display. Another innovation has been portable virtual reality theaters [46] or equipment such as the Elumens VisionStation [47] (see Figure 18.2). These seek to combine a greater sense of immersion than is provided by a laptop or standard projector display with the flexibility to take such facilities directly to different audiences.

Real-time visualization has long been seen as the logical next step from still images and animations. Arguably, the sense of control provided by a real-time model should enhance the viewing experience (e.g., allowing users to select landscape perspectives that are most relevant to them) and so increase public engagement in decision making, but empirical evaluation of such potential benefits is currently quite limited [46,48]. There is also the issue that the creation of a real-time model often involves some element of trade-off between interactivity (particularly smoothness of movement) and the detail with which features are represented. This is an important consideration because several studies have found that the level of detail, particularly for foreground features, is a key influence on how people relate to such computer-generated visualizations [49–51].

FIGURE 18.1 Multiprojector display system and 125-degree curved screen at UEA.

FIGURE 18.2   Portable visualization display for real-time landscape models.

As the graphics performance of PCs has improved, the degree of compromise required between interactivity and feature detail has declined. Nevertheless, it is still a real issue in certain contexts, examples including urban areas with detailed photographs displaying textures on buildings or heavily vegetated landscapes with many thousands of plants or trees. There are several methods for representing vegetation in landscape visualizations including terrain textures, billboard images, geometric solids, or hybrids of the latter two [21,52]. Techniques also exist for producing plant models through procedural geometry or rule-based methods, and for generating realistic ecosystems of plants to cover defined geographical areas [53–55].

Two key issues in the visualization of landscapes with many plants or trees are the ability of software to mix several of the different representation methods, and to dynamically control levels of detail so that foreground features appear as realistic as possible, while those in the distance are simplified to reduce the processing load. An excellent example of the current state of such capabilities is provided by the Lenné 3D player [56,57]. Nevertheless, there is undoubtedly room for further work on the relevant data structures and algorithms to improve the interactivity and feature detail of real-time displays.

The following sections expand on some of the issues regarding data and software by presenting two recent applications of GIS-based landscape visualization methods. These examples also serve to illustrate the types of output currently possible. The first study involved the creation of photorealistic still images to represent a long-term vision for the creation of ecological corridors in part of the Norfolk Broads and the discussion focuses particularly on some of the data integration issues involved in the research. In the second example the emphasis is on real-time visualization, and several different software tools are compared in terms of their ability to show a rural landscape with increased planting of a biomass crop for energy generation.

18.3 PHOTOREALISTIC IMAGES OF LANDSCAPE CHANGE IN BROADLAND

18.3.1 BACKGROUND TO THE VISUALIZATION PROJECT

The Norfolk and Suffolk Broads is a unique area of water, grazing marshes, fen, and woodland in the United Kingdom. Originally created as a result of the uncontrolled flooding of medieval peat diggings, an area of approximately 300 km² is now designated as a national park [58]. Broadland is home to numerous species of flora and fauna that are not found elsewhere in the United Kingdom and contains 28 Sites of Special Scientific Interest (SSSIs), amounting to 7,000 ha in total, which benefit from protection either as Special Protection Areas (SPAs) or Special Areas of Conservation (SACs) under European law. However, economic, environmental, and social changes pose considerable challenges for the future management of such sites and it is recognized that there is a need to move toward more integrated strategies at the catchment or landscape scale [59].

One area of Broadland that has changed considerably as a result of historical land use change is the region in the northern Broads around the Upper Thurne, Ant, and Bure SSSIs. These unique habitats, protected by their SSSI status, were once joined by areas of heath, grassland, and woodland. Today the sites are separated by tracts of arable land through which run a number of busy roads, most notably the A149. One option for the future management of this area would be to convert sections of the present arable land into a mixture of heath, grassland, or woodland, thus creating a series of ecological corridors that would facilitate species movement between the protected sites. Furthermore, the busiest roads could be bridged by a number of ecobridges. However, if such interventions were implemented, there would be considerable changes to the existing landscape.

In 2006 the Broads Authority asked researchers at UEA to prepare a set of visualizations showing some hypothetical changes to the landscape of the area on a timescale of 15 to 20 years in the future (i.e., the 2020s). These visualizations were designed so that they could be used to help inform and stimulate discussion at a stakeholder workshop discussing the possibility for such long-term managed change in the Broads.

18.3.2 DATA SOURCES AND METHODS

Staff at the Broads Authority provided some sketch maps depicting the land use changes that could potentially take place in the study area. They also supplied a set of digital spatial information for a rectangular area (some 6 km by 9 km in extent) covering the region of interest. These data included:

- Ordnance Survey MasterMap topographic data [29] (polygons)
- Land cover details dated 1998 from the Rural Development Service (RDS, now Natural England [60]; polygons)

- Intermap Technologies [61] digital elevation model (DEM) and digital surface model (DSM; 5 m raster)
- LiDAR elevation information supplied by the Environment Agency [62]
  (1 m points)
- BlueSky color aerial imagery [28] (25cm, acquired 2004)

Additional information was obtained from the MAGIC geodata portal [63] (SSSI boundaries) and Edina Digimap service [64] (raster imagery of Ordnance Survey 1:50,000 and 1:25,000 mappings).

A polygon database of existing land cover/use information was prepared in ESRI ArcGIS 9 [65]. The OS MasterMap was used as an initial framework, but the attribute details in this are not good at distinguishing some key types of rural land use (e.g., arable and grassland), so the RDS data was used to supplement it and create an extended classification. This exercise involved a significant amount of editing in the GIS and a number of feature boundaries did not coincide, so a variety of polygon-to-point, point-in-polygon, and spatial join operations were used to transfer the RDS land use/cover codes to the MasterMap polygons.

Hedges and trees are important features providing structure in the landscape, but they were not included in the information provided by the Broads Authority or other sources. Hedges were therefore digitized from the aerial imagery and then densified (1 vertex per 2 m) to meet the requirements of the visualization software. Single trees are numerous and so a procedural method was devised rather than relying on hand-digitizing each position individually. As both a DSM and DEM were available, the latter was subtracted from the former to identify all areas where there were landscape features above ground. Areas of significant difference (>5 m) were extracted and the resulting raster was converted to points, which were then given the attribute of the land use/cover at their location. Points in areas where there were already trees (e.g., woodland or scrub) or where trees would not be present (e.g., roads, buildings, or water), were removed from the data set, and those in wet areas were given an identifying attribute so as to be able to use appropriate trees in the visualizations. As there were still almost 250,000 points after these operations, 10% of them were randomly selected for use in the visualizations (visual comparison with the aerial imagery was used to arrive at 10% as an appropriate proportion). The random 10% was chosen with a tool from the Hawth's Analysis Tools extension for ArcGIS [66].

Following the stage of data preparation, a series of still image visualizations were created in Visual Nature Studio (VNS) [40]. This software reads GIS data files in a variety of formats and projections, and allows the user to build a virtual representation of the landscape in question based upon the imported information. The form of the landscape is taken from the DEM, and other features are laid on top according to point, line, polygon, or raster information. These features may comprise the color and texture of the ground surface, two- or three-dimensional representations of buildings, vegetation, people, and objects, and the simulation of water, as well as physical modifications to the terrain surface itself such as banks and ditches. Global attributes such as light and shadow, atmospheric haze, and sky features can also be adjusted.

The GIS data were structured to take account of some of VNS's inbuilt tools. In this study the most important were the abilities to read in attributes from a GIS database, perform simple queries to identify particular database items, and associate database items and visual properties with individual scenario/viewpoint combinations. For example, splitting the current land use/cover data into "change" and "no change" areas allowed relevant features to be easily selected or disabled when working with a view of the future state. Furthermore, when particular database items were enabled, the relevant visual parameters were automatically applied to them based on their land use/cover classification. These sorts of capabilities are especially useful for generating multiple scenario visualizations from a database.

An "ecosystem" (a collection of visual parameters including ground cover and vegetation images) was created in VNS for each land use/cover category in the database. Appropriate colors and textures were defined with reference to the aerial imagery and through feedback from Broads Authority staff; many textures were computer-generated variations on a random fractal noise pattern, but for the areas of farmland or unknown land use/cover, the aerial imagery (downsampled to 1 m) was applied as an image texture.

Billboard vegetation items were applied from the library that comes with VNS, from images developed for previous visualization projects, or from fieldwork photographs taken in the area in July 2006. Photographs were edited to black out unwanted background, leaving the required tree or other feature for the software to place within the landscape. Due to limitations within the range of available images, the species used were not always 100% accurate, but were used to achieve an appearance (e.g., of deciduous woodland) that was generally correct. Other features shown as billboards included people, animals, and small watercraft; again, the images either came from previous projects or from royalty-free images found online at Geograph [67]. Some objects within the visualizations (e.g., bridges and larger boats) were created as simple 3-D objects in Wings 3D [68]. Others (including all buildings) were created as simple extrusions of their footprint in the MasterMap data, given a brick-tone wall and a gray roof. Creating more detailed building models (e.g., with photo textures) is quite possible but was not considered important in the context of this project.

Final still images for eight viewpoints were rendered at high resolution (3360 × 2520 pixels) to allow printing at large sizes if required. Memory requirements for these images meant that they were rendered in several parts by VNS and were then merged to make a final image. Rendering time was between approximately two and six hours per image (largely dependent on the amounts of high-resolution photography and vegetation visible, with realistic volumetric clouds and water reflections also adding to the time taken). Two 60-second (1200 frame) animations were also generated to provide flyovers of the key landscape changes in the north and south of the study area.

18.3.3 EXAMPLE VISUALIZATIONS AND STAKEHOLDER RESPONSE

Figures 18.3 and 18.4 show examples of the types of overview landscape visualizations produced. Both images show a view looking northeast, with part of the wetland around the River Ant in the foreground, the village of Catfield in the mid distance, and the western end of Hickling Broad visible in the top-right corner. Figure 18.3 shows the existing landscape, while Figure 18.4 illustrates how this could change with corridors of heath and woodland across the arable fields.

FIGURE 18.3   Overview VNS rendering of existing landscape in the northern Broads.

An illustration of a possible future landscape at a site slightly further to the south is shown in Figure 18.5. This image looks northwest towards the confluence of the Rivers Bure (on the left) and Ant (on the right). The potential landscape changes shown are: restored heathland between the two rivers, and extended wetland (including a new broad) on the eastern side of the River Ant. Ground-level views of these areas of change are presented in Figures 18.6 and 18.7. The former shows a possible view on a footpath through the heathland and the latter an illustration of how the new broad could be used for recreation purposes.

Taken together, these examples illustrate the high degree of feature detail that is now possible to include in still image visualizations. There is obviously an element of artistic interpretation in producing such visualizations (especially the ground-level views in Figures 18.6 and 18.7), but one important strength of the GIS-based approach is the transparency provided by basing representations on actual map information. Other advantages are the variety of data that can be incorporated into a visualization and, with software such as VNS, the flexibility to generate output depicting a range of different scenarios, options, or viewpoints in an efficient manner.

A selection of the visualizations were shown to some 60 participants who attended a meeting of the Broads Research Advisory Panel on the theme of "Designing Land Use Change in the Broads" in October 2006. During subsequent discussion [69], the value of the visualizations as a means of generating interest in a future landscape vision was widely recognized, but it was also argued that such images needed to be supplemented with information on other possible impacts (e.g., with respect to biodiversity, the agricultural economy, and water resources).

FIGURE 18.4   Overview VNS rendering of possible northern Broads landscape with ecological corridors.

Other comments concerned the level of financial resources that would be required to implement such land use changes over a large area. It was also suggested that care would be needed if the visualizations were to be presented to individuals whose livelihoods would be directly impacted (e.g., farmers). The level of detail in the visualizations was seen as a positive asset, but it was also noted that this could be both distracting (i.e., attracting attention to specific details when the general principle involved was more important) and deceptive (i.e., providing a sense of definiteness when there were many uncertainties involved). Similar opinions have been expressed in several other studies involving the use of landscape visualizations to support decision making [70,71].

18.4   REAL-TIME VISUALIZATION OF BIOMASS CROPS

18.4.1   Background to the Visualizations

Energy crops are becoming increasingly recognized as an important, if sometimes controversial, means by which many countries can increase renewable energy generation and help achieve targets of reduced greenhouse gas emissions [72]. There are several types of energy crops, including cereals, maize, oilseed rape, and sugar beet, that can be used to produce transport fuels, and other forms of grasses and trees that supply biomass that can be processed to generate heat or electricity. One type of perennial biomass crop, miscanthus grass, is currently planted on over 5,800 ha of agricultural land in England. A recent government report [73] estimates that the area of such crops in the United Kingdom might expand to 350,000 ha by 2020, a total that

FIGURE 18.5   VNS rendering of possible future landscape at the confluence of the Rivers Ant and Bure.

FIGURE 18.6   Ground-level VNS rendering of potential restored heathland.

FIGURE 18.7   Ground-level VNS rendering of potential new wetland and broad.

would equate to planting on about 10% of arable land in some regions. However, these perennial energy crops are physically different than most current rural land uses; they are in place for 7 to 25 years, harvest is normally early spring (February–March), and they are dense and tall (3–4 m). These factors mean that expansion in planting has the potential to modify the rural landscape, with particular implications for visual appearance, cultural heritage, tourism, farm incomes, hydrology, and biodiversity [74]. Such issues are of interest to a range of government agencies and nongovernmental organizations, and have stimulated discussion regarding the development of planning policies and tools to maximize the benefits of planting [75,76].

   The social, economic, and environmental implications of increased UK rural land under perennial energy crops are currently being investigated as part of the RELU-Biomass project [77]. One component of the project is examining public attitudes about energy crops and it is planned to use landscape visualizations to illustrate the potential impacts of different scales of planting to focus group audiences. The ability of participants to select their own viewpoints can be important in such a context and, consequently, real-time visualizations are potentially useful. An issue, however, is the capability of such tools to display large amounts of vegetation in a realistic manner. Therefore, as a prelude to the focus groups, a pilot assessment was conducted to compare the real-time landscape modeling capabilities of (1) the 3D Analyst extension of ArcGIS 39, (2) Visual Nature Studio 40 with the Scene Express extension and the NatureView Express viewer, and (3) the Lenné 3D player [78]. This selection of software was chosen to represent a range from a general-purpose GIS, through a commercial landscape visualization product capable of generating a wide variety of outputs, to research tools designed specifically for real-time capabilities. Further details of the assessment are given by Lovett et al. [79].

18.4.2 Data Sources and Methods

A suitable pilot area was identified near the village of Dunholme in Lincolnshire, eastern England. Lincolnshire is part of one of the major current concentrations of miscanthus cultivation in England and the site near Dunholme was the subject of a successful application for a miscanthus planting grant under the energy crops scheme [80]. However, as of late 2007, no planting had taken place. As a consequence, it was possible to collect information on the existing land uses and visualize how the area could change under different miscanthus planting scenarios.

For the purposes of comparing the different real-time modeling tools it was decided to focus on creating visualizations for an area covering 1,400 m by 1,400 m on the western side of Dunholme. Digital data assembled for this area included a 5 m resolution NEXTMap Britain DEM [81], a 25 cm BlueSky orthophoto [28] (acquired in June 2003), and Ordnance Survey MasterMap vector mapping [29]. Attributes from the MasterMap data were combined in ArcGIS 9 [65] to produce a polygon land cover data set, to which information on specific crops grown in 2007 was added. Details of tree positions and hedgerows were digitized from the orthophoto, and photographs taken during site visits were used to create billboard and texture images for relevant vegetation. Other visits were made to sites where miscanthus is already grown to obtain photos of the crop.

A variety of 3-D building models for the area were created by combining polygon footprints extracted from the MasterMap data with textures and height details from photos taken during field visits. The building models were generated in ModelBuilder 3D (a subset of MultiGen-Paradigm Creator [42]) and then converted from OpenFlight to 3D Studio format using Deep Exploration [82] and Polytrans [83].

The real-time models were all created on PCs with at least a 1 GHz processor, 1 Gb RAM, and a graphics card with 256 Mb memory. Several publications have discussed the general procedure involved in creating real-time models with the software tools used [19,41,57], so this information will not be repeated here. Two points that should be noted, however, are that the more specialized visualization software was much better than ArcGIS in allowing feature attributes to be efficiently linked to billboard images or 3-D models of vegetation, and also in assigning these representations to vertices along a line (e.g., to represent a hedge) or filling a polygon (e.g., to depict a field of miscanthus). During the research it was found that the real-time viewers for ArcGIS (ArcGlobe) and VNS (NatureView Express, NVE) could not cope with the large number of miscanthus models involved (some 200,000 plants in a 10 ha field) and therefore compromises has to be made in terms of either the density shown or area covered. The Lenné 3D player performed much better in this respect, since it has a more efficient level of detail control to reduce scene complexity and can automatically switch between textured polygon mesh plant models for foreground detail and billboards for mid or background representation. This meant that it was possible to include several million separate plant models in the Dunholme scene.

18.4.3 EXAMPLE VISUALIZATIONS AND INTERACTIVITY

Figure 18.8 presents four close-up visualizations of a field with miscanthus. The images were generated from the same viewpoint, but use four different software tools. A standard in terms of feature detail is provided by Figure 18.8a, which is a still image rendered from VNS, whereas the other three are screenshots from real-time models. There are significant contrasts in visual appearance, highlighting both the differences between the systems and the subjectivities involved in creating such visualizations. With respect to the representation of vegetation, one clear difference is that the 3D player output is the only one that can display plants in the buffer strip around the miscanthus with similar detail to the still render. In the other two views this area has a rather bare appearance, which varies according to how the ground texture image is handled.

There are also differences in how the miscanthus crop is shown. With ArcGlobe the density of plants is lower, the models are quite regularly spaced and sized, and there is a halo effect around their edges, which probably stems from how transparency information (in the alpha channel) is handled. The other real-time outputs do not show this problem and there is a more plausible variation in the appearance and height of the miscanthus plants.

Another important characteristic of real-time visualizations is the ease of interactively moving through a landscape model. All of the viewing environments used provided a good range of navigation modes, but had problems with slower frame rates when moving in proximity to the miscanthus plants. For example, NVE achieved a rate of 12 frames per second (fps) with a distant view, an average of ~4 fps when approaching the miscanthus plot, and ~2 fps when moving around or through the plants and viewing at close range. At the latter speed, control tends to become difficult as it is easy to issue multiple consecutive navigation commands before the first one has completed. One solution to this problem is to predefine sets of viewpoints and move between them on a tour sequence. However, such routes need to be planned with care to avoid disorientating jumps (e.g., passing quickly through a hill or woodland) during transit from one viewpoint to another. An alternative option is provided by the 3D player, which uses a preview facility to assist with navigation when frame rates are low.

Several other aspects of interactivity merit comment because of their potential value in decision-making contexts. For instance, one strength of ArcGlobe is the integration of the display capability within a GIS, which provides query/analysis facilities and a ready ability to change the displayed layers on the fly. Making such alterations is difficult in the 3D player, and in NVE alterations are essentially restricted to switching the display of entire classes of features (e.g., building or vegetation) on or off.

Further work is currently in progress using VNS, NVE, and the Lenné 3D player to produce visualizations of planting scenarios for perennial energy crops at several sites in the East Midlands and South West regions of England. Subsequently, both rendered stills and real-time visualizations will be shown to focus group audiences (e.g., using the type of portable display shown in Figure 18.2) as part of research to examine public attitudes about different scales of energy crop planting.

(a)

(b)

FIGURE 18.8   Views of miscanthus crop produced using four different software tools.
a) VNS rendered still image b) Nature View Express screenshot c) Lenné 3D-Player screenshot
d) ArcGlobe screenshot.

(c)

(d)

18.5 CONCLUSIONS: CURRENT ISSUES AND FUTURE CHALLENGES

Generating 3-D landscape visualizations from GIS databases is now much more straightforward than was the case a decade ago. Still images can be rendered with a high degree of feature detail and some real-time systems are beginning to match them in terms of photorealism. It is still the case, however, that considerable data integration and editing is often necessary to produce a GIS database suitable for landscape visualization purposes. Furthermore, the argument of Orland et al. in 2001 [84] concerning the tendency for technical advances in visualization to occur faster than improvements in understanding about how to best use them for planning or decision support is still very relevant. Studies are starting to appear that assess the benefits and ethics of employing different visualization techniques in practical decision-making contexts [85–89], but undoubtedly more research of this type is required to develop robust guidance concerning the appropriate design and use of visualizations.

Current developments suggest that photorealistic real-time visualization systems will soon become quite commonplace. A key technical challenge for the future will be to extend other aspects of interactivity, particularly the ability of audience members to change the content of the environment they are examining. Such a capability to make on-the-fly changes has been identified as an important capability for stakeholder engagement [87] and examples that make use of personal digital assistants or other handheld devices are beginning to appear in the research literature [46,90]. This type of what-if ability requires efficient integration of GIS databases, modeling, and visualization tools, and is something that ongoing advances in distributed GIS and grid computing are likely to facilitate [91]. Other technical challenges concern the representation of uncertainty or nonvisual phenomena in landscape visualizations, and these are areas where insights from the wider geographic information science literature may prove useful [92]. Overall, it is fair to say that the field of GIS-based landscape visualization has made some significant advances in the past decade and there are good prospects for further technical innovations and improvements in understanding regarding applications that will help enhance decision making and planning with respect to the natural environment.

ACKNOWLEDGMENTS

Funding support for the research discussed in this chapter was provided by the Broads Authority, the Economic and Social Research Council (ESRC) Programme on Environmental Decision Making, the Rural Economy and Land Use (RELU) program of the UK Research Councils, and a British Council/German Academic Exchange Service ARC grant. We would like to thank Philip Paar (Lenné 3D), Lutz Ross (Technical University of Berlin), and Gilla Sünnenberg (UEA) for their contributions to the biomass crop visualizations. Helpful comments from an anonymous reviewer are also appreciated. The NEXTMap DEM data were supplied by the NERC Earth Observation Data Centre, and the Ordnance Survey MasterMap data are Crown copyright/database right 2008. An Ordnance Survey/Digimap supplied service.

REFERENCES

1. Hudson-Smith, A. and Evans, S., Virtual cities: From CAD to 3-D GIS, in *Advanced Spatial Analysis: The CASA Book of GIS*, Longley, P. A. and Batty, M., Eds., ESRI Press, Redlands, CA, 2003, 41.
2. Bishop, I. and Lange, E., Eds., *Visualization in Landscape and Environmental Planning*, Taylor & Francis, London, 2005.
3. Lovett, A. and Appleton, K., Eds., *GIS for Environmental Decision-Making*, CRC Press, Boca Raton, FL, 2008.
4. Hobbs, R., Future landscapes and the future of landscape ecology, *Landscape and Urban Planning*, 37, 1, 1997.
5. Selman, P., *Planning at the Landscape Scale*, Routledge, London, 2006.
6. Council of Europe, The European Landscape Convention, http://www.coe.int/t/dg4/cultureheritage/ Conventions/Landscape/default_en.asp, accessed April 2008.
7. European Commission, The EU Water Framework Directive – integrated river basin management for Europe, http://ec.europa.eu/environment/water/water-framework/index_en.html, accessed April 2008.
8. Dwyer, J., The state of the rural environment in Europe: What challenges and opportunities for future polices, Background paper for the Land Use Policy Group and Bundesamt für Naturschutz conference, Future Policies for Rural Europe 2013 and Beyond— Delivering Sustainable Rural Land Management in a Changing Europe, 19–20 September 2007, Brussels, available at: http://www.lupg.org.uk/Default.aspx?page=29, accessed February 2008.
9. Foresight, Land use futures project, http://www.foresight.gov.uk/LandUse/LandUse.html, accessed April 2008.
10. Cobb, D., Dolman, P., and O'Riordan, T., Interpretations of sustainable agriculture in the UK, *Progress in Human Geography*, 23, 209, 1999.
11. MacFarlane, R., Multi-functional landscapes: Conceptual and planning issues for the countryside, in *Landscape and Sustainability* (2nd ed.), Benson, J. F. and Roe, M., Eds., Routledge, London, 2007, 138.
12. Bishop, I., The role of visual realism in communicating and understanding spatial change and process, in *Visualization in Geographic Information Systems*, Unwin, D. and Hearnshaw, H., Eds., Belhaven Press, London, 1994, 60.
13. Tress, G. and Tress B., Scenario visualization for participatory landscape planning—a study from Denmark, *Landscape and Urban Planning*, 64, 161, 2003.
14. Meitner, M. J., Sheppard, S. R. J., Cavens, D., Gandy, R., Picard, P., Harshaw, H., and Harrison, D., The multiple roles of environmental data visualization in evaluating alternative forest management strategies, *Computers and Electronics in Agriculture*, 49, 192, 2005.
15. United Nations Economic Commission for Europe, Introducing the Aarhus Convention, http://www.unce.org/env/pp/, accessed April 2008.
16. Dunn, C. E., Participatory GIS—a people's GIS? *Progress in Human Geography*, 31, 616, 2007.
17. Bishop, I., Developments in public participation and collaborative environmental decision making, in *GIS for Environmental Decision-Making*, Lovett, A. and Appleton, K., Eds., CRC Press, Boca Raton, FL, 2008, 181.
18. Scottish Executive Development Department, Community Engagement—Planning with People, Planning Advice Note 81, Scottish Executive, Edinburgh, 2007.
19. Appleton, K., Lovett, A., Sünnenberg, G., and Dockerty, T., Rural landscape visualization from GIS databases: A comparison of approaches, options and problems, *Computers, Environment and Urban Systems*, 26, 141, 2002.
20. Lovett, A., Futurescapes, *Computers, Environment and Urban Systems*, 29, 249, 2005.

21. Ervin, S. M. and Hasbrouck, H. H., *Landscape Modeling: Digital Techniques for Landscape Visualization*, McGraw-Hill, New York, 2001.

22. NASA Jet Propulsion Laboratory, Shuttle Rader Topography Mission, http://www2.jpl. nasa.gov/srtm/, accessed February 2008.

23. Brown, I., Jude, S. R., Koukoulas, S., Nicholls, R., Dickson, M., and Walkden, M., Using virtual reality to simulate coastal erosion: A participative decision tool?, in *GIS for Environmental Decision-Making*, Lovett, A. and Appleton, K., Eds., CRC Press, Boca Raton, FL, 2008, 193.

24. Smith, S. L., Urban remote sensing: The use of LiDAR in the creation of physical urban models, in *Advanced Spatial Analysis: The CASA Book of GIS*, Longley, P. A. and Batty, M., Eds., ESRI Press, Redlands, CA, 2003, 171.

25. Balzter, H., Luckman, A., Skinner, L., Rowland, C., and Dawson, T., Observations of forest stand top height and mean height from interferometric SAR and LiDAR over a conifer plantation at Thetford Forest, UK, *International Journal of Remote Sensing*, 28, 1173, 2007.

26. GeoEye, GeoEye products, http://www.geoeye.com/products/default.htm, accessed February 2008.

27. DigitalGlobe, http://www.digitalglobe.com, accessed February 2008.

28. BlueSky, Aerial photography products, http://www.bluesky-world.com/products/, accessed February 2008.

29. Ordnance Survey, OS MasterMap, http://www.ordnancesurvey.co.uk/oswebsite/products/ osmastermap/, accessed February 2008.

30. Lees, B. G., Remote sensing, in *The Handbook of Geographic Information Science*, Wilson, J. P. and Fotheringham, A. S., Eds., Blackwell, Oxford, 2008, 49.

31. Tao, V. C., Empowering the web with location, CD Proceedings of Virtual Geographic Environments: An International Conference on Developments in Visualization and Virtual Environments in Geographic Information Science, Chinese University of Hong Kong, January 2008.

32. Google Earth, http://earth.google.com, accessed February 2008.

33. Microsoft, Microsoft Virtual Earth, http://www.microsoft.com/virtualearth/, accessed February 2008.

34. Google SketchUp Home, http://www.sketchup.com, accessed February 2008.

35. Goodchild, M. F., Virtual geographic environments as collective constructions, CD Proceedings of Virtual Geographic Environments: An International Conference on Developments in Visualization and Virtual Environments in Geographic Information Science, Chinese University of Hong Kong, January 2008.

36. Simonite, T., Virtual Earths let researchers "mash up" data, *Newscientist.com*, 2 May 2007, Available at: http://technology.newscientist.com/channel/tech/dn11773-virtual-earths-let-researchers-mash-up-data-.html, accessed February 2008.

37. Lange, E. and Bishop, I., Communication, perception and visualization, in *Visualization in Landscape and Environmental Planning*, Bishop, I. and Lange, E., Eds., Taylor & Francis, London, 2005, 3.

38. Fisher, P. and Unwin, D., Eds., *Virtual Reality in Geography*, Taylor & Francis, London, 2002.

39. ESRI, ArcGIS 3D Analyst, http://www.esri.com/software/arcgis/extensions/3danalyst/ index.html, accessed February 2008.

40. 3D Nature, Visual Nature Studio, http://3dnature.com/vnsinfo.html, accessed February 2008.

41. Appleton, K. and Lovett, A., Visualizing rural landscapes from GIS databases in real-time—a comparison of software and some future prospects, in *The ASPRS Manual of GIS*, Madden, M., Ed., American Society for Photogrammetry and Remote Sensing, Maryland, in press.

42. Presagis, Creator overview, http://www.presagis.com/products/multigen_paradigm/ details/creator/, accessed February 2008.

43. Presagis, Terra Vista overview, http://www.presagis.com/products/terrex/details/terra_ vista/, accessed February 2008.

44. Presagis, Vega Prime overview, http://www.presagis.com/products/multigen_paradigm/ details/vegaprime/, accessed February 2008.

45. Quantum 3D, Mantis image generator, http://www.quantum3d.com/products/Software/ mantis.html, accessed February 2008.

46. Miller, D., Morrice, J., Coleby, A., and Messager, P., Visualization techniques to support planning of renewable energy developments, in *GIS for Environmental Decision-Making*, Lovett, A. and Appleton, K., Eds., CRC Press, Boca Raton, FL, 2008, 227.

47. Appleton, K. and Lovett, A., Display methods for real-time landscape models—an initial comparison, in *Trends in Real-Time Landscape Visualization and Participation*, Buhmann, E., Paar, P., Bishop, I., and Lange, E., Eds., Wichmann, Heidelberg, 2005, 246.

48. Bishop, I., Visualization for participation—the advantages of real-time?, in *Trends in Real-Time Landscape Visualization and Participation*, Buhmann, E., Paar, P., Bishop, I., and Lange, E., Eds., Wichmann, Heidelberg, 2005, 2.

49. Lange, E., The limits of realism: perceptions of virtual landscapes, *Landscape and Urban Planning*, 54, 163, 2001.

50. Appleton, K. and Lovett, A., GIS-based visualization of rural landscapes: Defining "sufficient" realism for environmental decision-making, *Landscape and Urban Planning*, 65, 117, 2003.

51. Paar, P., Landscape visualizations: Applications and requirements of 3D visualization software for environmental planning, *Computers, Environment and Urban Systems*, 30, 815, 2006.

52. Muhar, A., Three-dimensional modeling and visualization of vegetation for landscape simulation, *Landscape and Urban Planning*, 54, 5, 2001.

53. Perrin, L., Beauvais, N., and Puppo, M., Procedural landscape modeling with geographic information: The IMAGIS approach, *Landscape and Urban Planning*, 54, 33, 2001.

54. Deussen, O., Colditz, C., Coconu, L., and Hege, H., Efficient modeling and rendering of landscapes, in *Visualization in Landscape and Environmental Planning*, Bishop, I. and Lange, E., Eds., Taylor & Francis, London, 2005, 56.

55. Röhricht, W., oik–nulla vita sine dispensatio—vegetation modeling for landscape planning, in *Trends in Real-Time Landscape Visualization and Participation*, Buhmann, E., Paar, P., Bishop, I., and Lange, E., Eds., Wichmann, Heidelberg, 2005, 256.

56. Paar, P. and Rekittke, J., Lenné 3D—walk-through visualization of planned landscapes, in *Visualization in Landscape and Environmental Planning*, Bishop, I. and Lange, E., Eds., Taylor & Francis, London, 2005, 152.

57. Werner, A., Deussen, O., Döllner, J., Hege, H.-C., Paar, P., and Rekittke, J., Lenné 3D—walking through landscape plans, in *Trends in Real-Time Landscape Visualization and Participation*, Buhmann, E., Paar, P., Bishop, I., and Lange, E., Eds., Wichmann, Heidelberg, 2005, 48.

58. Broads Authority, http://www.broads-authority.gov.uk/, accessed February 2008.

59. Broads Authority, Broads Plan 2004: A Strategic Plan to Manage the Norfolk and Suffolk Broads, Broads Authority, Norwich, 2004.

60. Natural England, What is Natural England?, http://www.naturalengland.org.uk/, accessed February 2008.

61. Intermap Technologies, http://www.intermap.com/, accessed February 2008.

62. Environment Agency, LiDAR, http://www.environment-agency.gov.uk/science/monitoring/ 131047/, accessed February 2008.

63. MAGIC, http://www.magic.gov.uk/default.htm, accessed February 2008.

64. Edina, Digimap collections, http://edina.ac.uk/digimap/, accessed February 2008.

65. ESRI, ArcGIS Desktop, http://www.esri.com/software/arcgis/about/desktop_gis.html, accessed February 2008.

66. SpatialEcology.com, Hawth's Analysis Tools for ArcGIS, http://www.spatialecology.com/htools/index.php, accessed February 2008.

67. Geograph, Welcome to Geograph British Isles, http://www.geograph.org.uk/, accessed February 2008.

68. Wings 3D, http://www.wings3d.com/, accessed February 2008.

69. Broads Authority, Broads Research Advisory Panel: Proceedings of Seminar/Workshop on Designing Land Use Change in The Broads, Broads Authority, Norwich, 2006.

70. Appleton, K. and Lovett, A., GIS-based visualization of development proposals: Reactions from planning and related professionals, *Computers, Environment and Urban Systems*, 29, 321, 2005.

71. Perkins, N. H. and Barnhart, S., Visualization and participatory decision making, in *Visualization in Landscape and Environmental Planning*, Bishop, I. and Lange, E., Eds., Taylor & Francis, London, 2005, 241.

72. UN-Energy, Sustainable Bioenergy: A Framework for Decision Makers, United Nations, New York, 2007.

73. Defra, DTI and DfT, UK Biomass Strategy, Department for Environment, Food and Rural Affairs, London, 2007.

74. Rowe, R., Street, N., and Taylor, G., Identifying potential environmental impacts of large-scale deployment of dedicated energy crops in the UK, *Renewable and Sustainable Energy Reviews*, in press.

75. English Heritage, *Biomass Energy and the Historic Environment*, English Heritage, London, 2006.

76. Defra, Planting and Growing Miscanthus: Best Practice Guidelines for Applicants to Defra's Energy Crops Scheme, Department for Environment, Food and Rural Affairs, London, 2007.

77. RELU-Biomass, Welcome to RELU-Biomass, http://www.relu-biomass.org.uk, accessed February 2008.

78. Lenné 3D, Lenné 3D—digital botany/virtual landscapes, http://www.lenne3d.com, accessed February 2008.

79. Lovett, A., Appleton, K., Paar, P., and Ross, L., Evaluating real-time landscape visualization techniques for public communication of energy crop planting scenarios, CD Proceedings of Virtual Geographic Environments: An International Conference on Developments in Visualization and Virtual Environments in Geographic Information Science, Chinese University of Hong Kong, January 2008.

80. Natural England, *Energy Crops Scheme—Establishment Grants Handbook*, Natural England, Sheffield, 2007.

81. BlueSky, NEXTMap Britain, http://www.bluesky-world.com/products/height_data/nextmap.php, accessed February 2008.

82. Right Hemisphere, Deep Exploration Standard Edition, http://www.righthemisphere.com/products/dexp/ index.html, accessed February 2008.

83. Okino Computer Graphics, Polytrans, http://www.okino.com/conv/conv.htm, accessed February 2008.

84. Orland, B., Budthimedhee, K., and Uusitalo, J., Considering virtual worlds as representations of landscape realities and as tools for landscape planning, *Landscape and Urban Planning*, 54, 139, 2001.

85. Sheppard, S. R. J., Lewis, J. L., and Akai, C., *Landscape Visualization: An Extension Guide for First Nations and Rural Communities*, Sustainable Forest Management Network, Edmonton, Alberta, 2004.

86. Sheppard, S. R. J., Landscape visualization and climate change: The potential for influencing perceptions and behaviour, *Environmental Science and Policy*, 8, 637, 2005.
87. Von Haaren, C. and Warren-Kretzschmar, B., The interactive landscape plan: Use and benefits of new technologies in landscape planning and discussion of the interactive landscape plan in Koenigslutter am Elm, Germany, *Landscape Research*, 31, 83, 2006.
88. Lange, E. and Hehl-Lange, S., Combing a participatory planning approach with a virtual landscape model for the siting of wind turbines, *Journal of Environmental Planning and Management*, 48, 833, 2005.
89. Bishop, I. D. and Miller, D., Visual assessment of offshore wind turbines: The influence of distance, contrast, movement and social variables, *Renewable Energy*, 32, 814, 2007.
90. Stock, C., Bishop, I., and Green, R., Exploring landscape changes using an envisioning system in rural community workshops, *Landscape and Urban Planning*, 79, 229, 2007.
91. Jarvis, C., Grid-enabled GIS: Opportunities and challenges, in *GIS for Environmental Decision-Making*, Lovett, A. and Appleton, K., Eds., CRC Press, Boca Raton, FL, 2008, 165.
92. MacEachren, A. M., Robinson, A., Hopper, S., Gardner, S., Murray, R., Gahegan, M., and Hetzler, E., Visualizing geospatial information uncertainty: What we know and what we need to know, *Cartography and Geographic Information Science*, 32, 139, 2005.

19 Visualizing Species Distributions
The Role of Geostatistics and GIS in Understanding Large-Scale Spatial Variation in Breeding Birds

David J. Lieske and Darren J. Bender

CONTENTS

OVERVIEW

The geographical element of conservation planning is fundamental to conservation decision making, with maps constituting a critical communication and visualization tool. However, their constraints and assumptions are not immediately obvious, and the impact of the choice of method or scale is rarely considered or discussed. To elucidate the utility of geostatistical (semivariography and kriging) and geographic information systems (GIS)-based visualization (inverse-distance weighting [IDW] and three-dimensional [3-D] vertical extrusion plots), we used these approaches to visualize broad-scale breeding distributions for four bird species. Our findings suggested that each of the methods revealed different but complementary aspects of spatial distribution. IDW was somewhat sensitive to the size of the spatial neighborhood and variation in this parameter highlighted the potential role of different environmental factors operating at different spatial scales. Kriging assumptions were, at best, only weakly met, but in the case of the American robin (*Turdus migratorius*) it highlighted similar patterns to IDW. Despite a limited ability to convey underlying trends, 3-D extrusion plots allowed for the easy identification of locations of unusual abundance or geographic isolation. In summary, we expect that conservation planners will be especially interested in applying these methods to identify core areas of peak abundance, thereby assisting conservation prioritization. Additionally, these methods have a critical role to play in hypothesis generation, with the potential to identify important driving factors.

19.1 INTRODUCTION

A key element of conservation planning for any species is to understand where that species occurs [1]. When monitoring provides additional information, such as the observation of abundance, it becomes possible to better evaluate the geographic pattern of actual resource usage. This is important, as resources in nature are not typically uniformly or randomly distributed in space [2], but rather tend to occur in the form of patches or gradients. Species dependent upon those resources—whether for breeding, foraging or dispersal—can be expected to be distributed in ways that reflect this pattern.

The geographical element of conservation planning, whether it involves the identification of gaps in the protection of species [3] or the visualization of the spatial pattern of changes in abundance [4], is fundamental to conservation decision making. The importance of geographic analyses has been widely accepted and endorsed [5,6], and is now commonly conducted as part of many ecological and environmental studies (e.g., to identify gaps in conservation coverage; see Merrill et al. [7]). Thanks to their persuasive power, map products constitute a critical communication and visualization tool, but the constraints and assumptions needed to develop the visualization output are not immediately obvious [8]. Furthermore, the impact of factors like choice of method or the scale of the visualization are rarely considered or discussed. As a consequence, cases where particular visualizations worked well are rarely evaluated or compared to alternative methods that may provide complementary information not readily discernable.

The scarcity of comparative studies leaves little to guide conservation planners and scientists in choosing a visualization approach, and compounds the difficulties already facing managers and conservation planners. This chapter addresses the shortage of analyses comparing the use of different geostatistical and geographic information systems (GIS) methods to visualize broad-scale species distributions. Our goal was one of exploratory data analysis (EDA), with the prime motivation to understand the prevailing properties of species abundance distributions while making the fewest number of *a priori* assumptions. We first produced semivariograms to explore broad-scale patterns, followed by simulation experiments to better understand the relationship between idealized distributional patterns (random, trended, and patchy) and the resulting semivariograms. We also examined a number of interpolation methods (inverse-distance weighted and kriging), as well as three-dimensional (3-D) visualizations of "extruded" sample points. We anticipated that these methods would differ in their ability to reveal species-specific trends in relative abundance, and concentrated our attention on the loss or gain of information from employing these different approaches. Data for this analysis consisted of abundance information for four species—American robin (*Turdus migratorius*), purple finch (*Carpodacus purpureus*), black-throated green warbler (*Dendroica virens*), and least flycatcher (*Empidonax minimus*)—averaged over seven years of observations for the North American Breeding Bird Survey (BBS) [9]. We focused on the utility, limitations, and potential complementarities of these methods, and considered species-specific responses as part of our evaluation. This chapter provides an evaluation of these methods as well as recommendations for applying these approaches to conservation planning.

19.2 REVIEW OF INTERPOLATION APPROACHES

All interpolators share the common objective of predicting values at unsampled locations, using a smaller set of sampled locations within the same area or region [10,11]. Clearly, this is of importance to applied conservation planning, where biodiversity status (species occurrence or abundance) can only be assessed at a limited number of sample locations, but decisions must be applied to the entire area. A well-established deterministic method is inverse-distance weighting (IDW), which is simple to implement [12], is known to work well with noisy data [13], and has been shown to yield predictions comparable to more sophisticated methods (such as kriging; see Valley et al. [14]). However, IDW is sensitive to the distribution of the sampled locations [14]. Other methods, such as polynomial modeling, make use of covariates (e.g., locational coordinates, elevation, slope, aspect) to construct global stochastic models for predicting values at locations within the region [11,15]. Such methods are sensitive to edge effects and outliers [10], and have been shown to perform poorly [16]. Locally fit models, such as local polynomials or splines, base their predictions on the relationships observed for measurements within a neighborhood surrounding the prediction point and can yield very accurate predictions [16] (but see Kurtzman and Kadmon [15] for exceptions). Although flexible, local models are highly data-dependent and prone to overfitting, which can limit their generalizability outside the study area [13].

Geostatistical methods (kriging) make use of the autocorrelation structure between locations as a function of the distance between them, capitalizing on the tendency for points near to one another to be more alike than points farther apart [17]. Interpolated values at any point are estimated as a linear combination of the observed values in that neighborhood [18]. In addition to assuming that the spatial process is constant throughout the region (stationarity), kriging must apply a parametric model to estimate the form of the autocovariance structure [13,19]. Under optimal conditions, kriging yields the best linear unbiased predictions (see Ishida and Kawashima [18] and Luo et al. [20]) as well as estimates of uncertainty. However, IDW has been shown to outperform kriging when sampling is poor or there is an absence of clear spatial structure [11,19,21]. As a final note, some authors have combined methods, for example, regression modeling followed by IDW interpolation of the predictions [19] or kriging interpolation of the residuals [22].

On the whole, the relative performance of any of these methods depends heavily on the variable under study, the spatial configuration of the data, and the level of agreement with model assumptions [19,23]. For this reason, optimal interpolation methods may only be optimal for a restricted range of conditions [24], and a number of approaches should be considered and compared whenever possible [19].

19.3 METHODS

19.3.1 STUDY AREA AND SPECIES OBSERVATION DATA

The study area (approximately 617,675 km^2) constituted the entirety of the boreal-hardwood transition zone centered on the Great Lakes of North America, a physiographic region compiled by Bystrak [25] and adopted by the Partners in Flight to assist its conservation planning process [26]. Species abundance data was obtained from the North American Breeding Bird Survey (BBS). Both the outline of the study area and the spatial arrangement of the survey routes are indicated in Figure 19.1. Although primarily intended to detect long-term trends in species abundance (e.g., Robbins et al. [27]), individual surveys consist of 50 three-minute sample points spaced 0.8 km apart along a defined route. Precisely georeferenced sample points were only available for seven routes in the study area, so we were forced to subdivide the remaining routes using a linear referencing operation in ArcGIS [28]. This procedure divided intact route segments into 50 individual points spaced 0.8 km apart, resulting in 56 survey routes and 2,799 individual sample points (one stop point was missing from the data for the georeferenced survey routes). To assess the positional accuracy of the interpolated points, we compared 75 randomly selected, known sample locations to the positions assigned by linear referencing. We found an average discrepancy of 1.8 ± 1.9 km (SD), which was in close agreement with an average of 1.5 km reported in a previous study [29]. Using data available from 1997 to 2003, species abundances were averaged for each sample point over the seven-year period so as to minimize the risk of falsely classifying sample locations as unused when, in fact, they were suitable but unoccupied due to chance events or failed detection. It should be noted that not all routes were surveyed for each of the seven years. Aggregating data in this way also permitted the estimation of standard errors for each survey point, which we later used to estimate confidence intervals for the semivariograms.

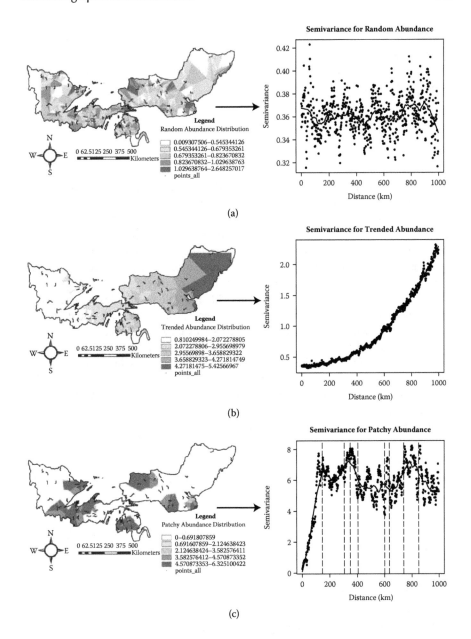

(a)

(b)

(c)

FIGURE 19.1 Results of the simulation experiment for (a) random, (b) trended, and (c) patchy abundances. Actual sample locations were used in each case, and the sample routes are indicated on each map. Randomly distributed abundances were generated by drawing absolute values from a population with a mean of zero and a variance of one, N(0,1). Spatial trend was created by adding a proportional increment (oriented east to west, and increasing toward the east) to the random normal deviates. The patchy distribution was generated by randomly seeding 10 cluster centroids throughout the study area, each with a zone of influence extending to all points within a 100-km radius. Interpatch distances are indicated as vertical hatched lines. Resulting semivariograms are indicated to the right.

19.3.2 Semivariogram Analysis

Semivariance (γ) was estimated for different sized spatial neighborhoods of radius h using the classical method implemented by Ribeiro and Diggle [30] for the R Statistical Package [31]:

$$\gamma(h) = \frac{1}{2N_h}\sum_{i=1}^{N_h}[Y(x_{(i+h)}) - Y(x_i)]^2 \tag{19.1}$$

where N_h is the number of pairs of points within radius h, and $Y(x_{(i+h)})$ and $Y(x_i)$ are the values of Y at locations $i+h$ and i. Semivariance (plotted as a semivariogram), captures the average degree of similarity between values as a function of the distance between them [32], and compared to covariance, is more robust to departures from assumptions [33,34]. Declines in semivariance for particular h distances indicate greater similarity (homogeneity) of the measured attribute at those distances, whereas increases in semivariance indicate less similarity (greater heterogeneity). We determined confidence envelopes at the 5th and 95th percentiles for each semivariogram by randomly drawing mean abundances, using either the standard error for that specific sample point when sampling was conducted over more than one year, or the overall mean standard error when sampling was conducted only once. Finally, the possibility that the directional orientation of the points under comparison impacted the measurement of semivariance (anisotropy) was also considered.

Prediction error was assessed using the root mean square error (RMSE), which provides a measure of discrepancy between measurements determined by the visualization method and known values gathered at sampled locations [34]:

$$RMSE = \sqrt{\frac{1}{N}\sum_{i=1}^{N}[Y(x_i) - \hat{Y}(x_i)]^2} \tag{19.2}$$

where $Y(x_i)$ and $\hat{Y}(x_i)$ are the observed and predicted values, respectively, at test location i.

19.3.3 Simulation Experiment

The purpose of this experiment was to generate spatial patterns for a set of simplified and hypothetical species distributions, and to investigate the semivariogram patterns that resulted. The results of this experiment were then used to make inferences about the much more complicated species abundance distributions. We simulated three different abundance distributions—random, trended, and patchy (clustered)—using the actual survey points. Randomly distributed abundances were generated by drawing absolute values from a population with a mean of zero and a variance of one, N(0,1). Spatial trend was created by adding a proportional increment (oriented east to west, and increasing toward the east) to the random normal deviates. The patchy distribution was generated by randomly seeding 10 cluster centroids throughout the study area, each with a zone of influence extending to all points within a 100-km radius.

19.3.4 GIS VISUALIZATION METHODS

We employed three tools in ArcGIS 9.0 [26] to visualize abundance distributions: (1) the IDW function and (2) universal kriging in the Geostatistical Analyst Extension, and (3) ArcScene. As a simple deterministic interpolator, the IDW function estimates no parameters and requires very few assumptions to be made about the process under examination:

$$\hat{Z} = \frac{\sum_{i=1}^{n} \frac{1}{d_i^k} * Z_i}{\sum_{i=1}^{n} \frac{1}{d_i^k}} \tag{19.3}$$

where d is the distance between the interpolation point and a neighboring sample point, n is the number of nearest neighbors (or neighborhood radius), and k is the power, commonly defined as 2. Use of the IDW requires the definition of the size of the neighborhood (search radius, or "bandwidth") around each point to be estimated, and as pointed out by Chainey et al. [35], it can be difficult to determine the optimal size. As we had no *a priori* reason to choose any particular neighborhood size, we used the following set and examined the resulting mean square error for each: 1, 5, 10, 15, 25, 50, 100, 250, and 500 nearest neighbors. We expected that increasing neighborhood size would result in greater smoothing, but also consequentially a loss of information. However, we also expected that this would be offset by greater clarity in detecting major spatial variation or trends. By varying the neighborhood size we were able to evaluate both the impact on the root mean square error as well as the visual interpretability of the resulting map.

Unlike IDW, universal kriging is a statistical interpolation method that incorporates spatial dependence by applying theoretical functions to the semivariogram pattern. Through the use of regionalized variable theory [10], kriging distinguishes the broad-scale deterministic variation $m(x)$ from that attributable to spatial dependence (ε', estimated directly from the semivariance function):

$$\hat{Z} = m(x) + \varepsilon' + \varepsilon'' \tag{19.4}$$

Remaining variation (ε'') is assumed to be random, spatially uncorrelated noise. Although a sophisticated technique for interpolation, it makes more assumptions about the process under study than IDW, in particular: (1) that the measurement is distributed in a Gaussian (normal) fashion [36]; (2) that spatial dependency is present and warrants modeling using regionalized variable theory; and (3) that the process exhibits a constant mean and variance throughout the study area (the so-called stationarity assumption, see Bailey and Gatrell [33]). Assumption 3 can be alleviated within the framework of universal kriging by fitting regression models, for instance, as a function of the geocoordinates, thereby eliminating broad-scale trend (drift) in the mean. We examined the utility of kriging as a means to perform interpolation when these assumptions were adequately met.

The ArcScene extension of ArcGIS 9.0 [28] was used to produce all 3-D visualizations. In the case of vertically exaggerated, "draped" surfaces, a preexisting prediction layer (e.g., an IDW prediction surface) was used to determine the height of the surface at each point. For 3-D bar plots, individual sample locations were vertically exaggerated without any interpolation.

19.4 RESULTS AND DISCUSSION

19.4.1 GEOSTATISTICAL ANALYSIS

19.4.1.1 Simulation Experiment

We evaluated the relationship between hypothetical spatial distributions and semivariogram patterns using a simulation experiment, the results of which are shown in Figure 19.1. In the first case (Figure 19.1a) we simulated a purely random spatial distribution, visible as a loose jumble of values with no discernable pattern. We expected this to be representative of a hypothetical species that uses resources that are randomly and haphazardly distributed throughout the landscape: a scenario that although possible seems unlikely given the general observation that resources in nature tend to show trends (gradients) or patches (clusters). The simulated random distribution produced a characteristic semivariogram pattern: a flat line at an elevation (or sill) equivalent to the global variance. Any deviation of the line from this variance value is purely attributable to the stochasticity inherent in the random number generator. It is also noteworthy how the lack of spatial dependence is reflected in the lack of a rapid increase in semivariance from 0 m (compare Figures 19.1a and 19.1c).

As argued by Legendre [37], random spatial distributions are uncommon in nature, suggesting that random processes are themselves unusual. According to Legendre, spatial trends or gradients are far more likely given the tendency for many processes to exhibit gradual change over space. Species that show peak abundance in some portion of their range and tend to gradually decrease in abundance toward their range peripheries might be expected to reveal a spatial trend or gradient of abundance distribution. The simulated spatial trend (Figure 19.1b) mimicked this scenario by producing abundances that increased toward the east side of the study area. Consideration of Figure 19.1 reveals that systematic changes in abundance values along a directional gradient increase variability in a predictable way, with points farthest apart on the spatial gradient showing the greatest differences in values. This was reflected in an exponential semivariogram pattern that was clearly dominated by the spatial trend.

Our final simulation case generated a patchy distribution by grouping unusually high values in close proximity. Species that display peak abundances in the form of many widely distributed patches, and that show no clear tendency to become gradually more or less abundant away from these patches, could be expected to exhibit this form of spatial distribution. The simulated spatial trend (Figure 19.1c) brought about a distinctive semivariogram, showing an initial rise and fall in semivariance (resembling a hump) and subsequent undulations (resembling peaks and troughs). The hump in the semivariogram, at the shortest distances, reflects the comparison of points within and outside the patches, whereas the subsequent troughs in semivariance illustrate what is referred to as the "hole effect" [32,38]. In this case, troughs of semivariance indicated the average distance between patches (localized peaks) of like values (see Burrough [39]). Average interpatch distances are indicated in Figure 19.1c by hatched lines.

19.4.1.2 Semivariogram Analysis

Figure 19.2 presents actual semivariograms for a wide range of spatial lag distances (*h*) for each of the four species. In each case, a running line "smoother" [40] was used to average 10% of the immediate neighbors to highlight underlying patterns. The

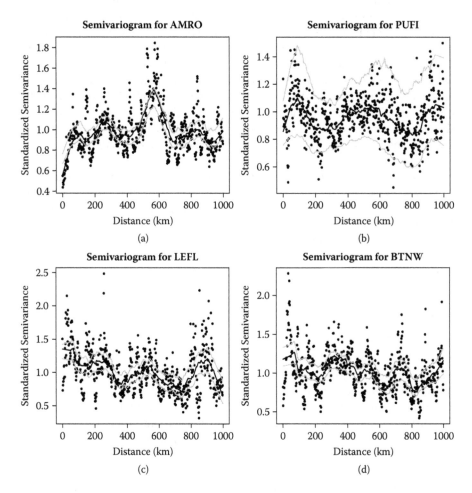

FIGURE 19.2   Semivariograms for the (a) American robin (AMRO), (b) purple finch (PUFI), (c) least flycatcher (LEFL), and (d) black-throated green warbler (BTNW). Semivariance was estimated using the classical method implemented by Ribeiro and Diggle [30]. Also indicated are the lower and upper bounds (5% and 95% percentiles) of semivariance derived from 100 simulations of the mean abundance values at each sampling location.

American robin (Figure 19.2a) and purple finch (Figure 19.2b) exhibited oscillating patterns of peak and valley semivariance, prominently illustrating the hole effect. Semivariance for both species increased rapidly as the distance intervals widened beyond 0 m, suggesting the effect of positive spatial autocorrelation. This pattern was also consistent with the simulated patchy distribution discussed in Section 19.4.1.1 (also see Figure 19.1c). In this case, peaks indicated maximum heterogeneity in relative abundance values, and likely corresponded to mean interpatch distances (as with the simulated patchy distributions of Figure 19.1c). Semivariogram patterns for the least flycatcher (Figure 19.2c) and black-throated green warbler (Figure 19.2d) were somewhat more difficult to interpret, as neither species exhibited the same initial rapid increase in semivariance as for the other two species. Nevertheless, the hole effect appeared in their semivariograms, as well as a tendency for declining semivariance.

Also shown in Figure 19.2 are the 5th and 95th percentile confidence bands (gray lines) based on random sampling of the mean abundance values. In most cases (Figures 19.2a, 19.2c, and 19.2d) variation in mean abundances served merely to raise or lower the relative variability, and did not alter the underlying patterns. However, the confidence intervals of the semivariogram for the purple finch (Figure 19.2b) were substantially wider than for the other species, signaling greater variability in abundance over the seven-year observation period. This is consistent with this species' tendency to show quasi-cyclical trends in abundance [41] and, hence, greater variation over time.

Anisotropy, or the tendency for directional differences in semivariogram patterns, was investigated for the American robin (Figure 19.3) on account of its strong response to a north–south latitudinal gradient (revealed by preliminary analysis). The hole effect appeared reasonably consistent for different orientations—indicating a tendency for patchiness—except the north–south (0°) bearing for which a gradually ascending semivariogram illustrated a long-range trend in response to latitude.

19.4.2 GIS Visualization Analysis

19.4.2.1 IDW Prediction Surfaces

We anticipated that IDW prediction surfaces might be sensitive to the definition of the neighborhood size, as well as the k parameter (Equation 19.3; see Attorre et al. [13]), so prior to generating the maps we examined the influence of these factors on root mean square error (Figures 19.4 and 19.5). Error (lack of fit) for the IDW prediction surfaces changed most dramatically up to about 10 km (Figure 19.4). Beyond that distance all IDW prediction surfaces performed similarly. With regard to k, errors in the IDW prediction surfaces seemed largely unaffected by changes in the value of this parameter (Figure 19.5).

On the basis of root mean square error, any distance weighting exceeding a 10 km fixed neighborhood size performed equally well for these species. However, visual inspection of the resulting IDW prediction surfaces tells a somewhat different story. For example, we produced 3-D plots of IDW prediction surfaces for the American robin at four different combinations of fixed-distance neighborhood size and k parameter (100 km and $k=2$; 400 km and $k = 2$; 900 km and $k = 2$; and

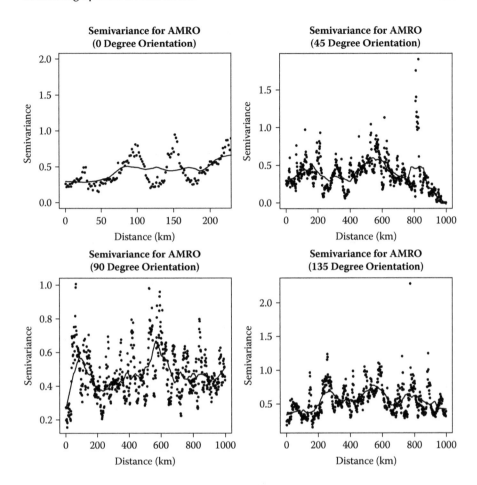

FIGURE 19.3 Directional semivariograms for the American robin (AMRO) for axes oriented at 0, 45, 90, and 135 degrees.

900 km and $k = 1$; Figures 19.6 a–d). We found pronounced differences in the amount of noise visible in each of the plots and, consequently, differences in the perceptibility of the signal of spatial pattern in species abundance. At the scale of 100 km, the signal is dominated by variation in species abundances for individual BBS survey routes, and as the BBS protocol calls for individual observers to complete entire routes, this represents the variation in the abundances recorded by individual observers. Because the identity of the observer is constant during the completion of the survey route, variation within routes (i.e., between sample points) represent responses to local environmental conditions, land cover, and so forth.

By 900 km, multiple survey routes and, hence, observations from multiple observers are aggregated during the production of IDW surfaces to produce interpolated values. At this definition of neighborhood size, spatial variation in abundance values are reflective of effects that are operating at broad, regional scales and could include such influences as climate or geology. Therefore, not only does the shape and pattern of distribution change with response to neighborhood size,

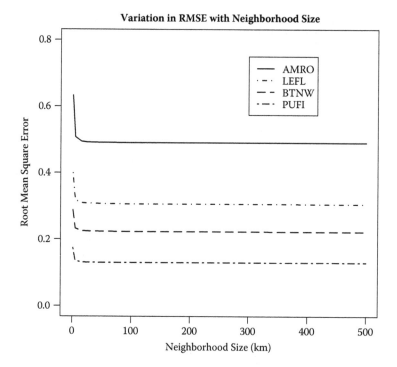

FIGURE 19.4   Root mean square error for the IDW prediction surfaces, as a function of nearest neighborhood size, for all four species: American robin (AMRO), purple finch (PUFI), least flycatcher (LEFL), and black-throated green warbler (BTNW).

becoming more or less smooth, but different driving processes are highlighted. This is because neighborhood size encompasses scale effects, a subject of key importance in geography and the environmental sciences [42,43]. We argue that factors that are slow acting and change only gradually over large distances are probably best studied using larger neighborhood sizes, whereas for fast acting factors the opposite is true.

Turning to prediction surfaces derived from the IDW method, we used two definitions of neighborhood: one based on a fixed neighborhood size of 300 km (Figure 19.7), approximately corresponding to the asymptotic semivariance for the American robin and purple finch (see Figures 19.2a and 19.2b); and another based on nearest neighbors (in this case, 15; Figure 19.8). The use of nearest neighbors to define the interpolation neighborhood (Figure 19.8) controls for uneven sample sizes across the sampling frame [44], as this method guarantees that predictions at any location are based on the same number of points. This has the benefit of helping to reduce the impact of unusual values in poorly sampled regions, but it also has the potential to obscure the spatial scale over which interpolation occurs. Qualitatively, patches of peak abundance based on this neighborhood definition are shardlike in appearance and lack the smoothness of neighborhoods defined using a fixed distance. The prediction surfaces for the American robin (Figures 19.7a and 19.8a) and purple finch (Figures 19.7b and 19.8b) confirmed our previous

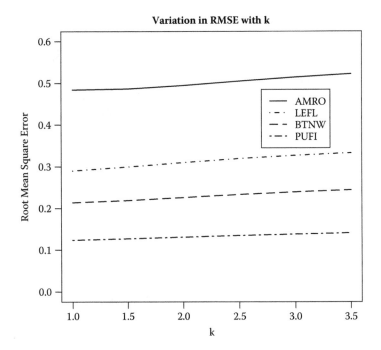

FIGURE 19.5 Root mean square error for the IDW prediction surfaces, as a function of varying k (see Section 19.3.4), but a constant nearest neighborhood size of 15, for all four species: American robin (AMRO), purple finch (PUFI), least flycatcher (LEFL), and black-throated green warbler (BTNW).

interpretations of the semivariograms (Figures 19.2a and 19.2b) in that localized (but widely separated) peaks of abundance were observable as patches or clusters surrounded by a matrix of lower predicted abundance. For the least flycatcher and black-throated green warbler, the tendency for declining variation in abundance at wider spatial scales was visible in the semivariograms (see Figures 19.2c and 19.2d) but not in the IDW surfaces (Figures 19.7c and 19.7d, and Figures 19.8c and 19.8d). For the most part, IDW surfaces for these two species resembled the patchiness exhibited by the first two.

19.4.2.2 Universal Kriging

Although not strictly an exploratory visualization technique, universal kriging is a powerful tool for generating prediction surfaces when data are Gaussian distributed, spatial dependence is strongly operating, and a constant mean and variance can be assumed (see Section 19.3.4). Assumptions also need to be made about the form of the spatial dependence through the choice of a fitted function [34]. We employed universal kriging for the American robin, with a first-order trend surface to remove the effects of an underlying north–south gradient in abundance ("drift"; see Ishida and Kawashima [18]), and a spherical model of the semivariogram. This resulted in both prediction (Figure 19.9a) and error (Figure 19.9b) surfaces, although we

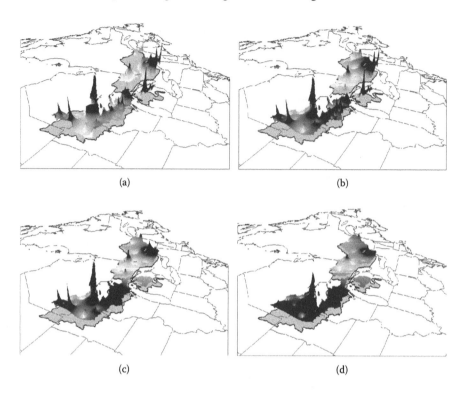

(a)   (b)

(c)   (d)

FIGURE 19.6 Impact of the choice of neighborhood size and magnitude of the k parameter (see Section 19.3.4) on smoothness of IDW prediction surfaces for the American robin (AMRO). On the basis of the spacing of semivariogram troughs (Figure 19.2), neighborhood sizes were set to (a) 100 km, (b) 400 km, (c) 900 km and (d) 900 km. The k parameter was set to 2 in all cases except for (d), in which k was defined as 1.

note that abundance data was decidedly nonnormal due to a large number of survey points without counts for this species. The lower prevalence of the other species (and a concomitantly large number of zero abundance values) resulted in abundance distributions that were even more skewed than that for the American robin, so we limited our universal kriging to an exploratory comparison with the IDW surfaces in this case alone. We found distributional patterns (Figure 19.9a) to be similar to that obtained using IDW surfaces, whether based on nearest neighbors (Figure 19.8) or fixed neighborhood sizes (Figure 19.7). It is not surprising that prediction variance (Figure 19.9b) is highest for points at the periphery of the study area where data was unavailable for predicting abundance.

19.4.2.3 Three-Dimensional Extrusion Plots

Three-dimensional views of the distribution of species abundances were generated (Figure 19.10). By virtue of making no *a priori* assumptions about the data, this method provided a literal (pure) view of the process under study. However, because no smoothing was applied, it was somewhat difficult to perceive spatial pattern from these figures. For example, the abundance distribution for the American robin

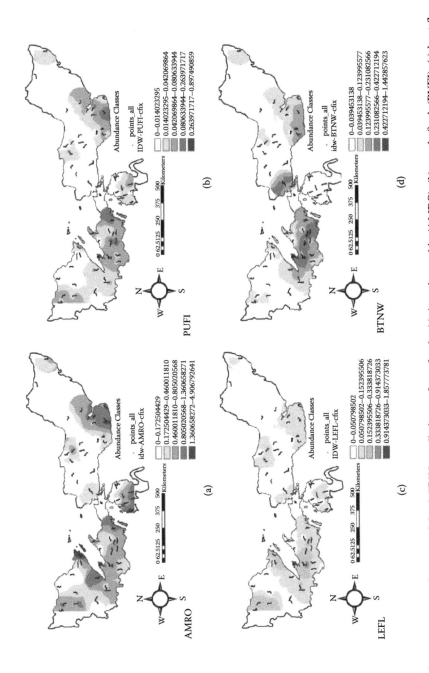

FIGURE 19.7 Inverse–distance weighted (IDW) prediction surfaces for the (a) American robin (AMRO), (b) purple finch (PUFI), (c) least flycatcher (LEFL), and (d) black-throated green warbler (BTNW) categorized into relative abundance categories using standard deviations (white = lowest relative abundance, black = highest relative abundance). In all cases, a fixed neighborhood of 300-km radius was used to produce the weighted estimate.

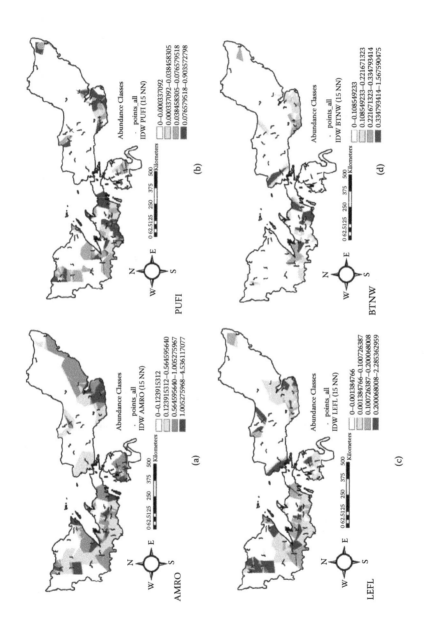

FIGURE 19.8 Inverse-distance weighted (IDW) prediction surfaces for the (a) American robin (AMRO), (b) purple finch (PUFI), (c) least flycatcher (LEFL) and (d) black-throated green warbler (BTNW) categorized into relative abundance categories using standard deviations (white = lowest relative abundance, black = highest relative abundance). In all cases a 15-nearest neighborhood size was used to produce the weighted estimate.

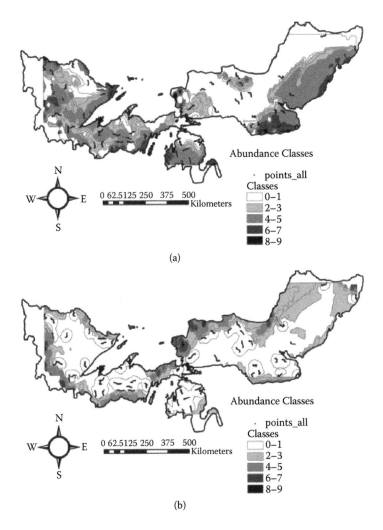

FIGURE 19.9   Kriging interpolation for the American robin (AMRO). The prediction surface (a) was generated using universal kriging and a spherical model to represent spatial dependence (range = 53.5 km, partial sill = 0.27474, nugget variance = 0.1904). Prediction uncertainty is indicated in (b).

(Figure 19.10a) appeared as a thick forest of extruded points. Nevertheless, it was still possible to observe the centers of major abundance patches, as they appeared as the highest columns in the plot. We found this information particularly useful for interpreting the IDW prediction surfaces generated in Section 19.4.2.1. For instance, the highly variable central region of the 3-D extrusion plot for the least flycatcher (visible from a particularly tall spike of abundance immediately south of Lake Superior; Figure 19.10c) was accentuated in this figure. Similarly isolated clusters of peak abundance were readily identifiable for the same region in the case of the black-throated green warbler (Figure 19.10d).

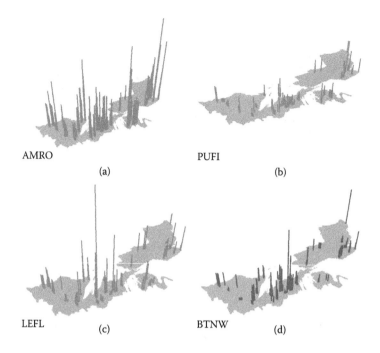

FIGURE 19.10 Mapped 3-D bar graphs of average abundances (vertically exaggerated by a factor of 150,000×) for the (a) American robin (AMRO), (b) purple finch (PUFI), (c) least flycatcher (LEFL) and (d) black-throated green warbler (BTNW).

Despite a limited ability to convey underlying trends, this method allowed for the easy identification of locations with measures of unusual magnitude (outliers) or which were geographically isolated. For this reason we anticipate that this visualization method would be usefully applied to very rare or elusive species, for which observations are few in number but critical to identify geographically due to their conservation value.

19.4.3 THE CONTINENTAL PERSPECTIVE

To help place our results within a continental context, we compared our findings to those provided by the continent-wide BBS (Figure 19.11). Although we used a subset of BBS survey routes for the Partners in Flight Boreal Conservation Region 12 (BCR 12)(see Section 19.3.1, [26]) and at a finer scale (the level of the individual survey point), it still constitutes the same core data. For this reason it is a somewhat circular comparison, but no other relevant, broad-scale distributional data exists for North America. The evaluation reinforced a number of distributional trends revealed at the finer resolutions of our study: peak abundances for the least flycatcher and black-throated green warbler (Figures 19.11c and 19.11d) south of Lake Superior, and a northward decline in abundance for the American robin (Figure 19.11a).

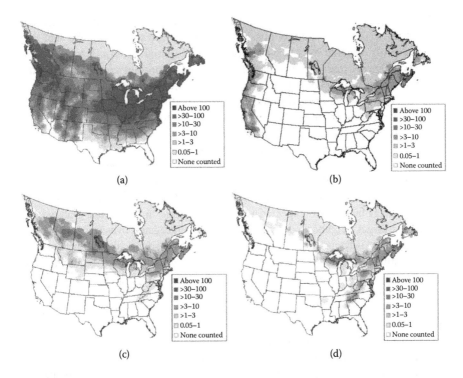

(a)   (b)

(c)   (d)

FIGURE 19.11 Breeding bird survey (BBS) summer distribution maps for the (a) American robin, (b) purple finch, (c) least flycatcher, and (d) black-throated green warbler, aggregated over the period 1994–2003 [49]. The maps were based on the starting points of all North American BBS survey routes and were not subdivided into each of 50 individual survey points as done in this study. Maps were produced using an inverse-distance weighting (IDW) interpolation.

19.5 CONCLUSIONS

Given that the goal of this study was to provide an initial exploratory visualization of abundance patterns while making the fewest assumptions about the distributional pattern of species habitat usage, we found each visualization method provided a different but complementary view of the underlying spatial pattern. The semivariogram was a very effective data interpretation tool for providing a simultaneous view of spatial variability at varying scales. Major discontinuities, indicating changes in the underlying process or trends, were detectable from these plots. Directional semivariograms were also useful for identifying directional trends in abundance. In terms of theoretical expectation, the results of our simulation experiment confirmed that patchy abundance distributions were universal for all species. Nevertheless, semivariograms are geostatistical plots without a geographic context, so while suggesting distances at which major discontinuities (e.g., patches of dissimilar abundance) occur, they could not be used to infer where they occurred.

IDW prediction surfaces were a simple yet effective means of interpolation and provided the geographic context missing from the semivariograms. We feel they were especially suitable as a first-pass glimpse of spatial pattern. Evidence from our study suggests that neighborhood size was an important consideration when using this tool, and, depending upon the choice of scale, different spatial patterns could either be isolated or obscured. But this method makes very few assumptions about the process under observation and produces readily interpretable maps. It quite clearly revealed patches of peak abundance for the least flycatcher and American robin west and/or south of Lake Superior, for instance.

We recommend that neighborhood size be chosen to match the scale of interest. In the case of species abundance distributions, variation within individual survey routes was revealed by smaller neighborhood sizes (e.g., 15 nearest neighbors). At this scale, local factors are expected to play a key role in determining abundance patterns. At larger neighborhood sizes, points are aggregated over larger distances, highlighting the influence of factors operating over broader scales. Visualizations, when overlain with land use or climatic maps, can be used as a heuristic tool to identify potential driving factors [36].

Three-dimensional extrusion plots were another valuable visualization tool that we found best suited for displaying raw data values and were very helpful for identifying exceptional peaks of abundance. For instance, the concentrations of abundance for the least flycatcher and black-throated green warbler in the Lake Superior region were clearly visible (through vertical exaggeration) as high points in the plot. The general utility of this data visualization technique was more limited in the case of common and widespread species such as the American robin, where the large number of presence points led to plots that were too dense to interpret. However, we anticipate that this would be an excellent choice of visualization tool for studies involving sparse or isolated data values, for example, for studies involving rare species.

For conservation planners, these methods allow for a complete visualization of species distribution patterns. They make a minimum number of assumptions about the process of species habitat usage and are therefore ideal for exploratory analysis. We expect that conservation planners will be especially interested in applying these methods to identify core areas of peak abundance and to use this information to guide the prioritization of areas for conservation. Additionally, these methods have a critical role to play in hypothesis generation, with the potential to assist in the identification of candidate factors that may influence the observed abundance patterns. Furthermore, we expect that particular hypotheses regarding, for example, dominant land-use patterns or the influence of climate, could be tested using the overlay of GIS prediction layers in combination with statistical models. In summary, visualization has a key role to play in all levels of conservation planning and scientific inquiry, both by highlighting patterns and suggesting the scales at which relevant factors are operating.

19.6 RECOMMENDATIONS FOR FUTURE WORK

An important area for future research is the assessment of the interpretability and acceptability of the various maps by a wider audience, some of whom may not have training in the physical sciences or be familiar with statistical methods (e.g., universal kriging). Elith et al. [45] also point out that an important factor limiting the acceptability of any geovisualization is the degree of confidence in the results. Prediction errors, undersampled regions, positional uncertainty in the case of data derived without precise GPS coordinates, and so forth can all combine to depreciate the decision-support value of the prediction surfaces [46]. There remains, then, an ongoing challenge of how best to quantify and visualize uncertainty (see Hunter and Goodchild [47]). We feel that the exploratory methods used in this chapter make the fewest assumptions about the data, and will be the most robust and widely applicable, but would be aided by visualizations of data uncertainty. Our use of simulation to construct confidence intervals for the semivariograms did illustrate how extra information about interannual variability could be simultaneously displayed, and reinforced that it should form a routine part of any uncertainty analysis. Finally, given that the four species unanimously showed some degree of patchiness in their distributions, a further area for exploration might involve an analysis of the properties of these patches. For example, delineating patches, determining the degree to which they are connected or isolated, and assessing their spatial arrangement [48]. A complementary analysis might also focus on the cold spots, or areas where species rarely occur (such as the outer portion of range limits).

ACKNOWLEDGMENTS

We acknowledge the efforts of the many American and Canadian BBS participants in the field, as well as the U.S. Geological Survey (USGS) and Canadian Wildlife Service (CWS) researchers and managers for providing this data. We also like to thank the Department of Geography, University of Calgary. for providing graduate support to D. J. L., and acknowledge D. Jacobson for providing advice and input into this analysis.

REFERENCES

1. Underhill, L. and Gibbons, D., Mapping and monitoring bird populations: Their conservation uses. In *Conserving Bird Biodiversity: General Principles and their Application*, Cambridge University Press, Cambridge, 2002, 34.
2. Legendre, P. and Fortin, M.-J., Spatial pattern and ecological analysis, *Vegetatio*, 80, 107, 1989.
3. Scott, J. M., Davis, F., Csuti, B., Noss, R., Butterfield, B., Groves, C., Anderson, H., et al., Gap analysis: A geographic approach to protection of biological diversity, *Wildlife Monographs*, 123, 1, 1993.
4. Villard, M.-A. and Maurer, B., Geostatistics as a tool for examining hypothesized declines in migratory songbirds, *Ecology*, 77, 59, 1996.
5. Maurer, B.A., *Geographical Population Analysis: Tools for the Analysis of Biodiversity*, Blackwell, Cambridge, MA, 1994.
6. Johnston, C. A., *Geographic Information Systems in Ecology*, Blackwell Publishing, Cambridge, MA, 1998.
7. Merrill, E. H., Kohley, T. W., Herdendorf, M. E., Reiners, W. A., Driese, K. L., Marrs, R. W., and Anderson, S. H., Wyoming gap analysis: A geographic analysis of biodiversity, Final report, Cooperative Fish and Wildlife Unit, University of Wyoming, Laramie, WY, 1996.
8. MacEachren, A. M., *How Maps Work: Representation, Visualization and Design*, Guilford Press, New York, 2004.

9. Robbins, C. S., Bystrack, D., and Geissler, P. H., *The breeding bird survey: Its first fifteen years, 1965–1979*, U.S. Fish and Wildlife Service Publication, 157, 1986.

10. Burrough, P. A. and McDonnell, R. A., *Principles of Geographic Information Systems*, Oxford University Press, 1998.

11. Ninyerola, M., Pons, X., and Roure, J. M., A methodological approach of climatological modelling of air temperature and precipitation through GIS techniques, *International Journal of Climatology*, 20, 1823, 2000.

12. Coley, A. R., and Clabburn, P., GIS visualisation and analysis of mobile hydroacoustic fisheries data: A practical example, *Fisheries Management and Ecology*, 12, 361, 2005.

13. Attorre, F., Alfo, F., De Sanctis, R., Francesconi, A., and Bruno, F., Comparison of interpolation methods for mapping climatic and bioclimatic variables at regional scale, *International Journal of Climatology*, 27, 1825, 2007.

14. Valley, R. D., Drake, M. T., and Anderson, C. S., Evaluation of alternative interpolation techniques for the mapping of remotely-sensed submersed vegetation abundance, *Aquatic Botany*, 81, 13, 2005.

15. Kurtzman, D. and Kadmon, R., Mapping of temperature variables in Israel: A comparison of different interpolation methods, *Climate Research*, 13, 33, 1999.

16. Lennon, J. J. and Turner, J. R. G., Predicting the spatial distribution of climate: Temperature in Great Britain, *Journal of Animal Ecology*, 64, 370, 1995.

17. Tobler, W. R., Cellular geography, in *Philosophy in Geography*, Gale, S. and Olsson, G., Eds., Reidel, Dordecht, 1979, 379.

18. Ishida, T., and Kawashima, S., Use of co-kriging to estimate surface air temperature from elevation, *Theoretical Applied Climatology*, 47, 147, 1993.

19. Nalder, I. A. and Wein, R. W., Spatial interpolation of climatic normals: Test of a new method in the Canadian boreal forest, *Agricultural and Forest Meteorology*, 92, 211, 1998.

20. Luo, W., Taylor, M. C., and Parker, S. R., A comparison of spatial interpolation methods to estimate continuous wind speed surfaces using irregularly distributed data from England and Wales, *International Journal of Climatology*, 28, 947, 2008.

21. Mueller, T. G., Pusuluri, N. B., Mathias, K. K., Cornelius, P. L., Barnhisel, R. I., and Shearer, S. A., Map quality for ordinary kriging and inverse distance weighted interpolation, *Soil Science Society of America Journal*, 68, 2042, 2004.

22. Hengl, T., Heuvelink, G. B. M., and Stein, A., A generic framework for spatial prediction of soil variables based on regression-kriging, *Geoderma*, 120, 75, 2004.

23. Martinez-Cob, A., Multivariate geostatistical analysis of evapotranspiration and precipitation in mountainous terrain, *Journal of Hydrology*, 174, 19, 1996.

24. Isaaks, E. H. and Srivastava, R. M., *An Introduction to Applied Geostatistics*, Oxford University Press, New York, 1989.

25. Bystrak, D., The North American Breeding Bird Survey, *Studies in Avian Biology*, 6, 34, 1981.

26. Williams, E. J. and Pashley, D. N., Designation of conservation planning units, in *Strategies for Bird Conservation: The Partners in Flight Planning Process*, Cornell Lab of Ornithology, Cornell, 1999.

27. Robbins, C. S., Sauer, J. R., Greenberg, R. S., and Droege, S., Population declines in North American birds that migrate to the neotropics, *Proceedings of the National Academy of Science (USA)*, 86, 7658, 1989.

28. Environmental Systems Research Institute, *ArcGIS version 9.0*, Environmental Systems Research Institute, Redlands, 2004.

29. Dobbyn, J. and Couturier, A., *A comparison of BBS stop locations in Ontario derived from GIS and GPS,* Canadian Wildlife Service, Ottawa, 1998.

30. Ribeiro Jr., P. J., and Diggle, P. J., geoR: A package for geostatistical analysis, *R-News*, 1, 15, 2001.

31. Ihaka, R. and Gentleman, R., R: A language for data analysis and graphics, *Journal of Computational and Graphical Statistics*, 5, 299, 1996.
32. Rossi, R. E., Mulla, D. J., Journel, A. G., and Franz, E. H., Geostatistical tools for modeling and interpreting ecological spatial dependence, *Ecological Monographs*, 62, 277, 1992.
33. Bailey, T. C. and Gatrell, A. C., *Interactive Spatial Data Analysis*, Prentice Hall, Upper Saddle River, NJ, 1995.
34. Webster, R. and Oliver, M. A., *Geostatistics for Environmental Scientists*, John Wiley & Sons, Toronto, 2001.
35. Chainey, S., Reid, S., and Stuart, N., When is a hotspot a hotspot? A procedure for creating statistically robust hotspot maps of crime, in *Innovations in GIS 9: Socio-economic applications of geographic information science,* Kidner, D., Higgs, G., and White, S., Eds., Taylor & Francis, New York, 2002, 21.
36. Walker, J. S., Balling, R. C., Briggs, J. M., Katti, M., Warren, P., and Wentz, E. M., Birds of a feather: Interpolating distribution patterns of urban birds, *Computers, Environment and Urban Systems*, 32, 19, 2008.
37. Legendre, P., Spatial autocorrelation: Trouble or new paradigm? *Ecology*, 74, 1659, 1993.
38. Fortin, M.-J. and Dale, M., *Spatial Analysis: A Guide for Ecologists*, Cambridge University Press, New York, 2005.
39. Burrough, P. A., Spatial aspects of ecological data, in *Data Analysis in Community and Landscape Ecology*, Jongman, R. H. G., Ter Braak, C. J. F., van Tongeren, O. F. R., Cambridge University Press, 1987, 213.
40. Friedman, J. H., *A variable span scatterplot smoother*, Laboratory for Computational Statistics, Stanford University, Technical Report No. 5, 1984.
41. Wootton, J. T., Purple Finch (Carpodacus purpureus), in *The birds of North America online,* Poole, A., Ed., Cornell Lab of Ornithology, Ithaca, 1996, Available at: http://bna.birds.cornell.edu.bnaproxy.birds.cornell.edu/bna/species/208doi:bna.208.
42. Meentemeyer, V., Geographical perspectives of space, time, and scale, *Landscape Ecology*, 2, 163, 1989.
43. Jelinski, D. E. and Wu, J., The modifiable areal unit problem and implications for landscape ecology, *Landscape Ecology*, 11, 129, 1996.
44. Anselin, L., *Data and spatial weights in spdep: Notes and illustrations*, 2003, available at: http://sal.agecon.uiuc.edu.
45. Elith, J., Burgman, M. A., and Regan, H. M., Mapping epistemic uncertainties and vague concepts in predictions of species distribution, *Ecological Modelling*, 157, 313, 2002.
46. Urban, O. and Janssen, R., Why are spatial decision support systems not used? Some experiences from the Netherlands, *Computers, Environment and Urban Systems*, 27, 511, 2003.
47. Hunter, G. J. and Goodchild, M. F., Communicating uncertainty in spatial databases, *Transactions in GIS*, 1, 13, 1996.
48. Jacquez, G. M, Maruca, S., and Fortin, M.-J., From fields to objects: A review of geographic boundary analysis, *Journal of Geographical Systems*, 2, 221, 2000.
49. USGS Patuxent Wildlife Research Centre, *The North American breeding bird survey results and analysis, 1966-2003*, 2008, available at: http://www.mbr-pwrc.usgs.gov/bbs/ramapin.html.

20 Visualizing Risk for Hill Walkers

Alastair Jardine and William Mackaness

CONTENTS

OVERVIEW

Digital mapping offers greater flexibility in the symbolization and composition of maps. Map designers have proved reluctant to reengineer their traditional series mapping, yet there is evidence to suggest that maps will support additional information. In the context of maps for outdoor enthusiasts, this chapter explores the addition of risk information to 1:25,000 scale mapping. After a discussion of risk, we present a methodology for combining land cover and slope, and visualizing this information by superimposing the information within the contour isoline. The chapter presents various styles, and tests them in a route selection task that shows that the visualization of risk information does indeed affect the quality of decision making, particularly among novice map users. Although the focus was on improving the design of static, paper-based products, it is clear that dynamic information (such as weather conditions) delivered over mobile devices offers the chance to convey a much richer description of risk—one that reflects the changing dynamics of risk.

20.1 INTRODUCTION

The notion of risk, of human susceptibility, and the consequences of taking risks are in themselves fascinating topics. In the context of outdoor pursuits we can readily identify risk associated with the environment, and the changing nature of that risk, either in response to decisions made or to changes occurring in the environment (such as changing weather or hours of daylight remaining). There are complex dependencies between risk (both perception and evaluation of that risk) and the decision making process [1–4]. As if to prove this point, Ralston [5] provides an amazing illustration of how people assess and choose to cope with risk. This research has as its case study the rambler or hill walker, and explores ideas of how certain types of risk might be visualized and incorporated into existing (or future forms of) mapping. In essence, it explores ways in which we might visualize risk to support the decision making associated with a number of hill walking activities (namely, route planning and execution).

Whereas some people accept risk as a necessary condition, others remain totally unaware of the risk element until they experience it by chance [6]. This brings into focus the issue of how explicitly risk information should be conveyed to the hill walker (in map or any other form)—keeping in mind that given the large number of dynamic interdependencies, it is never possible to show *all* of the components of risk. Issues of liability and the misplaced belief associated with digital map products [7] may deter map producers from explicitly conveying risk—preferring instead to let the map reader, with their own skills and experience, interpret risk from the information typically found on a map.

There is a counterargument that suggests we should do more to show risk. Johnston [6] notes that over 28% of close-call experiences (situations that very nearly could have caused accidents) were from humans falling or rock or ice falls (the rest falling under such headings as exposure, avalanche, drowning, lost, shooting, or other). We would suggest that such evidence points to the need for information on land cover type and steepness to be made more explicit on maps.

20.2 MODELING AND VISUALIZING RISK

What is intriguing about risks typically associated with rambling is that we can infer a great deal about risk by viewing maps and drawing upon our experiences, for example, in the United Kingdom via the Ordnance Survey's 1:25,000 scale OS Explorer products, or the 1:50,000 scale Ordnance Survey Landranger® product [8]. The interpretive process very much depends on (1) the users' experiences; (2) their cartographic knowledge and ability to interpret the map (particularly slope and form); and (3) their understanding of the interaction between the morphology of the landscape, prevailing climatic conditions, and forecast. There is a complex interplay between spatial decision making, remoteness, resources and equipment, and changing conditions that can alter the situation from one of being relatively risk-free to one of great danger and exposure [9–11]. It is acknowledged that there is a real dynamic to risk, but it was not the goal of this research to integrate temporal variables (weather, season, or changing level of the fitness of the walker), though one could envisage incorporating these more dynamic aspects in a handheld device.

20.2.1 Maps Tailored for Outdoor Use

There are a set of variables related to risk that hold constant, and maps do exist that explicitly convey component parts of risk, such as maps tailored for the outdoor enthusiast. For example, Harvey Maps [12], in partnership with the British Mountaineering Council, have produced an award-winning set of 1:40,000 scale British mountain maps intended for hill walkers. The maps do not show risk per se, but use strong colors and hill shapes to show height, and convey slope via contour information. They also include a limited land classification (forest, boulder fields, and scree).

Orienteering maps are produced at a range of large scales (often 1:10,000 scale) [13] and contain very detailed information (such as ditches, earth banks, and gullies), as well as information on the type of land cover and its density (whether the going is easy or tough). The maps explicitly state out-of-bounds zones, but they do not attempt to convey risk and are only able to show detailed information over the region covering the course. There is no attempt to systematically map such fine detail at the national scale.

One example of risk mapping is a range of maps produced by Swiss Topo (the Swiss Federal Office of Topography), which has produced a Back Country Ski Edition version with risk information overlaid on the standard topographic mapped product. This 1:50,000 scale product displays areas of elevated risk using a red wash. The red wash simply highlights any region with a slope greater than 30°. It does not take into account aspect or ground cover type. Figure 20.1a shows a small region from sheet 268 (normal edition) around the Juliapass, west of St. Moritz. Figure 20.1b shows the same region for the Ski Edition.

The Ski Edition communicates risk clearly, using a limited set of criteria. The *dark line* are ski routes or lifts. Some routes deemed riskier are shown as a *dark dashed line* symbol. However, it is not clear what decision-making process was undertaken to classify these trails as being high risk. The map is not able to contain temporal information, even though it is difficult to prescribe risk in a snowy environment due to the seasonal effects of avalanche risk. It is clear that in this case, the onus is on the skier to make a judgment in conjunction with the map to assess the safety of tackling a particular route.

The challenge of this research was to examine ways of making static risk information more explicit (combining land cover and slope), to model risk, and to devise ways of including this information in mapping (in this case the focus being on 1:25,000 scale mapping) [14]. Such work is relevant in teaching, and in the context of changing access to land [15], where the public has the right to ramble freely across areas where paths may be nonexistent or very poorly defined. Land that does not contain marked paths requires more careful assessment given the breadth of choice in where one can wander (for example, over open moorland).

20.3 METHODOLOGY

There were two distinct components to this research. The first step was to devise and implement a model of risk, and then explore different ways of visualizing that risk. The model and the form of visualization were iteratively improved via feedback—both from users tasked with using the map, and from comments from cartographers at the Ordnance Survey (Figure 20.2). The risk model is described in Section 20.4,

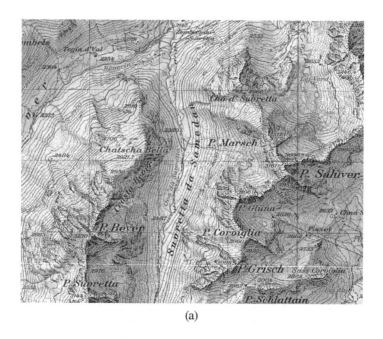

(a)

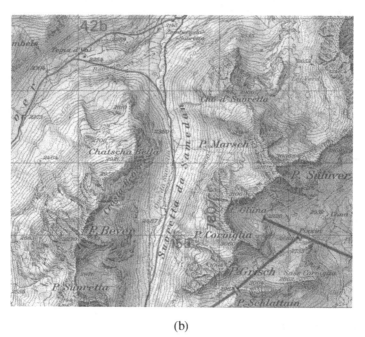

(b)

FIGURE 20.1   (a) 1:50,000 scale map (Sheet 268) and (b) Back Country Ski Edition, same area. © Federal Mapping Agency of Switzerland (Swiss Topo).

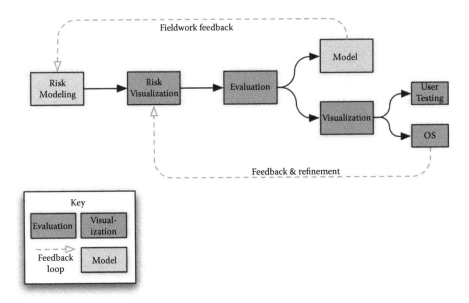

FIGURE 20.2 An overview of the methodology showing the key components of risk modeling and visualization.

and the visualization of risk is described in Section 20.5. The test area for the project was a 12 km² area in the Lake District National Park in the United Kingdom. The results from the visualization of risk were evaluated in a variety of contexts.

20.4 MODELING RISK

By definition, mountains offer exceptional risk either because of natural disaster (avalanche, earthquake, etc.) or because of human activity (climbing, scrambling, walking). Many factors govern the degree to which risk can be mitigated (e.g., a high level of fitness, an early start, summertime, good preparation, plenty of mountain experience, and a familiarity with the area). It goes without saying that visitors to mountainous areas do not desire to have accidents—even though we might acknowledge that risk can be a motivation for recreation [16]. The understanding of risk and risk taking is affected by the cultural environment in which it develops as well as by personal experience [17]. The notion of vulnerability and perception of risk affect our response to it. Risk is multifarious—something we might try and define in terms of isolation, duration, or exposure. One might consider there to be a strong correlation between risk and remoteness, but this relationship can be defused and confused by factors such as mobile reception and the proximity of an air ambulance service. Rather than attempt to model the dynamics of risk and its spatiotemporal interdependencies, we have restricted our visualization to a static view—of absolute values rather than relative ones (thus our model does not attempt to model remoteness or proximity to rescue services)—drawing from information more readily available.

There are three aspects that need to be developed as part of modeling risk: how can risk be (1) defined, (2) experienced by individuals in a walking context, and (3) assessed by individuals in different ways. *Hazard* can be defined as a situation or landform that may cause harm or has the potential to do so. Risk is the likelihood of harm caused by the hazard. Risk assessment is the identification and quantification of the risks associated with those hazards [18]. This research used a fairly recently devised subjective risk ranking assessment technique [19]. This approach offers a useful means of prioritizing components of risk by first asking an expert to define and rank the relevant risk attributes, and then asking other experts or lay persons to refine the ranking based on nontechnical descriptions of those attributes [20]. This helps handle the subjective task of ranking risk attributes. The technique enabled us to produce a quantitative model of walking risk, by first ranking hazards associated with walking risk [19], and then applying a scale to the ranked values. Once the land cover types have been ranked, a numerical risk value was assigned to each type. The scale used was based on a classification proposed by Malczewski [21] that links a textual importance scale to a number (Table 20.1).

Ranking allows a hierarchy of risk to be assigned to the categories in relation to one another, which then facilitates a greater clarity when assigning numerical risk values to categories individually. Ranking the risks is, of course, not value free and depends on the knowledge of an expert who might have previous biases or preconceptions. Establishing a list of variables that could affect the walker is essential, but from a pragmatic point of view, must relate to the data available. In the case of this research, it was OS MasterMap®, and the descriptiveTerm field. Initially it was assumed that the final output for the model would be the temporally fixed paper map. Therefore, temporal aspects were not considered, only those that can be defined in space.

OS MasterMap contains individual land cover types (Table 20.2) in the descriptiveTerm field. In this research only the nongeneric land cover types were utilized. These are the items with asterisks in Table 20.2. Only these nongeneric land cover types were used because no actual land cover can be inferred from the land cover types Multi Surface or General Surface. In accordance with the risk ranking methodology, these individual land cover (LC) types were ranked in risk order, and then Malzewski's scale (MV) was applied to the ranked land cover types (Table 20.3).

The risk values assigned in Table 20.3 were subjectively ranked, and it was assumed that the ground was flat, which is (of course) not realistic. The LC risk was

---

TABLE 20.1

Malczewski's Classification of Importance

1 – Equal importance

2 – Equal to moderate importance

3 – Moderate importance

4 – Moderate to strong importance

5 – Strong importance

6 – Strong to very strong importance

7 – Very strong importance

8 – Very to extremely strong importance

9 – Extreme importance

TABLE 20.2

Ordnance Survey's 20 Land Cover Types in OS MasterMaps

Boulders*	Rock (Scattered)*
Boulders (Scattered)*	Rough Grassland*
Cliff*	Scree*
Coniferous Trees*	Scrub*
Coppice or Osiers*	Slope*
Heath*	Coniferous Trees (Scattered)*
Marsh Reeds or Saltmarsh*	General Surface
Multi Surface	Nonconiferous Trees (Scattered)*
Nonconiferous Trees*	Step
Rock*	Track*

* Nongeneric land cover types.

TABLE 20.3

Linking Malczewski's Scale to Land Cover Type

Ranked Position	Land Cover Type	Malczewski Value
1	Track	1
2	Scrub	2
3	Coniferous Trees	3
4	Coppice or Osiers	3
5	Nonconiferous Trees	3
6	Coniferous Trees (Scattered)	3
7	Coppice or Osiers	3
8	Nonconiferous Trees (Scattered)	3
9	Heath	3
10	Boulders (Scattered)	4
11	Rock (Scattered)	4
12	Boulders	6
13	Rock	6
14	Rough Grassland	6
15	Marsh Reeds or Saltmarsh	7
16	Scree	8
17	Slope	8
18	Cliff	9

multiplied by a slope risk increase percentage (S%) to produce a model that incorporated the fact that as slope steepens, the risk increases. The scale used is illustrated in Table 20.4.

Slope information was derived from Ordnance Survey Land-Form PROFILE® data. A linear percentage increase was used, anticipating that these values would be

TABLE 20.4
Modeling Slope Risk in the Model

Degrees of Slope	% Weighting Increase
0	100
5	111
10	122
15	133
20	144
25	155
30	166
35	177
40	188
45	199
50	210
55	221
60	232
65	243
70	250
75	250
80	250
85	250

refined during the evaluation stage. This works in tandem with the established land cover risk. Thus, the risk value (RV) can be defined as being the product of the land cover Malczewski value (LCMV) and the slope percentage weighting increase:

$$RV = LCMV \times S\% \tag{20.1}$$

In practice, OS MasterMap contains concatenated categories. Thus, a single region might contain multiple single categories in an attempt to convey more precisely the nature of the ground. For example, a single region might be classified Coniferous Trees, Nonconiferous Trees, Rock, and Rough Grassland. To cope with this, equal weightings were applied for up to three categories, with the fourth ignored in all cases. Table 20.5 illustrates how the weighting system was applied.

For example, the equation for a descriptiveTerm containing three categories would be:

$$RV\ Overall = (33\% \times RV1) + (33\% \times RV2) + (33\% \times RV3)$$

To use the model with OS MasterMap data, slope was added into the attribute table, achieved by intersecting a TIN Feature class with OS MasterMap. This enabled the model to lookup the slope value together with the land cover type at any point within a given region. The risk model was constructed in a spreadsheet, and the risk values were calculated by importing OS MasterMap's attribute table into the spreadsheet and performing a lookup routine on the data. The table was then imported back

TABLE 20.5

Apportioning Weights According to the Number of Categories Describing Each Region

Number of Concatenated Categories	Percentage Weighting of Individual Categories			
	1	2	3	4
1	100%	—	—	—
2	50%	50%	—	—
3	33%	33%	33%	—
4	33%	33%	33%	Ignored

into ArcMap for visualization. The model is flexible such that any future changes in OS MasterMap could be reflected in changes to the weightings in the model.

20.4.1 RISK MODEL OUTPUT

The risk model produced a range of output values ranging from 0 (minimum risk) to 22.5 (maximum risk). Ground truthing indicated that modeled risk was accurately represented on the ground where the land cover type was correctly attributed. Ground truthing was undertaken by walking some areas of the landscape, and readjusting the weightings where they were deemed out of place. It is acknowledged that this readjustment was subjective, but this does not negate the underlying concept—and the idea that over time the weighting and ordering of land cover type and slope values could be refined. Figure 20.3 shows the associated risk values for various land cover types. It has been superimposed on a photograph to help illustrate the associated land cover. There are a range of values for each polygon, representing the minimum and maximum risks for the associated polygonal areas. The polygons have been sketched on with reference to the original OS MasterMap data.

20.5 VISUALIZING RISK

In the context of the research presented here, the key visualization requirements were that it be readily interpretable against the Ordnance Survey's current Explorer style (distinct from other map information) and be visually ranked (from low to high risk). Three maps styles were created, and the process of iterative design was applied to derive a suitable end product.

20.5.1 MAP STYLE 1

There are specific benefits to using a spectral color ramp to display all risk values (0–22; see Brewer [22] for detailed discussion). In response to comments from cartographers at the Ordnance Survey, a style was designed that was intended as a color wash that could be overlaid on the map (Figure 20.4). As one might expect, there is clear similarity between a map showing only slope (Figure 20.5a) and the risk

The model performs well in this situation. To gain foothold on an area with boulders, heath, rock and grassland would be easier than maintaining footing on an area of scree.

FIGURE 20.3 Superimposing the risk values over a photograph to assess the accuracy of the risk classification.

map (Figure 20.5b). But important differences are also evident because Figure 20.5b reflects additional information concerning the different land cover types. The combining of land cover and slope reflects the idea that in combination Figure 20.5b is more informative.

20.5.2 MAP STYLES 2 AND 3

A different design style, map style 2 (formed from a critique of the map style presented in Figure 20.4), was found to compete too much with the underlying information but again led to revisions that led to evaluation of a third style. Rather than color wash the entire map (all the information contained within the map), in map style 3 the risk information was embedded into the contour lines. In this instance, the contour lines were segmented according to their intersection with each risk polygon and each segment color coded with the risk value of the polygon in which it fell. This is illustrated in Figure 20.6 and detailed in Figure 20.7.

The same spectral color ramp was utilized from map style 1, but instead embedded within the contours. Thus in the case of map style 3, this means that there is no competition for space, only color. Due to the production method of map style 3, the contour values and spot heights are not present (not an ideal situation). Any final version of the map would need to contain this information. Technically, this symbology is a separable bivariate symbol [23], as it conveys risk and topography concurrently.

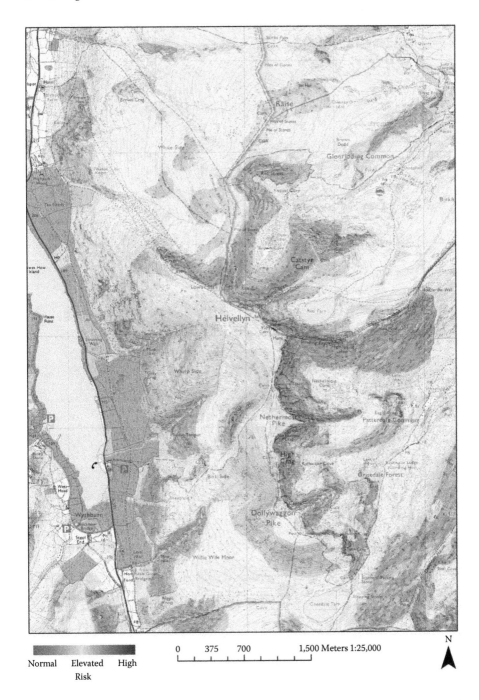

Normal Elevated High
Risk

0   375   700   1,500 Meters 1:25,000

N

FIGURE 20.4 Map style 1: intended as color wash overlaid on OS mapping (presented here in grayscale). © Crown copyright/database right 2005. An Ordnance Survey/EDINA supplied service.

(a)

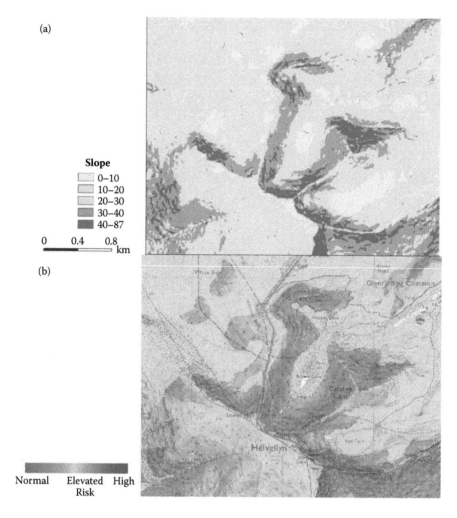

Slope

- 0–10
- 10–20
- 20–30
- 30–40
- 40–87

0   0.4   0.8
km

(b)

Normal   Elevated High
Risk

FIGURE 20.5 Detail from Figure 20.4: comparison with a map showing only slope. © Crown copyright/database right 2005. An Ordnance Survey/EDINA supplied service.

In this manner, map style 3 comprehensively displays risk for an area, leveraging existing content while maintaining overall coherency of the Ordnance Survey Explorer map.

20.6 EVALUATION AND DISCUSSION

To evaluate whether and how this risk information was being used, a test was undertaken comparing route selection using the Explorer map, and the Explorer map with the risk information superimposed (using map style 3). Subjects were asked initially to draw a route from A to B on the plain OS Explorer Map. The task was then repeated with the risk indicator map (RIM) and the results compared [14]. In both

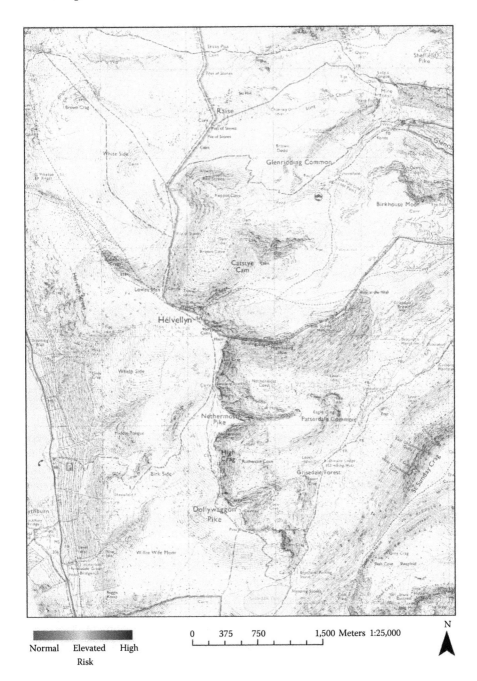

Normal   Elevated   High
      Risk

0   375   750   1,500 Meters 1:25,000

N

FIGURE 20.6 Map style 3 in which the risk value is color coded into various segments of each isoline (presented here in grayscale). © Crown copyright/database right 2005. An Ordnance Survey/EDINA supplied service.

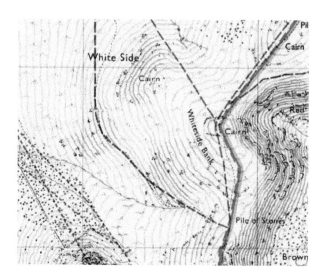

FIGURE 20.7   A detail from Figure 20.6 illustrating how risk information is embedded into the contour line. © Crown copyright/database right 2005. An Ordnance Survey/EDINA supplied service.

cases the task was to choose an efficient route that would get them from A to B and take in the summit of Helvellyn. To assess the difference that the risk information has made between the OS Explorer and the RIM, the routes were digitized and the average risk calculated for each path. To illustrate the various routes proposed, all routes are shown in Figure 20.8.

There were a number of interesting results. Two of the five candidates were inexperienced map users, two were experienced map users, and another had some experience of hill walking. The route choice for experienced users with the RIM was unaffected, but caused one to comment that it would be "useful for nonexperienced users." Conversely, the way inexperienced users utilized the information was of more relevance. Using the plain OS Explorer map, a beginner hiker struggled to quickly interpret the information on the map. When presented with the RIM, the user immediately commented "it just makes sense," and said it allowed her to "take a more scenic route" while also allowing a "more accurate predecision [*sic*]." Another inexperienced candidate recognized that bunching contours would generally mean higher risk, and thus aimed to stay "on rivers and valley bottoms ... it is flatter." The risk map allowed the user to "see right away what to avoid," speeding up the route selection process when compared with OS Explorer. The user with moderate experience said the RIM was "much more easy to read by [just] looking at it," and that it was "easier for someone who isn't an expert user."

It is considered that the RIM would be of great utility for teaching purposes, as it makes explicit relationships between spatially related features that combine to give an indication of risk. As a teaching tool, it could be introduced, then removed as students became more familiar with the map, and the intricacies of interpretation were learned.

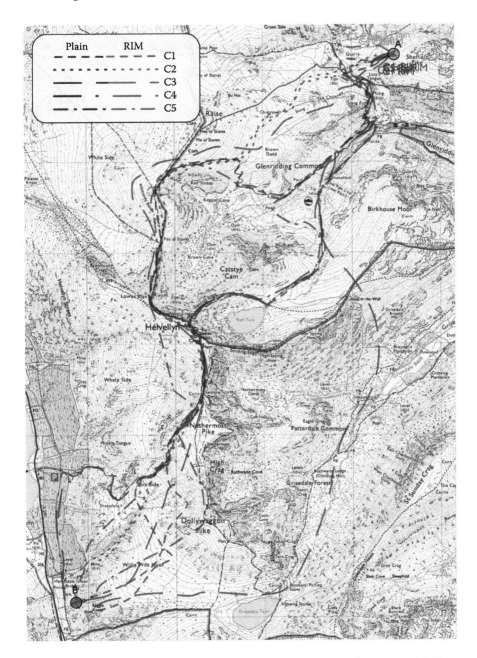

FIGURE 20.8   Showing route choices arising from user testing. © Crown copyright/database right 2005. An Ordnance Survey/EDINA supplied service.

OS Explorer maps contain a great deal of information; this is the nature of topographic maps since they seek to cater to as broad an audience as possible. Adding risk information makes the map busier still! Though cartographers expressed concern over the volume of information and the user's ability to interpret the information, the route testing conducted with novice users indicated that with some training they could cope with this additional information, and that the additional information did influence their choice of route.

20.7 CONCLUSION

The quantification of risk is subject to a number of assumptions and simplifications that are open to challenge. Risk is multifarious, affected by changing conditions and the dynamics of decision making in the hills. But the focus here was to model risk using fixed variables known for national coverage (combining digital elevation model information with land cover type), and not to use human dependent, temporally variable factors. If this type of risk mapping was available over mobile devices, then access to such dynamic information may well afford a more truthful and current representation of risk (for example, including weather conditions).

The initial weightings were set by best guess and adjusted slightly after fieldwork. The ordering of some categories was also adjusted in response to the fieldwork. The sensitivity of the model was assessed by varying the weightings against each of the variables, but because of the spatial autocorrelation, the changes in the map were slight.

The evaluation demonstrated that the more explicit presentation of risk information did indeed modify people's decision making. From a graphical point of view, it is clear that OS Explorer and OS Landranger products are already information rich. Therefore, any design solution needs to avoid adding information to the point that the challenges of interpretation play against the benefits of this additional information. To that end there is a clear need for further investigation and evaluation of the risk model outputs and how they are visualized.

ACKNOWLEDGMENTS

Figure 20.1 is a copyright of the Federal Mapping Agency of Switzerland (Swiss Topo). Figures 20.4, 20.5, 20.6, 20.7, and 20.8 are Crown copyright/database right 2005. An Ordnance Survey/EDINA supplied service. The authors are very grateful for the support of the Ordnance Survey and the invaluable advice of their cartographers.

REFERENCES

1. Slovic, P., Finucane, M. L., Peters, E., and Macgregor, D. G., Risk as analysis and risk as feelings: Some thoughts about affect, reason, risk and rationality, *Risk Analysis*, 24(2), 311, 2004.
2. Parkin, R. T. and Balbus, J. M., Variations in concepts of "susceptibility" in risk assessment, *Risk Analysis*, 20(5), 603, 2000.
3. Husdal, J., Can it really be that dangerous? Issues in visualization of risk and vulnerability, 2001, available at: http://www.husdal.com/gis/visirisk.htm, accessed June 2005.
4. Sjöberg, L., Factors in risk perception, *Risk Analysis*, 20(1), 1, 2000.
5. Ralston, A., *Between a Rock and a Hard Place*, Atria Books, New York, 2004.
6. Johnston, M., Accidents in mountain recreation: The experiences of international and domestic visitors in New Zealand, *GeoJournal*, 19, 323, 1989.

7. Cho, G., *Geographic Information Systems and the Law: Mapping the Legal Frontiers*, Wiley, Chichester, 1998.
8. Ordnance Survey, Ordnance Survey Leisure Map Shop, 2005, http://www.ordnancesurvey.co.uk/oswebsite/mapshop/, accessed 17 August 2005.
9. Saku, J., Map use teaching and experience, *Cartographica*, 29(3-4), 38, 1992.
10. Walker, K., *Mountain Hazards*, Constable and Company, London, 1988.
11. Carter, J. R., The Many Dimensions of Map Use, International Cartographic Conference, A Coruna, Spain, July 2005 CD-ROM Proceedings, 2005.
12. Harvey Maps, http://www.harveymaps.co.uk, accessed April 2008.
13. British Orienteering, http://www.britishorienteering.org.uk, accessed April 2008.
14. Jardine, A. J., Modeling and visualizing the risk associated with hill walking to support route planning and execution for ramblers, unpublished masters dissertation, Institute of Geography, School of GeoSciences, University of Edinburgh, 2005.
15. Countryside Agency, Leaflet CA65: Countryside access and the new rights, 2004, available at: http://cms.countrysideaccess.gov.uk/var/csa/storage/original/application/php-PQaMPb.pdf, accessed 24 June 2005.
16. Ewert, A. W., Playing the edge: Motivation and risk taking in a high altitude wilderness like environment, *Environment and Behavior*, 26(1), 3, 1994.
17. Seigneur, V., The problems of the defining the risk: The case of mountaineering, *Forum: Qualitative Social Research*, 7(1), Article 14, 2006 http://www.qualitative-research.net/fqs-texte/1-06/06-1-14-e.pdf.
18. Barrow, C., Risk assessment and crisis management, in *The Royal Geographical Society Expedition Handbook*, Winser, S., Ed., Royal Geographical Society, London, 2004.
19. Morgan, M. G., Florig, H. K., Dekay, M. L., and Fischbeck, P., Categorizing risk for risk ranking, *Risk Analysis*, 20(1), 49, 2000.
20. Florig, H. K., Morgan, M. G., Morgan, K. M., Jenni, K. E., Fischhoff, B., Fischbeck, P. S., and DeKay, M. L., A deliberative method for ranking risks (I): Overview and test bed development, *Risk Analysis* 21(5), 913, 2001.
21. Malczewski, J., *GIS and Multicriteria Decision Analysis*, John Wiley & Sons, New York, 1999.
22. Brewer, C. A., Spectral schemes: Controversial color use on maps, *Cartography and Geographical Information Systems*, 24(4), 203, 1997.
23. Nelson, E. S., Designing effective bivariate symbols: The influence of perceptual grouping processes, *Cartography and Geographic Information Science*, 27(4), 261, 2000.

21 Using Web-Based 3-D Visualization for Planning Hikes Virtually
An Evaluation

Susanne Bleisch and Jason Dykes

CONTENTS

OVERVIEW

The use and usefulness of an interactive three-dimensional (3-D) desktop application is explored in the context of evidence that landform may be understood more intuitively through realistic 3-D visualization than in traditional maps. A Web-based digital environment that combines elevation models and photorealistic imagery with additional abstract information is tested for planning hikes in the foothills of the Swiss Alps. Participants' efforts in fulfilling standard hike planning tasks are recorded through questionnaires and interviews and subsequently analyzed. The 3-D visualization is preferred over the 2-D (two-dimensional) map for tasks associated with getting an overview, but participants requested more abstract information or additional 2-D maps for tasks involving extracting exact information. The application is more useful to those who spend a longer time using it. Problematically, participants underestimated the scale of the relief depicted. Additionally, their confidence in solving the tasks involving extracting exact information from the 3-D visualization was low, even when tasks were solved correctly. We conclude that there is scope for using 3-D visualization for effective hike planning among tech-savvy hikers, especially in combination with additional information or 2-D maps, but that this may involve some risk of underestimating the scale of the topography involved.

21.1 INTRODUCTION

A range of recently developed technologies has made it increasingly straightforward to digitally visualize information in interactive three-dimensional (3-D) forms. Realistic 3-D visualizations may be understood more intuitively than traditional maps [1,2], and distributing them over the Internet makes them accessible to a wide audience [3]. They are also popular, as demonstrated by the rapid advances and widespread uptake associated with desktop virtual reality environments or geobrowsers such as Google Earth [4] or NASA's World Wind [5]. But are these forms of representation effective when used for particular activities to help support successful task completion? And does their use lead to any particular biases or behaviors?

Geovisualization applications are frequently technology driven, but while we continue to build exciting and impressive applications the need for evaluating them in supporting particular tasks is paramount [6]. Here we address a particular need and test a desktop 3-D visualization technology in a specific context and with a predetermined and real task set. Hiking is the most popular recreational activity among the Swiss [7], and the tourism organizations and other agencies are looking for ways to promote their region and support hiking [8]. We explore the usefulness of an interactive 3-D application that provides Web-based realistic visualization in the context of planning hikes in the foothills of the Swiss Alps. Our 3-D maps are photorealistic but enhanced with abstract information that is valid to the task at hand, such as a hiking route and spot heights. The assessment of usefulness includes aspects of both utility (the tool's performance regarding a specific task) and usability (the users efficiency and satisfaction with a tool), as defined in the field of human–computer interaction design [9]. Although many different sources may be consulted when planning hikes, we test the 3-D visualization as the sole tool used in this study. Doing so enables us to evaluate its usefulness in isolation, and make inferences about its potential contribution to hike planning and the possible effects of using it as part of this process.

21.1.1 Visual Realism in a 3-D Environment: Background, Data, and Visualization

Hiking tours in Switzerland are generally planned using the Swiss National Maps 1:25,000 (LK25) which are highly detailed and reliable. Several researchers suggest that people have difficulties interpreting the kind of 3-D information that is essential for planning hikes from 2-D maps, where the information is encoded with 2-D visual variables through contour lines or hill shading [10,11]. Thus, new types of maps and maplike representations have been developed that look more similar to the real world and may be easier to decode and to understand (e.g., [12,13]). Recent advances in computer processing capabilities, graphics hardware, and data availability have led to widespread digital 3-D terrain visualization as found in Google Earth [4] or Memory-Map [14]. Dickmann [15] compares the interpretation of spatial information from a Web-based 3-D visualization and a print-based traditional map, concluding that the 3-D visualization is generally more effective in communicating the information presented. However, studies also indicate that the acceptance and utilization of 3-D cartography is relatively low among certain groups, and the effort required to learn to use it effectively can be very high when compared with the increase in knowledge achieved [2].

This investigation focuses on a digital environment that combines digital elevation models and photorealistic imagery with additional abstract information to support hike planning. We use the virtual reality (VR) technologies from Geonova AG—software tools that allow highly interactive 3-D visualizations to be generated from digital terrain models and orthoimagery, and subsequently distributed over the Internet. The elevation data used for the 3-D visualization in this study is the DHM25, the most frequently used digital height model of Switzerland with a resolution of 25 m. Our study focuses on an area in the heart of Switzerland that permits moderate hiking between 430 and 2400 m above sea level. The area is established but far from notorious among hikers, and representative of many similar regions, enabling us to compare users who know the region with those who are new to it and to generalize our results to other areas of similar geography. To facilitate the comparison, the 3-D representation contains approximately the same information as the traditional Swiss National Maps (LK25). There are some key differences, however. When on the map, the third dimension is encoded using contour lines and spot heights; this information is represented by perspective views of the digital elevation model in the 3-D scene. Lange [16] shows that draping orthoimages over terrain is the most important element for generating virtual environments that are perceived as realistic. Following Lange, we drape orthoimagery with a resolution of 0.5 m over the elevation model (Figure 21.1). This approach is typical of many of the current wave of geobrowsers. Another key difference thus relates to the way that land cover is represented. On the map it is preinterpreted and represented through coded symbols, while the 3-D visualization employs the uninterpreted orthoimagery to show land cover. MacEachren et al. [18] recommend cartographic designs for virtual environments that combine abstract and realistic representations of geographic phenomena. For this study, abstract vector information showing a hiking route, spot heights, and place name labels were added to the 3-D visualization (Figure 21.1). Similar information is available in traditional panoramic representations of tourism regions (e.g., [19]). The addition of only one specific hiking route ensures that all participants in the study refer to the same route when attempting the tasks. One significant difference relating to information content is that place names and height labels are less dense in the 3-D representation than in the 2-D LK25 so that the 3-D scene is not overcrowded with labels.

FIGURE 21.1   Monochrome reproduction of 3-D visualization used in the experiments. This example view includes a marked hiking route, place names, and height labels.

21.1.2 PLANNING OF HIKES

The Swiss Federal Office of Sports has a program called Youth & Sport (J&S) with guidelines describing the sequence of events involved in planning a hiking tour [20]. The J&S instructions are used as guidelines for measuring the utility of the 3-D visualization for planning hikes. The map-related parts of the guidelines can coarsely be divided into three categories: getting an overview of a hiking region, which means studying maps and getting a feeling for the region and its topography; selecting a suitable route; and planning the hike in detail, for example, by checking the heights, steepness of slope, and distances. This is often an iterative process.

21.2 DATA COLLECTION AND ANALYSIS

To capture information on the way in which the 3-D visualization is used in a virtual hike planning activity and to measure task performance, we developed an experimental framework based upon established methodologies for the evaluation of visualization techniques and of virtual environments [21,22]. This involves developing plausible user tasks (see Section 21.2.1) and performing a formative evaluation by asking a representative sample of likely users (hikers and mountaineers) to complete the tasks. Their efforts are recorded through questionnaires and interviews (see Section 21.2.2), and analyzed with a focus on overall patterns and trends within particular groups. These methods of evaluation have been employed in a number of user-centered geovisualization studies (e.g., [17,23]).

21.2.1 TASKS

A number of specific tasks were designed to represent hike-planning activities by converting the guidelines from each of the J&S categories into questions. Tasks that had been designed for paper-map hike planning [20,24] were analyzed according to their information content and translated into questions that required the same information to be acquired by interacting with the 3-D visualization. For example, a height profile is created during the paper-map planning stage because this information is regarded as being difficult to interpret from contour lines in the 2-D map. The contention that realistic 3-D visualizations are more intuitive than traditional maps [1,2] suggests that information about steep sections or the highest and lowest points of a route may be visually interpreted directly from the oblique views of the 3-D visualization by interacting with the model. To fulfill the tasks, the users were encouraged to navigate around the 3-D visualization. Use of the mouse to control viewing position and to fly across the landscape was explained to all participants.

21.2.2 QUESTIONNAIRE AND INTERVIEWS

A series of questions about the specific tasks were designed to record activities undertaken and levels of task completion (Figure 21.3 and Figure 21.4 show task titles that summarize the questions used here). Users were also asked to record their feelings about the ease of task completion and their confidence in the results. Details about prior map reading skills, knowledge of the region, prior contact with 3-D visualization applications, and demographic status were also recorded [25]. Multiple choice questions were employed in some cases with one or more answer possibilities. Where appropriate, a "Don't know" or "Other" category was added with opportunities to provide further explanation to allow for omitted answers [26]. Short answer (closed) questions that could also be quantitatively analyzed were used for some of the tasks, for example, finding the steepest part of the hike. Open-ended questions allowed the participants to express their thoughts and feelings about using the visualization. To generate more qualitative information and to follow up on written statements, interviews were conducted with a sample of respondents [27]. This allowed us to validate our interpretations of the responses from those participants and explore some of the reasons behind them.

21.2.3 THE SAMPLE

The population sampled for the purpose of this study consisted of all people interested in hiking in the foothills of the Swiss Alps. The application and questionnaire were publicized through the Internet, through hiking Web sites, and communicated to hiking contacts via e-mail. All hikers informed about the study were asked to further distribute the information to acquaintances interested in hiking. As both the 3-D visualization and the questionnaire were made accessible over the Internet, only hikers with some IT knowledge were able to participate in the study. A subset of those who completed the questionnaire was interviewed. Producing an unbiased sample of this diverse population is an extremely difficult task [25]. The sample of the hikers reached through the methods described in this study may not be wholly representative of the background population. The findings are thus applicable only to this sample, but due to the similarities between the characteristics of those used in the sample and a section of the background population, they may be more widely relevant to hikers who are digitally aware and empowered by the Internet. These are likely to be those who use geobrowsers and to whom tourism agencies may want to promote hiking through interactive 3-D cartography.

21.2.4 DATA ANALYSIS

Quantitative data were analyzed graphically and, where appropriate, with statistical comparison techniques to investigate possible differences in successful task completion among gender, occupation, age, prior technical experience with 3-D visualizations, self-reported levels of map reading ability, previous knowledge of the study area, and time spent on the tasks. The data from personal feedback, open-ended questions and the interviews were analyzed in part through quantitative techniques, but predominantly through qualitative methods including content analysis [28]. The qualitative data was used to contextualize and interpret the quantitative findings.

21.3 RESULTS AND FINDINGS

Approximately 600 people were informed about the study and 99 participants completed the questionnaire. A further 68 people did not fill in the questionnaire completely or provided other forms of feedback, such as not being able to access the 3-D visualization or feeling motion sickness when navigating. The questionnaire reached people from a full range of ages and about a quarter of the participants were women. Notably, most participants claimed to have very good or good map reading skills. More than two-thirds of the participants stated that they had come into contact with 3-D visualizations prior to the study, perhaps suggesting a bias in the sample that may be associated with the Internet-based nature of both the software used and the dissemination techniques employed. More than half of the participants (58%) did not know the region. Qualitative statements from the open-ended questions in the questionnaire and interviews were used to triangulate the quantitative data obtained through the questionnaire.

21.3.1 GETTING AN OVERVIEW OF A REGION

Responses to tasks associated with the first of the J&S hike-planning categories are summarized in Figure 21.2. They show that 56% of the participants preferred the realistic 3-D visualization over the map for the task of getting an overview or an

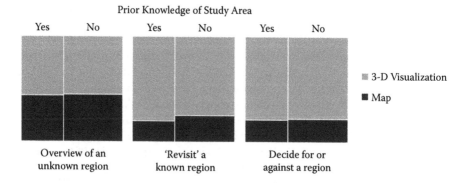

FIGURE 21.2 Three-dimensional visualization or map preferences subdivided by prior knowledge of the study region (graphic area per task = 100%).

impression of an unknown region (41% of these with and 59% without prior knowledge of the region). For virtually "revisiting" a known region, 76% of the participants preferred the 3-D visualization (43% of these with and 57% without prior knowledge of the region). Also, 78% preferred the 3-D visualization for helping with the decision as to whether to visit the region for hiking (41% with and 59% without prior knowledge of the region). These results suggest that users without prior knowledge of the study area have a stronger preference of using the 3-D visualization for any of the three tasks. However, employing a chi-square test reveals that these differences are not significant at the 0.05 level. The possible advantages of the 3-D visualization over the 2-D map for this stage of hike planning are evidenced through the content analysis, which highlighted a sizable number of positive statements about hike planning using 3-D: "It is faster to get an overview of a hike (where does it lead through) or region with the 3-D visualization"; "It is possible to know what the views from a specific point on the hike will be"; "It is more intuitive to comprehend the topography of the region."

21.3.2 SELECTING A SUITABLE ROUTE

The second stage of hike planning involves selecting a suitable hiking route by detecting possible tracks on the orthophoto draped over the digital elevation model (DEM). About 56% of the participants were able to locate routes along streets and clearly visible paths. Another 15% located concealed or less visible paths such as those across pastures or through woodland. For selecting routes, 68% of the participants prefer to use the traditional map. However, 12 participants mentioned that this would change in favor of the 3-D visualization if more than one hiking route was included in the visualization. This coincides with participants' statements indicating that the orthoimagery is difficult to interpret by itself for exact information extraction tasks as, in contrast with most traditional 2-D maps, the realistic content is not preinterpreted and highlighted by cartographers. This supports the contention of MacEachren et al. [18].

21.3.3 EXTRACTION OF EXACT INFORMATION

The participants were asked to gather exact information, such as heights, lengths, steepness of slope, and information about the nature of particular features from the 3-D visualization. These tasks are representative of the third stage of the J&S guidelines. Figure 21.3 shows that tasks that are completed using additional abstract information are solved most successfully. For example, the highest or lowest points can be extracted directly from height labels in the 3-D visualization. This corresponds with qualitative feedback on the importance of the height labels in the 3-D scene.

Tasks like estimating the length of a hike section or finding the steepest section of the hike were rarely solved satisfactorily in the 3-D visualization. Only 22% of the participants estimated the length of a 3 km section of the hike to within ±500 m.[1] On

---

[1] This is not a level of error that is generally regarded as being acceptable. We were unable to find a source that provided a standard distance. We derived this figure by calculating hiking time (based on the formula of 4 LKm/h [25], 1 LKm is either 1 Km distance or 100m height difference upwards) and deciding that spending 7.5 minutes (equivalent to 500m) more or less time to hike the section was acceptable. Some sensitivity analysis of the effects of this assumption would be appropriate.

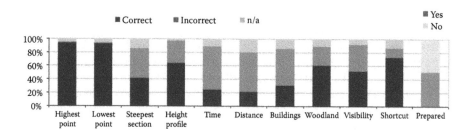

FIGURE 21.3 Extraction of information. Spot heights were labeled in the 3-D visualization; the questionnaire contained four graphical height profiles that could be compared with the visualization.

average they overestimated the length of the section by 700 m (23%) with a high standard deviation of 1.5 km. One or both of the two steepest sections of the hike were identified by 41% of the participants. These findings further support the suggestion that ancillary information, such as distance or height information or measuring tools, is needed when using realistic environments to help the successful completion of some tasks that are deemed essential in planning hikes. One clear trend was identified: 58% of the incorrect answers to the question about the steepest section selected the longest steep section, but this did not contain the steepest slope as measured between adjacent cells at a scale equivalent to the resolution of the DEM. The finding that a minority of participants was able to identify the steepest section at this scale may be surprising as 3-D maps are believed to be understood more intuitively. Such judgments may relate to the fact that slope is a scale-dependent property. This suggests that although the form of the landscape might be understood, the nature and scale of its variation may not be interpreted accurately or effectively from the 3-D visualization. Alternatively, participants may have used their experience of longer shallower sections being harder to hike than the short steeper sections in completing this task. The content analysis of our qualitative data identified three statements in support of this interpretation, noting that it is difficult to estimate the steepness of slopes in the 3-D visualization. Our data also show that those who solved the different tasks of extracting exact information from the 3-D visualization correctly were not confident that this was the case, associating considerable uncertainty with this task.

The conflicting overestimation of distance and likely misinterpretation of slope and the associated uncertainty suggest some difficulties in effectively planning hikes using the 3-D visualization software evaluated here. It is important in this context that even though almost all participants were unable to use the 3-D visualization to extract exact information with confidence, 51% of them felt sufficiently prepared to go hiking having used the 3-D representation. These possible misinterpretations could have serious consequences. More than half of those 51% who stated that they felt sufficiently prepared also indicated that they did not have prior knowledge of the hiking area.

The participants were asked what they thought was missing from the 3-D visualization. About 71% suggested additional abstract height, time, and/or distance information in forms such as labels, contour lines, reference grids, or measuring tools, highlighting the need for additional abstract information and functionality to support specific tasks. Others desired additional functionality such as hotel or travel information or the automatic generation of height profiles. As neither of these features can be found in the traditional map it seems that the new technology has resulted in demands for new functionality and possibilities in general. But as shown by the rather poor task performance (see Figure 21.4), it also seems that the participants were not able to satisfy their information needs for hike planning from a 3-D visualization environment consisting solely of a DEM with draped orthoimagery and limited abstract ancillary information. The widely accessed geobrowsers discussed in Section 21.1 consist predominantly of precisely this combination, and our findings therefore suggest that solely using such tools for hike planning may be problematic.

21.3.4 Differences between User Groups

The tasks were evaluated in the light of differences between subgroups of participants using chi-square tests at a 0.05 significance level. Tests for differences between gender, occupation, age, and prior technical experience with 3-D visualizations suggested some trends but revealed no significant differences in relation to any of the findings reported above. Significant differences were found for some tasks when users were categorized according to completion time. Figure 21.4 shows that participants who spent more time with the 3-D visualization solved many tasks more successfully. The group with prior technical experience of 3-D visualizations also

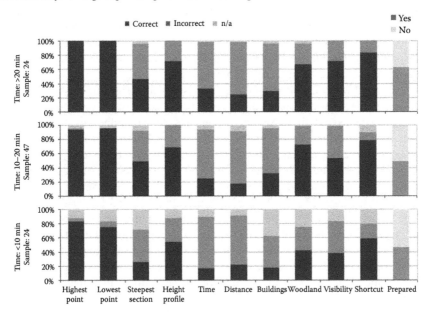

FIGURE 21.4 Hike-planning tasks by the time spent answering the questions. A trend toward higher proportions of correct answers among participants spending more time with the 3-D visualization is shown. This trend is significant for finding the lowest point, the steepest section, the number of buildings, the woodland, and the intervisilibity. Participants who spent more time with the 3-D visualization also feel better prepared for the hike.

solved some of the tasks with more success than the group without, though this was not a statistically significant difference. These results at least suggest that those who are more familiar or spend more time with 3-D visualizations (which may indicate a greater willingness to become familiar with the application) may find them more useful. Another trend relates to those who regarded themselves as having less accomplished map reading skills. This group of nine participants is a small sample of users, but their poor performance in the tasks warrants further investigation. Despite their inability to successfully complete the preparation tasks, this group felt best prepared for going hiking after completing the tasks in comparison to all other groups (e.g., in comparison with the group with more advanced map reading skills and also those groups relating to gender, age, completion time, etc.). It is interesting that the questionnaire data contained 13 statements from participants who mostly consider themselves as being proficient map readers suggesting that realistic 3-D visualizations might be most useful for moderate or poor map readers.

Evaluating the task performance of those with previous knowledge of the study area and those without yields almost no differences between these two groups. Only the task involving identifying the location of the steepest section seems to benefit slightly from prior knowledge of the area.

21.4 CONCLUSIONS AND OUTLOOK

Some sections of the public seem extremely interested in the new possibilities associated with planning hikes virtually. The findings reported here suggest that Web-based 3-D visualizations that combine realistic and abstract information may be helpful additions to the paper map for some when planning hikes, but as implemented for our tests in the Geonova AG software are not suitable replacements. We also identify some possible risks associated with hike planning using 3-D visualization software. Although use of the visualization software was not directly compared to use of the LK25 in this study, the experienced hiker seems to base their judgments of the usefulness of the 3-D visualization for planning hikes according to this comparison. This argument is supported by a number of related statements such as "Nothing will ever be able to replace the LK25" or "I would never plan a hike without the LK25." We note that the methods employed here, whereby participants were asked to self-report their task performance, may lead to statements that do not reflect the true value or usefulness of the 3-D visualization [25].

The 3-D visualization supported overview tasks more successfully than those involving route detection and planning, whereas the extraction of exact information relied upon abstract symbols and labels rather than realism. However, it appears that the usefulness of 3-D visualization may improve when users spend longer with the application (Figure 21.4). In combination, these findings also suggest that more intuitive understanding may conceal some important misinterpretations and false impressions and support the suggestions of MacEachren et al. [18] that realistic and abstract information should be combined. The qualitative information collected also indicates that the technologies employed can result in some frustrating impediments, such as not being able to install the software or experiencing computer crashes; the cited advantages are of course dependent upon being able to access the 3-D application and getting it to work (see Section 21.3).

Three-dimensional visualizations are likely to become more available, more realistic, and more important in the near future, and perhaps the obstacles will be reduced. However, to be usable by the public, careful design is required that is appropriate for specific tasks and the combination of realistic and abstract information is key. Effective interaction design, supported by usability evaluation, should be considered an equally important factor as traditional cartographic design [30]. The suggestion that participants in this experiment underestimated the scale of the relief depicted in the 3-D visualization, which appeared small and controllable, needs further research due to the potential dangers involved in underestimating hazards when planning hikes. Gradient is scale dependent and the question about steep slope made no assumptions about measurement scale. The selection of a less steep, larger-scale feature by many participants, rather than a steeper, smaller-scale slope adds weight to the argument that users of desktop VR may underestimate the scale of the features with which they are interacting. In this sense, photorealistic 3-D environments may be virtually unrealistic. We also identify some inconsistencies between the ways in which users think that tasks have been completed when using 3-D visualizations and actual successful task completion. This may be related to the trend noted by Appleton and Lovett [17] who found that people self-report being better able to imagine the landscape and relate to the 3-D visualization with increased levels of detail. These issues are important in applications of visualization that are associated with significant safety issues and should be of particular concern to those promoting hiking using the kind of 3-D landscape interfaces evaluated here.

The findings draw attention to a number of opportunities for further research. We may be able to identify groups of tasks or users for which/whom 3-D visualizations may be a better choice than 2-D representations, or usefully enhance 2-D displays in combination. The result that participants predominantly prefer the 3-D visualization over the 2-D map for tasks concerning getting an overview, but request more abstract information or the use of the 2-D map for extracting exact information, indicates that a combination of 2-D and 3-D representations may be sensible. Thus, each visualization type may be used for the tasks for which it is most useful. Our continuing research examines how abstract information can best be combined with and represented in 3-D environments that are designed to look realistic. Preliminary results show that users are not significantly influenced by topographic information such as distance or elevation differences when comparing the heights of abstract symbols in virtual environments. Further research may include the exploration of scale in 3-D desktop-based landscape visualizations. Using larger surveys with more varied terrain may allow exploring whether responses and task completion success varies not only with the familiarity or the time spent with the application, but rather with the environment (e.g., terrain variability, route length, height discrepancy).

The participant's requests for more abstract information and functionality could be taken into account (for example, by using more abstract or preinterpreted draped imagery instead of the orthoimagery). Related to this, other researchers are investigating the use of (photo)realism as a visual variable [31]. A different study [32] has shown that draping a digital terrain model with a 2-D map or leaving it undraped does not influence the estimation of walking times or of gradients. Employing more abstract information or combining 2-D and 3-D visualizations may result in users who are more confident in their findings. Research undertaken on the technical and human aspects of using 3-D visualizations with additional abstract information in the field (e.g., [33,34]) may also help in making 3-D visualizations more useful as a hike-planning tool. The assessment of the risks resulting from overestimating distances, underestimating the size or steepness of landscape features, or poor map readers feeling prepared for going hiking is an important area for additional research.

ACKNOWLEDGMENTS

The creation of the 3-D visualizations has been possible thanks to data from LIS Nidwalden AG and the free use of Geonova AG software. We appreciate the feedback provided by those who reviewed an extended abstract of this paper submitted to GISRUK 06 and those who participated in discussion during the subsequent visualization session at the meeting. Their comments and suggestions have informed our study and helped shape the work as presented here. The authors also thank the two anonymous reviewers of this chapter for their comments and suggestions, which have been used to improve the work.

REFERENCES

1. Meng, L., Missing Theories and Methods in Digital Cartography, in *Proceedings of the 21st International Cartographic Conference*, 10–16 August 2003, ICA, Durban, 2003.

2. Rase, W. D., Von 2D nach 3D—perspektivische Zeichnungen, Stereogramme, reale Modelle, in Kartographische Schriften, Band 7: Visualisierung und Erschliessung von Geodaten, Deutsche Gesellschaft für Kartographie, Hannover, 2003, 13.

3. Treinish, L. A., Web-Based Dissemination and Visualization of Operational 3D Mesoscale Weather Models, in *Exploring Geovisualization*, Dykes, J., MacEachren, A. M., and Kraak, M.-J., Eds., Elsevier, Amsterdam, 2005, 403.

4. Google, Google Earth—Explore, Search and Discover, http://earth.google.com/, accessed March 2008.

5. NASA, World Wind, http://worldwind.arc.nasa.gov/, accessed March 2008.

6. Fuhrman, S., Ahonen-Rainio, P., Edsall, R. M., Fabrikant, S. I., Koua, E. L., Tobon, C., Ware, C., and Wilson, S., Making Useful and Useable Geovisualization: Design and Evaluation Issues, in *Exploring Geovisualization*, Dykes, J., MacEachren, A. M., and Kraak, M.-J., Eds., Elsevier, Amsterdam, 2005, 553–566.

7. Kromer, F. K., Grandjean, M.J.P., Gschwend, P., Gloor, T., and Kaslin, T., *Leitbild Langsamverkehr Teilbereich Wandern*, Schweiz, Riehen, 2001.

8. Almer, A. and Stelzl, H., Multimedia Visualisation of Geoinformation for Tourism Regions Based on Remote Sensing Data, in *Proceedings of the Technical Commission IV/6*, ISPRS Congress, July 2002, Ottawa, 2002.

9. Rinner, C., A Geographic Visualization Approach to Multi-Criteria Evaluation of Urban Quality of Life, Working Paper, Workshop Visualization, Analytics & Spatial Decision Support, GIScience 06, 20 September 2006, Münster, 2006.

10. Peterson, M. P., Elements of Multimedia Cartography, in *Multimedia Cartography*, Cartwright, W., Peterson, M. P., and Gartner, G., Eds., Springer-Verlag, Berlin, 1999, 31.

11. Cartwright, W. and Peterson, M. P., Multimedia Cartography, in *Multimedia Cartography*, Cartwright, W., Peterson, M. P., and Gartner, G., Eds., Springer-Verlag, Berlin, 1999, 1.

12. Troyer, M., The World of H.C. Berann, available at: http://www.berann.com/, 2007.

13. Imhof, E,. Bern-Lötschberg-Simplon. Vogelschaukarte, available at: http://www.maps. ethz.ch/imhof/imhof7/, 1922.

14. Memory-Map, http://www.memory-map.co.uk/, accessed March 2008.

15. Dickmann, F., Mehr Schein als Sein? Die Wahrnehmung kartengestützer Rauminformationen aus dem Internet, *Kartographische Nachrichten*, 2, 61, 2004.

16. Lange, E., The Degree of Realism of GIS-Based Virtual Landscapes: Implications for Spatial Planning, in *Photogrammetric Week '99*, Fritsch, D. and Spiller, R., Eds., Wichmann, Heidelberg, 1999, 367.

17. Appleton, K. and Lovett, A., GIS-Based Visualisation of Rural Landscapes: Defining "Sufficient" Realism for Environmental Decision-Making, *Landscape and Urban Planning*, 65, 117, 2003.

18. MacEachren, A. M, Kraak, M.-J., and Verbree, E., Cartographic Issues in the Design and Application of Geospatial Virtual Environments, in *Proceedings of the 19th International Cartographic Conference*, August 1999, ICA, Ottawa, 1999.

19. Schatzalp Davos, Wanderkarte Schatzalp, available at: http://www.schatzalp.ch/down load/Karte2007.pdf, 2008.

20. Lehner, P. and Von Dach, J., *Lagersport/Trekking, Trekking—unterwegs sein*, Bundesamt für Sport, Magglingen, 2004.

21. Ware, C., *Information Visualization: Perception for Design* (2nd ed.), Morgan Kaufman, San Francisco, 2004.

22. Gabbard, J. L., Hix, D., and Swan, J. E. I., User-Centered Design and Evaluation of Virtual Environments, *IEEE Computer Graphics and Applications*, 19(6), 51, 1999.

23. Kirschenbauer, S., Applying "True 3D" Techniques to Geovisualization: An Empirical Study, in *Exploring Geovisualization*, Dykes, J., MacEachren, A. M., and Kraak, M.-J., Eds., Elsevier, Amsterdam, 2005, 363.

24. Gurtner, M., *Karten lesen: Handbuch zu den Landeskarten*, SAC-Verlag and Bundesamt für Landestopographie, Bern, 1995.

25. Elmes, D. G., Kantowitz, B. H., and Roediger, H. L., *Research Methods in Psychology* (8th ed.), Thomson Wadsworth, Belmont, CA, 2006.

26. Gendall, P. and Hoek, J., A Question of Wording, Available at: http://www.sysurvey. com/tips/whitepapers.asp, 2002.

27. Rudestam, K. E. and Newton, R. R., *Surviving Your Dissertation: A Comprehensive Guide to Content and Process* (2nd ed.), Sage, Thousand Oaks, CA, 2001.

28. Kitchin, R. and Tate, J. T., *Conducting Research into Human Geography—Theory, Methodology and Practice*, Pearson Education, Harlow, 2000.

29. Gurtner, M., *Karten lesen Handbuch zu den Landeskarten*, Schweiz, Bundesamt für Landestopographie and Schweizer Alpen Club, Bern, 1995.

30. Bodum, L., Modelling Virtual Environments for Geovisualization: A Focus on Representation, in *Exploring Geovisualization*, Dykes, J., MacEachren, A. M. and Kraak, M.-J., Eds., Elsevier, Amsterdam, 2005, 389.

31. Fabrikant, S. I. and Boughman, A., Communicating Data Quality through Realism, in Geographic Information Science, IfGIprints, Fourth International Conference, GIScience 2006, Münster, 269, 2006.

32. Wood, M., Pearson, D. G., and Calder, C., Comparing the Effects of Different 3D Representations on Human Wayfinding, in *Location Based Services and TeleCartography*, Gartner, G., Cartwright, W., and Peterson, M. P., Springer, Heidelberg, 2007, 345.

33. Coors, V., Elting, C., Kray, C., and Laasko, K., Presenting Route Instructions on Mobile Devices: From Textual Directions to 3D Visualization, in *Exploring Geovisualization*, Dykes, J., MacEachren, A. M., and Kraak, M.-J., Eds., Elsevier, Amsterdam, 2005, 529.

34. Edwardes, A., Burghardt, D., and Weibel R., Webpark—Location-Based Services for Species Search in Recreation Area, in *Proceedings of the 21st International Cartographic Conference*, August 2003, Durban, 2003.

22 PastureSim
A Visualization Tool for Pasture Management

Conrad E. S. Rider and Femke E. Reitsma

CONTENTS

OVERVIEW

The purpose of PastureSim was to provide a visualization tool for pasture management. The system allows the exploration of alternative pasture management strategies and their effect on the resulting grazing pattern. This is facilitated by an interactive pasture edit interface, an agent-based model of sheep grazing, and a spatially distributed model of pasture growth. Because it was designed for use by farmers and not researchers, there are more stringent constraints on complexity and computing resources used. This chapter discusses how the models and visualizations were implemented and describes some experiments carried out in testing.

22.1 INTRODUCTION

Traditionally passed down from generation to generation, the skill of pasture management is widely considered to be best learned from experience. Although this is certainly true, it can also be enhanced by the use of supplementary tools or methods. The purpose of this research was to investigate the potential for the application of a computational modeling and visualization tool to help farmers improve the management of their pastures. The aim was to produce a visualization system for pasture management. The tool would visualize the effects of pasture management regimes on sheep-grazing pressure and distribution. It would be achieved by implementing an exploratory spatial model of pasture growth and sheep grazing, usable by farmers in terms of simplicity and technologies available. To be useful to a farmer, such a tool must:

- Be easy to use
- Facilitate exploration
- Represent the user's farm
- Use technologies available on a typical home personal computer (PC)
- Be lightweight in terms of memory consumption and computational complexity

The overall approach was to produce a simple simulation system, allowing users to explore a set of what-if scenarios. There are, inevitably, a high number of choices available to a livestock farmer. Because of project constraints, it was decided to limit the scope to two areas primarily concerned with space: (1) field geometry and connectivity, and (2) water source distribution.

The simulation component of the system involved the construction of a spatial pasture model and an agent-based sheep model. Both individual-based models of foraging behavior [1,2] and models of pasture growth dynamics [3–5] have previously been the subjects of significant research. What makes this research unique is its focus on the end users—farmers. This increases the constraints on simplicity and interactivity when compared to an equivalent system designed specifically for scientists.

22.2 SYSTEM ARCHITECTURE

The overall system architecture consists of a biophysical model of pasture growth, an agent-based model of grazing sheep, and a physical model of pasture management options (including field boundaries, gates, drinking water sources, and stocking rate).

The system requires the following inputs:

1. An aerial photograph of the farmer's grazing areas. The photograph serves two purposes. First, it provides a representative visualization of the farmer's land, and second, it is used to calibrate the pasture growth model. To give a clear indication of pasture fertility, it is recommended that the photograph be taken in the summer season when the pastures are at maximum growth rate.
2. Climate parameters. These can be tuned to represent the climatic conditions in the farm's region. They are defaulted to the temperate conditions in England.

3. Pasture management choices. These are input into the system using a simple graphical edit interface. Field boundaries and water sources are defined by drawing polygons onto the pasture visualization. Gates (represented by points) can then be placed at any location along the edge of a field polygon. A sheep-stocking interface allows the farmer to add or remove groups of sheep agents at any point within the pasture area. In order to maximize interactivity, these activities may be carried out at any moment throughout the execution of the simulation.

Figure 22.1 shows the PastureSim user interface. General usage of the system involves the following workflow. Initially, the user creates a new simulation file by providing an aerial photograph and its geographic dimensions. Next, the pasture management options are input. This typically involves defining the location of field boundaries, gates, and water sources using the Pasture Edit interface, followed by the placement of sheep flocks at various positions within the pasture area using the Livestock Edit interface. Once the pasture management regime has been set up, it is then run through the execution interface. This causes both the pasture growth model and agent-based sheep model to execute simultaneously. The resulting animal movements and grazing patterns can be examined through a number of visualization interfaces. While running the simulation, its speed may be interactively changed from slow to allow close inspection of flock movements, up to the maximum computer capability to allow for quick access to final results. Additionally, previous states are saved, allowing the user to browse to previous moments in the simulation.

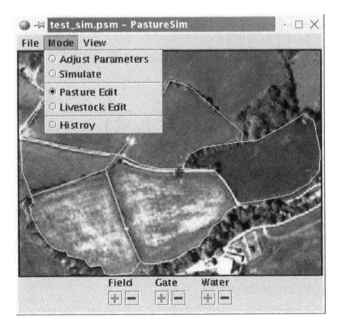

FIGURE 22.1   The PastureSim user interface.

22.3 MODEL OVERVIEW

22.3.1 PASTURE MODEL

The pasture simulation model was based on a previously developed and validated model: SEPATOU. This helps to maintain a good level of realism when simulating the effect of grazing on plant regeneration rate. The SEPATOU dairy farm simulator, developed by Cros et al. [3], was chosen on the basis of its simplicity and ease of calibration. Both of these criteria are essential for an interactive and user-friendly system [6]. The short computation time and daily time step of the model make it feasible to spatially distribute the simulation over a 2-D grid.

22.3.1.1 Variables

The pasture model is represented by a 2-D, m × n matrix of cells, where m is the number of rows and n is the number of columns. Each cell represents anything down to 1 square meter of the pasture area, depending on the chosen resolution. This allows the pasture growth model to be executed on each cell individually to reflect the specific growth conditions at its location and thus the natural spatial heterogeneity in the pasture area. Each cell in the pasture model corresponds to a single pixel in the aerial photograph, which is loaded into the system on initialization. Five state variables are maintained for every cell in the pasture model:

- SWC, soil water capacity
- N, soil nitrogen
- SM, soil moisture
- LAI, leaf area index
- DM, sward dry mass

This gives the model a total memory footprint of 5 mn.

22.3.1.2 Execution

Figure 22.2 shows a schematic diagram of how the model is executed with these five state variables. The SWC and N values are calibrated on initialization and are not altered by the model execution. The model proceeds as follows:

1. Soil water capacity (SWC) is used, along with the climate parameters to calculate soil moisture level (SM).
2. Soil moisture (SM), nitrogen levels (N) and the climate parameters are then used to calculate the leaf area index (LAI).
3. Leaf area index (LAI), soil nitrogen (N), soil moisture, (SM) and the climate parameters are then used to calculate sward dry mass level (DM).
4. The sheep agents then reduce sward dry mass (DM) and leaf area (LAI) when grazing over the cell.

Please see Cros et al. [3] for the model equations and their explanation.

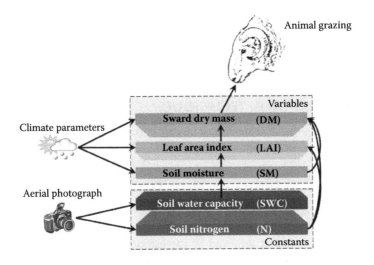

Animal grazing

Climate parameters

Aerial photograph

Variables

Sward dry mass (DM)

Leaf area index (LAI)

Soil moisture (SM)

Soil water capacity (SWC)

Soil nitrogen (N)

Constants

FIGURE 22.2 Pasture model execution.

22.3.1.3 Calibration

The accurate calibration of any biophysical model poses a major challenge to modelers. Because of the technical difficulty and apparatus involved, it would be inappropriate to expect farmers to manually calibrate the pasture model. Instead, the system uses an aerial photograph of the pasture resource to estimate the calibrated values. Although technically difficult, it may eventually be possible to use remotely sensed imagery to derive the soil and vegetation properties required [7–9]. The calibration algorithm developed applies various image filters to the original aerial photograph to estimate the SWC and N values. Although the calibration process is not completely accurate, it largely captures the heterogeneity between moist and arid, or fertile and infertile areas correctly. It does this by using simple rules (such as greener grass implies higher soil moisture). Although the inaccuracy of this method prevents any quantitative prediction of pasture yields, it does allow some degree of discrimination between areas of different quality.

22.3.2 Agent-Based Sheep

Agent-based modeling was used to capture the complex behavior of grazing ruminants. The technique simplifies the specification and coding of the grazing model. Instead of trying to generate complex rules to describe the grazing patterns resulting from collective foraging behavior, simpler rules can be specified for each sheep agent. A complex collective behavior then emerges from the execution of the individual-based model.

Four high-level behaviors were implemented in each sheep agent: grazing, drinking, flocking, and sleeping. For each of the behaviors, a utility function returns a floating point number ≥ 0 and ≤ 1, which represents the level of benefit the action is expected to achieve for the agent. For example, if a sheep becomes thirsty, it is more beneficial to drink than to eat, sleep, or flock. For every simulation loop iteration, the sheep executes the behavior of highest utility. Each of the four high-level behaviors is composed of simpler, lower-level behaviors such as following, foraging, and memorization.

22.4 IMPLEMENTATION TECHNIQUES

The construction of the aforementioned models into an executable computer simulation tool involved some specific challenges.

22.4.1 Pasture Data Storage and Manipulation

Storing and processing the $5 \times m \times n$ pasture state matrix efficiently poses a significant challenge, particularly where small PCs are concerned. Efficient processing of this large matrix is best achieved by the use of carefully designed, optimized algorithms. Building custom matrix processing algorithms in a high-level language such as JAVA is not the best way to achieve this. However, the fact that this data set can be split into five $m \times n$ rasters (or images) means it is possible to use the highly optimized image filtering APIs supplied by JAVA's standard libraries. The use of standard image processing libraries allows the execution of the pasture model to be allocated to the computer's graphics card, freeing up processing resources on the computer's main processor.

Storing the pasture state data as a series of images allows for very efficient processing but consumes a vast quantity of memory resources. For example, one 24-bit image uses 3 bytes for every pixel, so storing all five state variables would have a memory footprint of 15 bytes (5×3) for every pixel (or cell). It is possible to reduce this footprint by packing several of a cell's state variables into one image pixel. This is done by reducing the number of bits used to store each variable's value, so that they can be concatenated and stored in the single 24-bit pixel. Using this technique, we lose numerical precision in each variable value. By intelligent allocation of bits to variables it is possible to minimize the detrimental effects of loss in numerical precision. The idea is to allocate more bits to variables that must be capable of recording smaller changes in value. Analysis of the model's equations will give the minimal step in variable values for given inputs. Figure 22.3 shows the bit allocation scheme used.

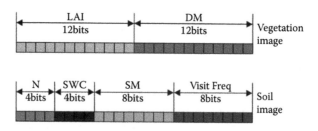

FIGURE 22.3 Pasture data allocation to two 24-bit image pixels.

22.4.2 Agent Navigation and Obstacle Avoidance

To develop sensible paths through gates and around obstacles, an agent requires knowledge of the environment's static objects and a means by which to interact with them. Using conventional geometric analysis to accomplish this can be problematic. Solving specific problems introduced by each object type simply serves to increase the computational complexity of the navigation algorithm, eventually to a point where it becomes unusable.

The problem of navigation is solved with much greater efficiency using the A* path finding algorithm [10]. The idea behind the algorithm is to build a matrix of directions to the resource. Each square describes the direction to a square closer to the resource or object. With this information an agent trying to reach the source can simply look up the matrix (or path map) to find the best direction to follow from that cell. This method is extremely fast because computation of the path map is done on initialization. During execution of the simulation, all direction lookups have a constant runtime, independent of pasture, size, and complexity [11].

So how is the path map constructed? The idea is fairly simple. We take all the cells in the matrix that are directly over the resource. These are initialized with a distance of zero and a direction of null. These cells are then all added to a queue (queue A). We then iterate through each cell in the queue. For every cell, all adjacent cells that have not been visited in the matrix are given the direction to the current cell and a distance of the current cell's distance plus one. All adjacent cells are then added to a second queue, B. When queue A is empty, then queue B is copied to queue A and the process is repeated. This will continue until all the reachable cells have been visited. The accuracy of the overall path map produced can be improved by inclusion of diagonal cells in the algorithm. If this is done, then all adjacent cells are visited first, and then the remaining diagonal ones can be dealt with before the next iteration. This ensures the growth of a regular path map. Figure 22.4 illustrates the path growth process.

22.4.3 Agent Flocking with the Boids Algorithm

A* provides an efficient means to solve navigation around static objects. However, because of its costly initialization process it cannot be used for interactions with dynamic objects, such as other agents. The solution to this problem takes advantage of the fact that the senses of a sheep only allow it to interact with objects within a limited distance. Using this fact, the location of the agents was tracked by a grid reference system, allowing the agent to quickly query neighboring grid cells for other agents. This made it possible to implement the core social behaviors of grazing sheep including flocking and following.

An adaptation of the well-known Boids algorithm by Reynolds [12] was used to simulate flocking. It is based on three basic principles, illustrated in Figure 22.5:

1. Cohesion: If an animal strays too far from the flock, it will move toward it.
2. Separation: If an animal is too crowded, it will move away from the flock.
3. Alignment: The animal will align its direction of movement with others in the flock.

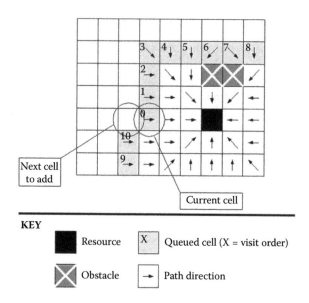

FIGURE 22.4 A* path map initialization process.

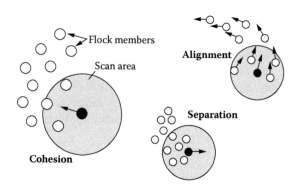

FIGURE 22.5 Three principles behind the Boids algorithm.

A problem with this is that to find the average location of the rest of the flock, each agent must determine the location of all the other animals. When it is considered that this needs to be done for every agent for each simulation loop iteration, it is clear that it is going to be a costly process.

If it is assumed that the rest of the flock is in the direction of the closest animal it is possible to greatly optimize the algorithm. This is not entirely unrealistic since a sheep's flocking decisions are more likely to be based on others close by rather than the entire flock. The optimization is made by simply scanning the immediate area for the closest animal. All flocking decisions are then made on the basis of the position and direction of the closest animal (within the scanning threshold distance). Although not exactly correct, it is a close enough approximation as it results in realistic flocking behavior.

22.4.4 VISUALIZATION INTERFACE

The system provides an interactive visualization interface that can be configured at any time during pasture setup or simulation. The key components to visualize were the sheep agents, the static pasture objects, and the pasture growth model. The static pasture objects including field boundaries, gates, and water sources were visualized using bold colors that were most likely to contrast the pasture visualization. Because these objects need not display any more information than simply their location in the area, their visualization is fairly straightforward. This is not the case, however, for the agents and raster-based layers of pasture state data.

22.4.4.1 Pasture Visualization

The variables representing pasture state are stored as five layers of 2-D matrices, with each matrix cell corresponding to a pixel in the aerial photograph. The visualization system must take these values and present them to the user in a meaningful way. The general technique used to visualize 2-D data of this nature is to shade or color each pixel according to the cell's value and display the resulting image. The choice of color or shade used for this method of visualization can have a very large impact on its success. The principle applied in this system was to visualize the data using colors that represent the substance's actual color, while ensuring that there is enough variation in shade to allow differentiation between areas of unequal value. Figure 22.6 shows the RGB (red, green, blue) color-mixing technique (shown in grayscale) used for each data set and the resulting image. Beside each visualization there are three formulas—describing the quantities of red, green, and blue used—in relation to the cell's value, x.

For all but one of the data sets, a linear relationship between the data value (x) and its corresponding RGB values is suitable. The visualization of leaf area, however, requires a little more processing on the data value to produce an effective visualization. This is because of the nature of the LAI values. When LAI is large, it is of much less consequence than when it begins to diminish to values less then 10% of the maximum. As grass grows, the increasing composite area of all grass blades allows the grass area as a whole to capture more of the sun's radiation. However, as the leaf area of the grass gets larger, leaf area at the canopy of the grass sward can inhibit the penetration of radiation to the leaf below. This means that the increasing leaf area has a progressively lower effect on the increase of solar energy capture. Using an inverse quadratic transformation on the leaf area value will enhance the level of variance in color near the lower end of the scale where small perturbations are important. The resulting visualization permits a more useful analysis of the leaf area data.

22.4.4.2 Agent Visualization

The system visualizes four of the most essential animal state variables to allow the user to keep track of the changing livestock state: location, direction, hunger, and thirst. Visualizing animal location is simply done by superimposing a filled circle onto the pasture data visualization. Representing direction is done by drawing a second filled circle of half the radius on a point projected in the animal's direction. The point is projected by the original circle's radius so that it sits on its edge (see

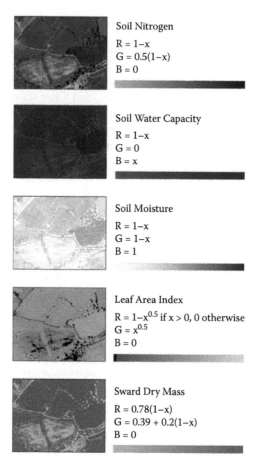

Soil Nitrogen

$R = 1-x$
$G = 0.5(1-x)$
$B = 0$

Soil Water Capacity

$R = 1-x$
$G = 0$
$B = x$

Soil Moisture

$R = 1-x$
$G = 1-x$
$B = 1$

Leaf Area Index

$R = 1-x^{0.5}$ if $x > 0$, 0 otherwise
$G = x^{0.5}$
$B = 0$

Sward Dry Mass

$R = 0.78(1-x)$
$G = 0.39 + 0.2(1-x)$
$B = 0$

FIGURE 22.6 RGB (red, green, blue; shown here in grayscale) mixes for the pasture state visualization.

Figure 22.7). This allows the user to establish the direction of the animal while making the object look slightly more like an animal by adding a headlike element.

Thirst and hunger are visualized by coloring the animal's body. Animal thirst (Figure 22.8) uses a scale from dark to light gray in figure, and hunger (Figure 22.9) uses a scale from light to dark gray in figure. The heads are colored blue for thirst and green for hunger. This allows differentiation between the animals displaying hunger data and those displaying thirst, while preserving the general meaning of red being bad. It also allows the user to establish whether the animal is eating or drinking, because if thirst or hunger is zero then the shade of the head will be exactly the same as the shade of the body.

22.5 EXPERIMENTS AND RESULTS

In the evaluation of research models a quantitative approach is often adopted. However, because the purpose of this model is to support exploratory analysis and learning, it does not provide any quantitatively accurate predictions. For this purpose,

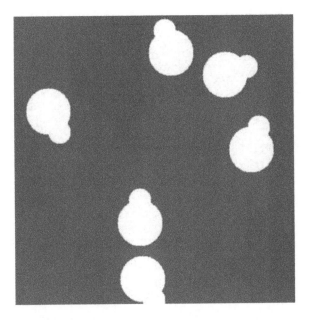

FIGURE 22.7   Basic sheep visualization.

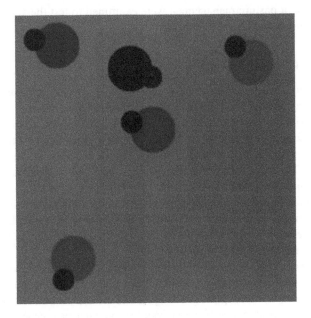

FIGURE 22.8   Sheep thirst visualization.

a useful system only needs to be capable of showing the relative effects of various pasture-management methods, for example, the difference between using many small fields and a few large ones. In this respect, PastureSim may be regarded as an exploratory model rather than a predictive one.

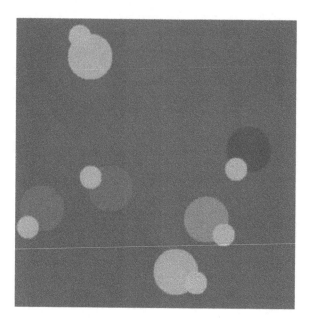

FIGURE 22.9   Sheep hunger visualization.

Two aspects of pasture management were examined to test the system: (1) water source distribution and (2) field geometry.

22.5.1 WATER SOURCE DISTRIBUTION

The use of a large number of drinking water sources and easily accessible placement is generally accepted as good pasture-management practice because it promotes an even and distributed grazing pattern. Poor choice of water source distribution can cause overgrazing and trampling, resulting in bare and useless pasture areas. This experiment was conducted to test the concept. Figure 22.10 shows the initial experimental setup. The same number of sheep, same field size and arrangement, same total water source area, and same pasture "growth power" (soil water capacity and soil nitrogen) were used for both experiments. The only variables are the number and location of drinking water sources. The first field contains 16 evenly distributed and well-accessible water sources. The second contains a single water source located in the bottom corner of the field, providing poor access.

The resulting grazing pattern is shown in Figure 22.11. The first field shows a generally even grazing pattern with no obvious bias for any part of the grazing area. The second field, however, shows great grazing pressure around the only water source, and the upper and rightmost regions (farthest from the source) appear generally untouched. This result shows that the model reproduces the expected behavior—with sheep preferring to graze closer to water sources. Using this experimental setup, the outcome may be obvious, but it illustrates the system's ability to highlight potential problems with a pasture-management regime. More subtle examples may not be as predictable from the outset, with potential problems only emerging after executing several simulation runs.

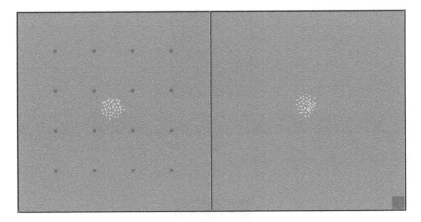

FIGURE 22.10 Initial experimental setup showing sheep (white), water (gray squares), and pasture visualization.

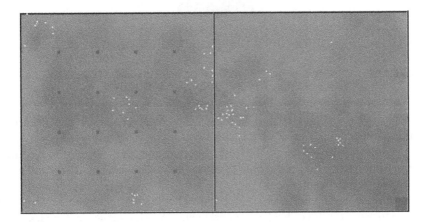

FIGURE 22.11 Experiment result visualizing final sward dry mass (DM).

22.5.2 FIELD GEOMETRY

This second experiment was carried out to find out whether and how field shape has an effect on sheep grazing efficiency. For the experimental setup (shown in Figure 22.12) two fields were used. They were identical in area, growth power, and water source distribution, with the only difference being their geometry. The first field was square and the second was an oblong rectangle.

Figure 22.13 shows the resulting grazing pattern. Looking at the darker, over-grazed patches in the square field, it is obvious that it has received more significant grazing pressure, but why? In the rectangular field the animals' movements are more constrained by the boundaries and the distance between the farthest extremes of the area are greater. It means there is greater difficulty in covering all areas while grazing. Flocking behavior also causes the sheep to cluster together, exacerbating the problem by slowing the movement of the group to better, ungrazed patches.

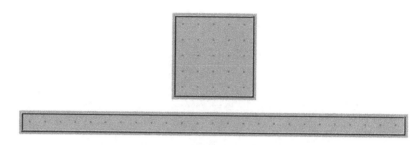

FIGURE 22.12   Initial experimental setup.

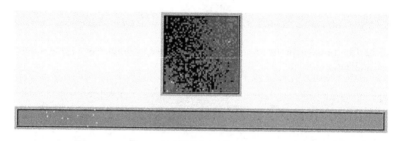

FIGURE 22.13   Experiment result visualizing final sward dry mass (DM).

The resulting uneven grazing profile means the animals are grazing less efficiently because a lower proportion of time is spent grazing on fresh, ungrazed grass. Even though both fields have been stocked with the same number of sheep with the same energy requirements, the greater grazing efficiency in the square field has resulted in greater overall grazing pressure.

So what are the implications of this result? Grazing efficiency can be improved by maximizing the ratio of field area to perimeter. As a consequence, the ideal field shape to maximize grazing efficiency is a circle and the optimal tessellating field shape is hexagonal. Unfortunately, it is usually not possible to implement this for practical reasons. Using square fields is a reasonable compromise. On the other hand, using elongated fields, minimizing field area to perimeter, has the effect of reducing grazing efficiency. It allows grazed pastures time to recover before they are revisited in a similar manner to rotational grazing.

This result is not as obvious as the previous one. It demonstrates how even a simple model can be used to generate or confirm new theories. Simplicity can eliminate the confusion caused by unnecessary model complexity to provide a better understanding of the main process at work.

22.6 CONCLUSIONS AND FUTURE WORK

There are numerous possible future directions the research project could take. Some examples are:

- A farmer usability survey to measure how well the system achieved its original goal. This would be the most important next step for the project as it is difficult to measure its effectiveness until it is tested with its target users. Results from this could also have implications for the interface design of research-based models.
- Validation of the sheep decision model with empirical measurement of movement and grazing patterns. This type of work might involve collaboration between computer scientists and ethologists to determine exactly how the movement patterns relate to sheep behavior and how these behaviors can be artificially recreated with realism.
- Better calibration of the pasture model using remote-sensed data or by direct measurement of soil and climate characteristics. With a more accurate pasture model, the system may be capable of producing better predictive results. Examples include average grazing efficiency and optimal stocking rates.
- Additions to the agent-based model to allow the simulation of other livestock, such as cows, horses, pigs, and poultry. Other animals not only graze pastures at a different rate, but also have a different grazing style. They also affect pasture growth by trampling and excretions. It is possible that there exists a combination of different livestock which results in optimal grazing efficiency or better pasture health.
- Further experimentation with pasture management options including water sources, field geometry, field connectivity, and field size. Because of time constraints it wasn't possible to conduct a comprehensive set of experiments on the existing model. Possible experiments that could be carried out in the existing model include:
    1. Comparing rotational grazing with free grazing.
    2. Finding the optimal field size for a given pasture.
    3. The effect of field connectivity. Do animals naturally rotate between fields?
    4. The relationship between the total number of sheep and the number of flocks that develop (or average flock size).
    5. Examining the effect of seasons and changing weather on grazing pressure and distribution.

Where improvements to the system are concerned, it is important that the original design goal of usability is maintained. The system's strength is as an interactive learning tool. Improvements to the model need to maintain the short processing times and low memory consumption, which permit user interaction.

By utilizing simple models, efficient algorithms, and interactive visualizations, it was possible to implement an exploratory tool for pasture management. Although its constituent models are not predictive, the system has been shown to allow the exploration of alternative pasture management methods. It helps to highlight subtle problems in potential management plans, as well as provide a highly visual and representative platform for the exploration of a pasture-farming regime.

REFERENCES

1. Beecham, J. A. and Farnsworth, K. D., Animal foraging from an individual perspective: An object orientated model, *Ecological Modelling*, 113, 141, 1998.
2. Dumont, B. and Hill, D. R. C., Multi-agent simulation of group foraging in sheep: Effects of spatial memory, conspecific attraction and plot size, *Ecological Modelling*, 141, 201, 2001.
3. Cros, M.-J., Duru, M., Garcia, F., and Martin-Clouaire, R., A biophysical dairy farm model to evaluate rotational grazing management strategies, *Agronomie*, 23, 105, 2003.
4. Riedo, M., Grub, A., Rosset, M., and Fuhrer, J., A pasture simulation model for dry matter production, and fluxes of carbon, nitrogen, water and energy, *Ecological Modelling*, 105, 141, 1998.
5. Thornley, J. H. M., Simulating grass-legume dynamics: A phenomenological submodel, *Annals of Botany*, 88, 905, 2001.
6. Engelen, G. Models in Policy Formulation and Assessment: The WADBOS decision support system, in *Environmental Modelling: Finding Simplicity in Complexity*, Wainwright, J. and Mulligan, M., Eds., John Wiley & Sons, Chichester, 2004, chap. 14.
7. Ustin, S. L., DiPietro, D., Olmstead, K., Underwood, E., and Scheer, G. J., Hyperspectral remote sensing for invasive species detection and mapping, IEEE International Geoscience and Remote Sensing Symposium, 2002.
8. Noborio, K., Measurement of soil water content and electrical conductivity by time domain reflectometry: A review, *Computers and Electronics in Agriculture*, 31, 213, 2001.
9. Selige, T. and Schmidhalter, U., Soil resource mapping for precision farming using remote sensing, *Geoscience and Remote Sensing Symposium*, 7, 3138, 2001.
10. Hart, P. E., Nilsson, N. J., and Raphael, B., A formal basis for heuristic determination of minimum cost paths, *IEEE Transactions on Systems Science and Cybernetics*, 4, 100, 1968.
11. Cormen, T. H., Leiserson, C. E., Rivest, R. L., and Stein, C., *Introduction to Algorithms*, MIT Press, Cambridge, MA, 2001.
12. Reynolds, C. W., Flocks, herds and schools: A distributed behavioral model, *SIGGRAPH '87: Proceedings of the 14th Annual Conference on Computer Graphics and Interactive Techniques*, 25, 1987.

Index